MARINE RADIOACTIVITY

RADIOACTIVITY IN THE ENVIRONMENT

A companion series to the Journal of Environmental Radioactivity

Series Editor
M.S. Baxter
Ampfield House
Clachan Seil
Argyll, Scotland, UK

Volume 1: Plutonium in the Environment
(A. Kudo, Editor)

Volume 2: Interactions of Microorganisms with Radionuclides
(F.R. Livens and M. Keith-Roach, Editors)

Volume 3: Radioactive Fallout after Nuclear Explosions and Accidents
(Yu.A. Izrael, Author)

Volume 4: Modelling Radioactivity in the Environment
(E.M. Scott, Editor)

Volume 5: Sedimentary Processes: Quantification Using Radionuclides
(J. Carroll and I. Lerche, Authors)

Volume 6: Marine Radioactivity
(H.D. Livingston, Editor)

MARINE RADIOACTIVITY

Editor

Hugh D. Livingston

International Atomic Energy Agency,
Marine Environment Laboratory,
Principality of Monaco

ELSEVIER

2004

AMSTERDAM – BOSTON – HEIDELBERG – LONDON – NEW YORK – PARIS
SAN DIEGO – SAN FRANCISCO – SINGAPORE – SYDNEY – TOKYO

ELSEVIER B.V.
Sara Burgerhartstraat 25
P.O. Box 211
1000 AE Amsterdam
The Netherlands

ELSEVIER Inc.
525 B Street, Suite 1900
San Diego
CA 92101-4495
USA

ELSEVIER Ltd
The Boulevard, Langford Lane
Kidlington
Oxford OX5 1GB
UK

ELSEVIER Ltd
84 Theobalds Road
London WC1X 8RR
UK

First edition 2004

Library of Congress Cataloging in Publication Data
A catalog record is available from the Library of Congress.

British Library Cataloguing in Publication Data
A catalogue record is available from the British Library.

ISBN: 0 08 043714 1

♾ The paper used in this publication meets the requirements of ANSI/NISO Z39.48-1992 (Permanence of Paper).

Printed in The Netherlands.

Contents

Foreword

Hugh D. Livingston

This series of books on *Radioactivity in the Environment* is ambitious in that it sets out to review and document the steady growth in knowledge in this young field. While it may date back a century or so in respect of the discovery of radioactivity, it really has been over only about fifty years that the application of the basic knowledge in nuclear physics and chemistry has been applied to understand the impact of radioactivity on the environment and, conversely, what can be learned about the terrestrial and marine environments using the unique new toolset of radionuclides. The oceans, covering 70% of the surface of the planet and containing the bulk of its water, form a major compartment of the global environment generally and one which holds a corresponding large share of radioactivity on our planet. In recent years, we have become increasingly aware of the importance of the oceans to life on the planet in diverse ways such as their influence on climate, their potential to provide food for an ever-expanding population and the vital role of the coastal ocean for mineral resources, transportation and a locale used by a very large fraction of the Earth's population.

This book on *Marine Radioactivity* sets out to cover most of the aspects of marine radioactivity which have been the focus of scientific study in recent decades. The authors and their reviews divide into topic areas which have defined the field over its history. They cover the suite of natural radioisotopes which have been present in the oceans since their formation and quantitatively dominate the inventory of radioactivity in the oceans. Also addressed are the suite of artificial radionuclides introduced to the oceans as a consequence of the use of the atom for development of nuclear energy, nuclear weapons and various applications of nuclear science. The major source of these continues to derive from the global fallout of atmospheric tests of nuclear weapons in the 1950s and 1960s but also includes both planned and accidental releases of radioactivity from both civilian and military nuclear technology. The other division of the major study direction depends on whether the objective is to use the radionuclides as powerful tools to study oceanic processes, to describe and understand the ocean distribution of the various natural or artificial radionuclides or to assess the different radionuclides' impact on and pathways to man or marine organisms.

The subject of natural radionuclides in the oceans has been often covered in general in books and reviews. This book features two very current and broad reviews of specific topics in this field. In Chapter 1 (Cochran & Masque), the uses of natural radionuclides to address practical problems in the vital coastal ocean are reviewed. In Chapter 5 (Bacon), the use of reactive natural radionuclides in the study of particle transport processes is covered.

Three chapters focus specifically on aspects of the nature of oceanic contamination with artificial radionuclides. Chapter 2 (Hamilton) provides a broad account of the history and evolution in the world oceans of the artificial radionuclides introduced both by accident and on purpose, from both military and civilian activities since the beginning of the nuclear age. More specifically, Chapter 3 (Leon Vintro et al.) reviews the parallel history of the long lasting transuranic radionuclides in the oceanic domain. Finally, in Chapter 4, Linsley et al. cover, in a very comprehensive fashion, the nature and distribution of all known locales where manmade radionuclides are located – the so-called point sources.

The chapter by Fowler and Fisher (Chapter 6) focuses on the interactions of both artificial and natural radionuclides with marine biota – a broad subject of wide interest and application in ocean science and very relevant to understanding the pathways back to man for effects evaluations. Such evaluations are fully described and reviewed in Chapter 7 (Hunt) where a complete assessment of the radiological impact of natural and artificial radionuclides is made and set in the context of the international standards and guidelines adopted to assess and control the effects of radioactivity on man and biota.

Finally, Chapter 8 (Povinec) documents the current 'state-of-the-art' critical analytical methodologies which are in use in both ocean sampling and analyses in connection with marine studies of radioactivity.

While the book thus deals with most salient aspects of marine radioactivity, coverage of tracer applications of radionuclides to characterize ocean current flows, advective and diffusive processes in the water column etc. is not the subject of a planned dedicated chapter since unfortunately at a late date the invited author was unable for personal reasons to deliver the manuscript. However, there are references throughout the book to such applications and so hopefully the reader will assimilate much on this topic if in indirect manner.

The authors are to be congratulated in their efforts to bring together a scholarly and topical selection of papers in this key area of environmental studies of radioactivity. I wish to thank them for their efforts knowing that they will be much valued by students in the field who will be carrying their efforts forward in the 21st century.

MARINE RADIOACTIVITY
Hugh D. Livingston (Editor)

Chapter 1

Natural radionuclides applied to coastal zone processes

J. Kirk Cochran, Pere Masqué[1]

Marine Sciences Research Center, Stony Brook University, Stony Brook, New York 11794-5000, USA

1. Introduction

The coastal zone represents the portion of the world ocean that is most directly affected by human activities. Development of the coastal zone, exploitation of its natural resources, and its use as a receptacle for societal wastes are resulting in problems ranging from contamination of sediments and living marine resources to eutrophication and the development of harmful or nuisance algal blooms. Radionuclides provide tracers for many of the processes related to coastal zone problems. Indeed, increasingly in the last 25 years, natural radionuclides have been used to quantify the rates of coastal ocean processes and many of these results are directly applicable to providing important information that may be used by managers tackling problems in the coastal zone.

This chapter had its inception in a request from the International Atomic Energy Agency to the senior author to provide an overview of some of the principal applications of natural radionuclides to studying processes in the coastal zone, with an emphasis on the management problems that these radionuclides are well placed to address. These applications are indeed numerous: virtually all are the focus of current research and application, and any one of them could serve as the basis for a review article. In preparing this chapter we have of necessity, therefore, been selective in the examples we have chosen to include.

It is fortuitous that the naturally occurring radionuclides, including those of the U and Th decay series and those produced by the interactions of cosmic rays with atmospheric gases (i.e. cosmogenic), comprise some of the most useful tracers for coastal zone processes. The choice of an appropriate radionuclide tracer is dependent on matching

(i) its distinctive geochemical behavior with the process being characterized, as well as
(ii) its half-life to the rate of the process.

[1]Present address: Institut de Ciència i Tecnologia Ambientals – Departament de Física, Universitat Autònoma de Barcelona, 08193 Bellaterra, Spain.

Indeed many coastal zone processes operate on time scales that are rapid relative to similar processes in the open or deep ocean, and thus require radionuclides with short half-life for proper characterization.

The suite of natural radionuclides used in studying coastal zone processes includes (half-lives in parentheses):

- ^{234}Th (24 d) – This ^{238}U series radionuclide, produced in solution from decay of dissolved ^{238}U, is rapidly scavenged onto particles and removed to bottom sediments. It is used to determine rates of scavenging, sediment mixing by organisms or physical processes (such as storms) or rates of sediment accumulation in cases of rapidly depositing sediments.
- ^{32}P (14.3 d) and ^{33}P (25.3 d) – These cosmogenic isotopes of phosphorus are beta emitters produced by the spallation of atmospheric argon by cosmic rays. They quickly become associated with aerosols and are input to the oceans, where they equilibrate with dissolved inorganic phosphorus. Only a small fraction of ^{32}P and ^{33}P is produced in situ in the surface ocean, and thus the principal source of both isotopes is wet deposition.
- ^{7}Be (53 d) – This cosmogenic radionuclide is produced by the interaction of cosmic rays with atmospheric gases (mainly N and O). It is delivered to the Earth's surface by wet and dry deposition and, where suspended particle concentrations are high, can be scavenged onto particles and deposited in bottom sediments. As with ^{234}Th, ^{7}Be can be used to determine mixing rates, or in certain instances, sediment accumulation rates.
- ^{210}Pb (22.3 y) – Lead-210 is produced from decay of ^{226}Ra via ^{222}Rn in the ^{238}U decay series. In the coastal ocean, the dominant supply of ^{210}Pb is typically via the atmosphere where it is produced from the decay of ^{222}Rn that has emanated from terrestrial rocks and soils. Owing to its long half-life, ^{210}Pb profiles in coastal sediments reflect both mixing and accumulation processes.
- ^{226}Ra (1622 y), ^{228}Ra (5.7 y), ^{223}Ra (11.4 d), ^{224}Ra (3.7 d), ^{222}Rn (3.8 d) – The element radium includes isotopes in the ^{238}U, ^{235}U and ^{232}Th decay series and ^{222}Rn is the immediate daughter of ^{226}Ra. All are the result of alpha decays and are produced dominantly in mineral structures. Recoil associated with the production of the nascent Ra or Rn atoms mobilizes these nuclides to sediment pore waters or to groundwater. Subsequently Ra and Rn can be released from sediments to overlying water. Recent interest in the distributions of Ra and Rn in coastal waters has focused on their use as tracers of the rate of groundwater inflow.
- ^{14}C (5730 y) – Although radiocarbon specific activities in the environment have been affected by the burning of fossil fuels as well as anthropogenic production associated with atmospheric nuclear testing, this radionuclide is also produced naturally as a cosmogenic radionuclide. Radiocarbon can be used, especially at depth in sediment cores, to determine long-term rates of accumulation. Such determinations provide an assessment of the accuracy of accumulation rate estimates produced using the shorter-lived radionuclides listed above. Recent advances in measurement using accelerator mass spectrometry (AMS) provide errors of only $\pm$50 years on the radiocarbon age for $\sim$2000 year old carbon, permitting sediment accumulation rates to be determined over time intervals of just a few hundred years.

2. Scavenging rates of particle-reactive contaminants in coastal waters

The simplest use of natural radionuclides to determine scavenging rates assumes a balance between the production of the radionuclide from decay of a dissolved parent and the decay and scavenging of the radionuclide itself (Matsumoto, 1975):

$$\lambda P = \lambda D + k_{\mathrm{d}} D, \tag{1}$$

where:

λ is the decay constant for the daughter radionuclide,
P and D are the activities of parent and daughter respectively in a water sample and
k_{d} is the scavenging rate constant (first-order) of the daughter radionuclide.

If the parent and daughter activities are measured, all quantities in equation (1) except k_{d} are known and k_{d} can be calculated; the inverse of k_{d} is τ_{scav}, the mean residence time of the daughter radionuclide with respect to scavenging. A variation of equation (1) includes separate measurements of the particulate and dissolved fractions and setting up analogous equations that lead to calculation of the residence time of D with respect to removal from solution onto particle surfaces and removal from the water column by sinking particles (e.g. Krishnaswami et al., 1976).

The thorium isotopes are perhaps the quintessential particle-reactive radionuclides, and both ^{234}Th and ^{228}Th have been used to measure scavenging rates of Th in coastal waters (see Moore, 1992, for a review). Indeed the high suspended particle concentrations and shallow water columns of coastal waters and the consequent intense benthic–pelagic coupling that results produces short residence times with respect to scavenging. For example, early measurements by Aller & Cochran (1976) showed that ^{234}Th had a mean residence time of about 1 day in the waters of Long Island Sound, USA. An extreme example of rapid scavenging of Th is seen in the waters of the Venice Lagoon (Italy) (Cochran et al., 1995). The shallow water column (<2 m) of the lagoon, coupled with tidally-mediated resuspension, produces residence times of dissolved Th of just a few hours.

The residence time of Th ranges from these extremely rapid values in shallow coastal environments to days – weeks offshore (Kaufman et al., 1981). It has been generally recognized that particle concentration and flux are fundamental controls on the removal of Th and other reactive chemical species from the water column (Santschi et al., 1979). As well, particle size and type play a role (Honeyman & Santschi, 1989; Quigley et al., 2002). These factors have been incorporated into progressively more complex models of scavenging and transfer of reactive radionuclides (especially Th isotopes) among the dissolved and particulate phases. A more detailed discussion of the application of natural radionuclides to scavenging models is given in Chapter 5, this volume.

The intensity of scavenging in continental shelf and estuarine waters makes these environments strong sinks for reactive radionuclides as well as contaminants. Natural radionuclide distributions in such environments provide useful information on the rate and extent of particle and contaminant transport (Aller et al., 1980; McKee et al., 1984; Gustafsson et al., 1998; Feng et al., 1999a).

3. Particle dynamics in coastal waters

After scavenging, particle-associated radionuclides serve as useful tracers of the fate of particles in coastal waters. Important applications include tracing the movement of particles through estuarine systems and determining chronologies of sediment deposition. The latter has many ramifications for coastal management, including determining rates of sediment accumulation in dredged areas or around piers or other structures, estimating accumulation chronologies of particle-associated contaminants, and characterizing the rate and intensity of reworking of sediments by the benthic fauna.

3.1. *Particle transport in estuaries*

Transport of suspended particles in estuaries is governed by river flow, tidal mixing and any residual estuarine circulation. Natural radionuclides are useful tracers of the movement of suspended particles in estuaries. The short-lived radionuclides ^{234}Th and ^{7}Be are especially useful in this context because their half-lives are sufficiently short to allow the distribution of these radionuclides to adjust to short-term fluctuations in river flow and tides. Significantly, ^{234}Th and ^{7}Be have different sources to estuaries: ^{7}Be is added directly from the atmosphere, and unless the estuarine system is very large, its supply can be considered to be spatially uniform, although temporally variable. On the other hand, ^{234}Th is produced in situ from decay of dissolved ^{238}U. Uranium behaves conservatively during mixing in many, although not all, estuaries (Toole et al., 1987; McKee et al., 1987), but dissolved U generally increases with increasing salinity. Thus ^{234}Th is produced preferentially in the seaward reach of the estuary. To the extent that Th and Be have similar geochemical behaviors under conditions of high suspended particle concentrations typically found in estuaries, the difference in sources of these two radionuclides ensures that particles that spend a significant amount of time in the high salinity portion of an estuary will have greater activity ratios of ^{234}Th/^{7}Be. Feng et al. (1999b) showed that this was indeed the case for the Hudson River Estuary (USA) (Fig. 1). Departures from theoretical ratios of ^{234}Th/^{7}Be predicted from a simple scavenging model indicated that, under conditions of low river flow, particles can be transported up-estuary over distances comparable to the tidal excursion. In contrast particles are displaced seaward during times of high flow.

Feng et al. (1999c) also used the ^{234}Th/^{7}Be ratio of suspended particles in the lower Hudson River Estuary to determine the contribution of locally resuspended sediment versus sediment transported from the seaward portion of the system to the turbidity maximum (TM). During a period of low river flow, box models of ^{234}Th and ^{7}Be in the system showed that about 30% of the suspended sediment in the TM was derived locally and the remainder advected in with the estuarine circulation. These results have implications for the transport of contaminants in the system. The Hudson (and many estuaries) show an urbanized pattern in which cities near river mouths produce a strong source of contaminants at the seaward end of the system. In the case of the Hudson River Estuary, New York City constitutes such a source of particle-reactive trace metals and organic contaminants. Indeed Feng et al. (1999a) showed that the ^{234}Th/^{7}Be ratio on suspended particles correlated well with the Ag/Fe ratio. Silver is a contaminant that is strongly linked to wastewater inputs to urban coastal waters (Sañudo-Wilhelmy & Flegal, 1992) and the good correlation shows the link between the lower

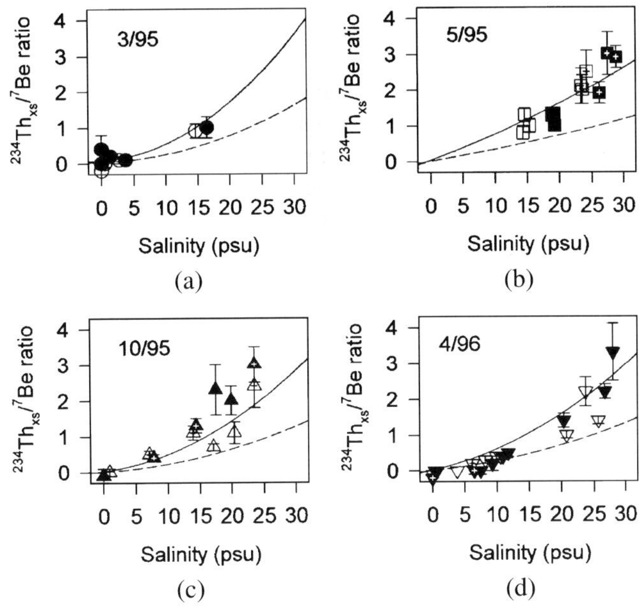

Fig. 1. The $^{234}Th/^{7}Be$ activity ratio in suspended particles versus salinity in the Hudson River Estuary. Open and filled symbols represent surface and bottom water, respectively. Samples collected on flood tide are designated with a +. The solid curves denote upper limits on $^{234}Th/^{7}Be$ activity ratios, estimated from a model assuming in situ production of ^{234}Th, atmospheric supply of ^{7}Be, rapid scavenging, and short residence time of suspended particles in water column. The dashed curves are estimated lower limits set by resuspension of local bottom sediments. (From Feng et al., 1999b, with permission from Elsevier.)

estuary source of ^{234}Th (from higher ^{238}U activities in the seaward portion of the system) and Ag (from urban wastewater inputs).

Beryllium-7 also serves as a useful tracer of temporal variability in the transport of sediment by river estuaries. In the Hudson, Hirschberg et al. (1996), Feng et al. (1998) and Woodruff et al. (2001) showed that the spring freshet transported large amounts of sediment through the system, depositing it temporarily in the lower Hudson. With time this material, labeled with high activities of ^{7}Be, was redistributed through the system. On a larger scale, Sommerfield et al. (1999) showed that the spring freshet of the Eel River, California (USA), transported river sediment labeled with ^{7}Be onto the adjacent shelf. The freshly supplied river sediment was distinguished from adjacent shelf sediments by its high ^{7}Be activities.

3.2. *Sediment accumulation*

Understanding the patterns of sediment accumulation in a coastal environment is critical for evaluating the long-term effects of dredging and the construction of piers, for example. Many of the particle reactive natural radionuclides mentioned above are useful in determining accumulation rates of coastal sediments. Indeed sediment accumulation rates in recently dredged areas can be quite rapid, necessitating the use of short-lived radionuclides such as ^{234}Th or ^{7}Be to determine chronologies. Once chronologies of sediment accumulation have been established, they can be used to construct chronologies of the fluxes of particle-associated contaminants to a site (e.g. Valette-Silver, 1993). This information has practical value in assisting coastal zone managers to determine whether contaminant inputs to a coastal system are increasing or decreasing. Finally, comparing sediment inventories of natural radionuclides, whose sources are commonly well constrained, to inventories of contaminants, whose sources are multiple and often poorly characterized, can help to sort out the relative importance of contaminant sources to coastal systems.

Since the early work by Koide et al. (1973) using ^{210}Pb as a chronometer for sediment accumulation, this nuclide has become the sine qua non for determining accumulation rates in coastal sediments. Lead-210 is present in sediments in excess of the activity of its grandparent ^{226}Ra and the decrease of excess ^{210}Pb with depth in the sediments can be used to extract the sediment accumulation rate. Using the familiar decay equation with time represented as (depth in the sediment)/(sediment accumulation rate), the change in excess ^{210}Pb with depth in a sediment column can be represented as:

$$A = A_0 \exp\left(-\frac{\lambda x}{S}\right), \tag{2}$$

where:

A is the activity at depth x in the sediment,
A_0 is the activity at the sediment water interface,
λ is the decay constant and
S is the sediment accumulation rate.

As applications of ^{210}Pb geochronometry to subtidal coastal sediments grew increasingly numerous in the 1970s and 1980s (e.g. Goldberg et al., 1977, 1978; Thomson et al., 1975; Benninger et al., 1979; Turekian et al., 1980), it became apparent that equation (2) was an oversimplification in that it neglected to take into account the effect of an important process, namely sediment mixing by infauna, on the radionuclide profiles. Animals living in sediments displace particles through feeding activities and enhance the transport of solutes across the sediment water interface through irrigation of burrows with overlying water. Particle mixing serves to modify depth gradients of radionuclides and causes the sediment accumulation rate determined from equation (2) to be erroneously high. Moreover, the intensity of particle mixing decreases with depth in the sediment column so that the effect on radionuclide profiles varies with depth. The mixing processes caused by surface deposit feeding organisms have been analogized to an eddy diffusion process, and the diagenetic equation pertaining to a particle-reactive radionuclide in sediments (with no compaction) is commonly represented as:

$$\frac{\delta A}{\delta t} = D_{\mathrm{B}} \frac{\delta^2 A}{\delta x^2} - S \frac{\delta A}{\delta x} - \lambda A, \tag{3}$$

where:

D_{B} is the particle mixing coefficient and
t is time.

(A)

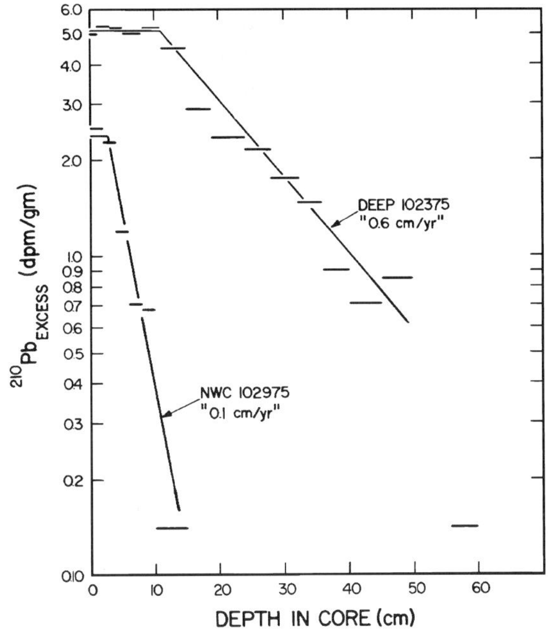

(B)

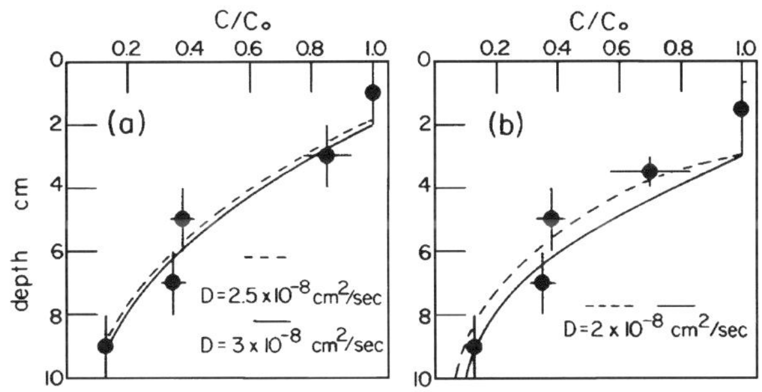

Fig. 2. (A) Excess ^{210}Pb activity profiles versus depth in two sediment cores from Long Island Sound. The indicated sediment accumulation rates are obtained assuming that no mixing is present below the upper 4 cm or 10 cm (cores NWC 102975 and NWC 102375, respectively). However, the presence of Pu (B) at similar depths as excess ^{210}Pb suggests that both deep mixing and sediment accumulation are important processes affecting the ^{210}Pb profile. In (B) best fits to normalized Pu activities (C/C_0) for core NWC 102975 are shown, obtaining a D_B below the surficial mixed zone that permits calculation of a sediment accumulation rate $\leqslant 0.05$ cm y^{-1}. (From Benninger et al., 1979; Turekian et al., 1980, with permission from Elsevier.)

It is apparent that equation (2) results from the steady state solution to equation (3) if there is no particle mixing ($D_B = 0$).

The effect of sediment mixing on a profile of excess ^{210}Pb is often to produce a well-mixed layer of nearly constant activity in the upper few centimeters (Fig. 2). Below the well-mixed layer is a zone of decreasing activity. Generally, inspection of the sediments visually or by X-radiography is sufficient to document the presence of an active benthic community that causes rapid mixing in the surficial sediments.

The short half-lives of both ^{234}Th and ^{7}Be make them well suited to determine mixing rates in surficial coastal sediments, because the rate of burial of the radionuclide by sediment accumulation is small compared with the rate of downward transport by mixing (Aller & Cochran, 1976; Krishnaswami et al., 1980; Canuel et al., 1990; McKee et al., 1984). A significant complication arises, however, in trying to estimate sediment accumulation rates from radionuclide profiles below the depth of rapid surficial mixing. A common assumption is that mixing is not occurring at depth in the core and ^{210}Pb excess is used (in equation (2)) to determine accumulation rates. However, excess ^{210}Pb is commonly present to only tens of centimeters in the coastal sediment column, a depth region that is affected by mixing events occurring over the time scale of the ^{210}Pb half-life. Benninger et al. (1979) articulated the effects of decreasing mixing with depth in a core on radionuclides with different half-life.

Thus, as emphasized by Smith (2001), a requisite step for validating accumulation rates from radionuclide profiles is to use multiple radionuclides of different half-life. Other potential chronometers include anthropogenic ^{137}Cs, added to the environment in association with atmospheric testing of nuclear weapons and with releases associated with the nuclear fuel cycle. Cesium-137 is used differently than ^{210}Pb as a chronometer, with reliance placed on identifying depth horizons corresponding to known pulse inputs of ^{137}Cs to the environment. These include the first large-scale appearance of this radionuclide from atmospheric testing of atomic weapons in the 1950s, a peak in global fallout 1963–1964 corresponding to the Nuclear Test Ban Treaty and, in some locations, a peak in 1986 resulting from the Chernobyl accident. Plutonium isotopes (^{239,240}Pu) or ^{241}Am can as well be used for the same purpose, and in some instances can be used with ^{210}Pb to resolve the effects of mixing and sediment accumulation on radionuclide profiles (Fig. 2).

Radiocarbon also has been used successfully to determine accretion rates in coastal sediments. Carbon-14 has the advantage of having a much longer half-life than ^{210}Pb. It is thus present deeper in the sediment column and is less subject to mixing effects (Fig. 3).

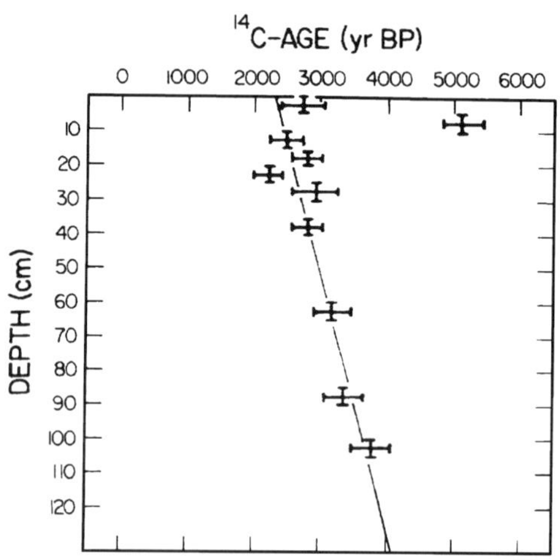

Fig. 3. Carbon-14 versus depth in organic fraction of core NWC 102975. The ^{14}C-based accumulation rate is ~0.07 cm y^{-1}, and can be compared to the ^{210}Pb results shown in Fig. 2(A) for the same core. (From Benoit et al., 1979, with permission from Elsevier.)

Moreover, recent advances in measurement of radiocarbon by accelerator mass spectrometry have improved precision such that age differences of a few hundred years can be determined. Comparison of sediment accumulation rates derived from ^{210}Pb (uncorrected for bioturbation) and radiocarbon often shows differences of factors of 2 or more, with the ^{210}Pb rates biased toward the high side, as expected from neglecting the effects of mixing (compare Figs 2 and 3; Benoit et al., 1979). Thus mixing must be taken into account to properly determine the sediment accumulation rate in many subtidal environments.

In some instances, however, the rate of accumulation in a coastal area is truly high, of the order several cm's to 10's of cm's per year. Sites where this is happening include recently dredged areas or those that have been created by activities such as mining sand in harbors to use as aggregate in concrete manufacture. Such areas are significantly out of 'equilibrium' with the hydrodynamics of the surrounding undisturbed sea floor and become loci of rapid sediment accumulation. The lack of significant mixing is supported by X-radiographs that show laminations and small opportunistic infauna (if any). In such cases, the short-lived radionuclide chronometers ^{234}Th and ^{7}Be are appropriate chronometers (Fig. 4).

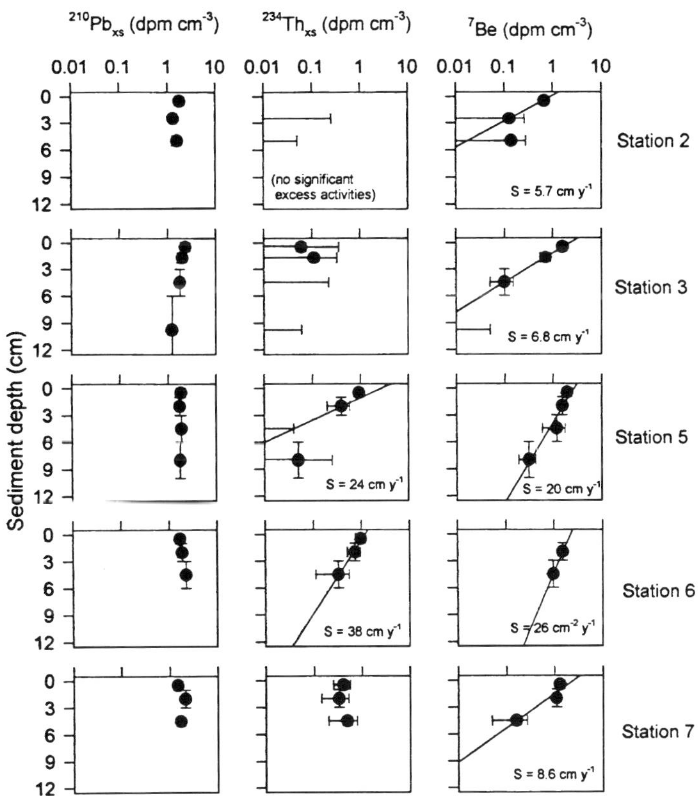

Fig. 4. Excess ^{210}Pb, excess ^{234}Th and ^{7}Be activity profiles versus depth in sediment cores from the western margin of the lower Hudson River Estuary. Sediments accumulate rapidly on the short term at this location and high sediment accumulation rates are obtained from excess ^{234}Th and ^{7}Be profiles. (From Feng et al., 1998, with permission from Estuarine Research Federation.)

4. Wetlands

4.1. *Accretion rates relative to forcing factors such as sea level rise*

Wetlands serve important functions as habitats, as nurseries for juvenile fish and other marine organisms, as filters for contaminants from land to oceans, as recyclers of organic material and as buffers against the effects of storms on coastal areas. Wetlands face many threats – from local development of coastal areas to the global increase in sea level. One important way of assessing the health of wetlands is through determination of their accretion rate. Wetlands such as salt marshes often accrete to keep pace with sea level rise. Wetlands that accrete at a rate less than sea level rise for a significant length of time run the risk of being lost by immersion. In many instances, development of land adjacent to the coast prevents the migration of wetlands inland as sea level rises, and this is a contributing factor to their net loss. Indeed, about two thirds of the wetlands in Europe and North America have been lost during the latter part of the 20th century, and 85% of those in Asia are threatened (GESAMP, 2001).

Accretion rates of wetland deposits can be determined using the same suite of natural and anthropogenic radionuclides as with subtidal sediments. Those used most commonly include ^{210}Pb (Armentano & Woodwell, 1975; McCaffrey & Thomson, 1980; Cochran et al., 1998a), ^{137}Cs (Kearney et al., 1994; Cochran et al., 1998b) and ^{14}C. Lead-210 provides chronologies over the past 100 years, an appropriate time scale because it matches the progressive industrialization of the 19th and 20th centuries and concurrent release of contaminants to the coastal marine environment (Figs 5 and 6). Radiocarbon is useful in extending salt marsh chronologies to the longer term (e.g. Allen & Rae, 1988; Varekamp et al., 1992). Unlike subtidal sediments, radionuclide profiles in salt marsh deposits are often less disturbed by mixing. The dense root structure of the marsh grasses tends to inhibit the turnover of surficial sediments by particle mixing that is common in subtidal sediments. Beryllium-7 and/or thorium-234 can provide further assessment of the presence or absence of mixing. If sampling sites are selected well away from drainage channels and without evidence of burrowing by crabs, a reliable accretion history of the marsh is likely to be preserved.

An interesting aspect of the application of ^{210}Pb to many wetland sites, particularly those poised near mean high tide, is that this radionuclide is supplied dominantly by atmospheric deposition. The atmospheric flux of ^{210}Pb to the Earth's surface is approximately constant with time at a given location (Turekian et al., 1977) and thus a model incorporating this fact can be used to determine ages of given depth horizons in a marsh core (Appleby & Oldfield, 1978). This 'constant flux' or 'constant rate of ^{210}Pb supply' model permits determination of variation in salt marsh accretion rate over time.

The accretion rate of a wetland, compared with the rate of sea level increase, is a simple measure of the health of the system. Wetlands that remain viable will, over the long term, keep pace with the sea level rise. Wetlands accreting faster than the sea level rise, if this is sustained for long periods of time, will be in transition from low marsh to high marsh or from high marsh to upland soil. The chronology of accretion is also useful in determining the record of contaminant fluxes to the marsh surface (see below).

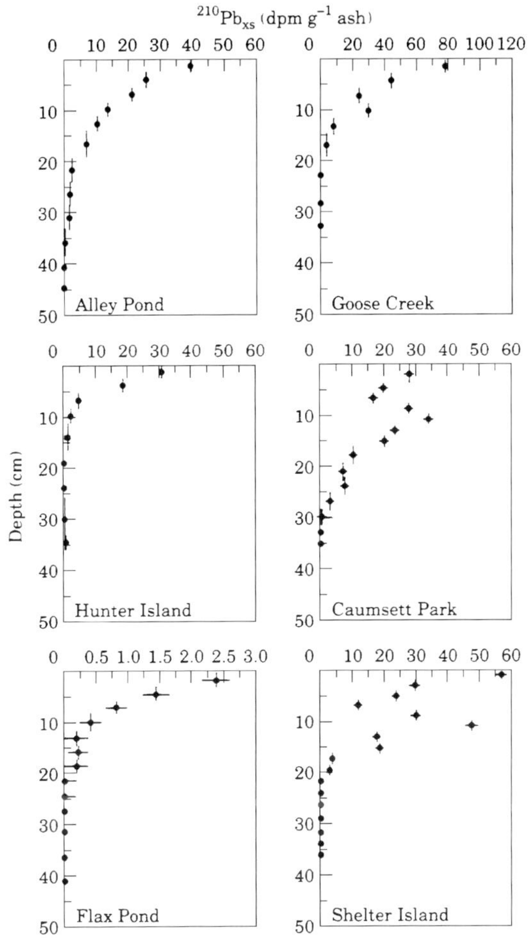

Fig. 5. Excess ^{210}Pb activity profiles versus depth in salt marsh cores from New York City and Long Island. Marsh accretion rates calculated with the constant flux model range from 0.1 cm y^{-1} to 0.4 cm y^{-1}. (From Cochran et al., 1998b, with permission from Elsevier.)

4.2. *Contaminant chronologies of coastal wetlands*

Wetlands act as filters for contaminants, and as a consequence the sediments record inputs of particle-associated organic and inorganic contaminants over time. Certain wetlands (e.g. high salt marshes) dominantly receive their burden of contaminants (as well as ^{210}Pb; see above) from the atmosphere, so the wetland sediments record the atmospheric deposition of contaminants to coastal waters. Knowing the relative fraction of contaminant input from the atmosphere is important in deciding on alternative management strategies for contaminant inputs. Thus contaminant chronologies in wetland deposits serve both a monitoring function (are inputs increasing or decreasing?) and a source indicator (what fraction of the contaminant burden in adjacent subtidal coastal sediments is atmospheric?).

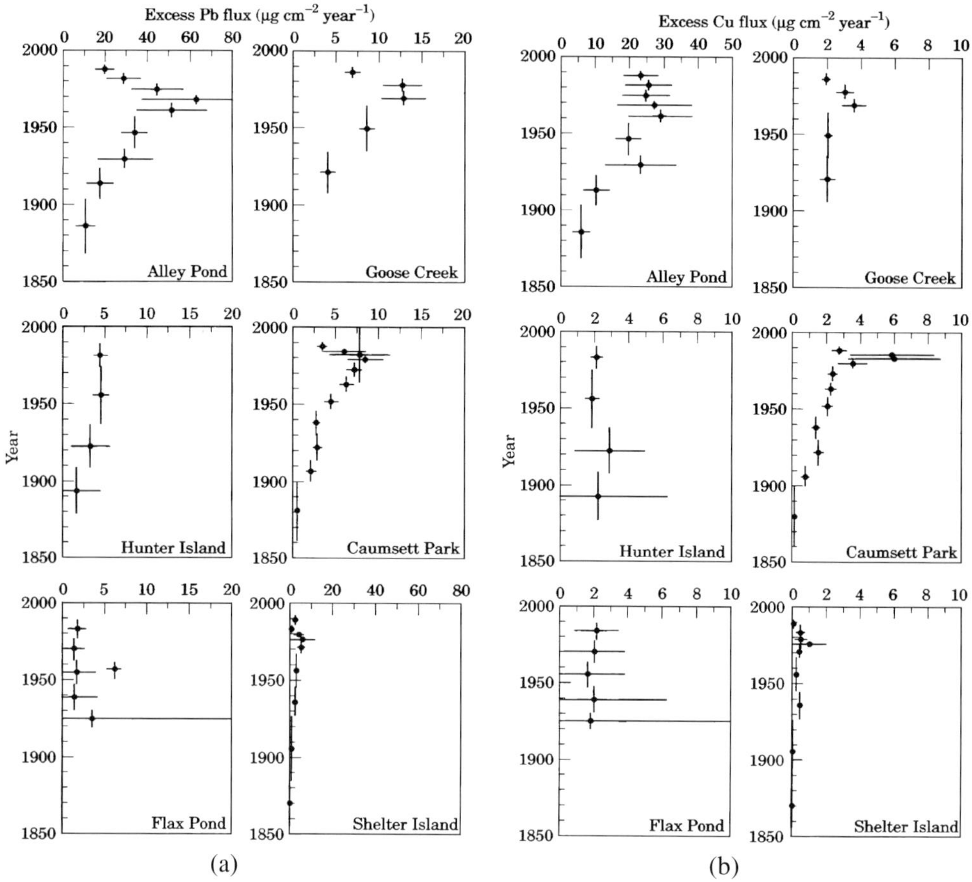

Fig. 6. Excess metal fluxes to salt marsh sites in New York, based on ^{210}Pb chronologies (see Fig. 5). (Excess metal concentrations are calculated by subtracting pre-industrial background values at depth in the cores from total concentrations.) (a) Pb, (b) Cu, (c) Zn and (d) ^{137}Cs. (From Cochran et al., 1998b, with permission from Elsevier.)

Determining contaminant chronologies in salt marsh deposits requires a valid chronology of accretion as outlined above, as well as measurements of organic and inorganic contaminant concentrations with depth in the core. Contaminant fluxes are derived by multiplying the accretion rate by the contaminant concentration (Fig. 6). The patterns of these fluxes over time often show sensible variations relative to the known inputs to the coastal environment.

A related issue is determination of the relative fraction of contaminants supplied to coastal waters by the atmosphere. One must first verify that the wetland receives its burden of contaminants from the atmosphere. This may be done by comparing the ^{210}Pb inventories in the wetland core with those expected from atmospheric deposition. If direct measurements of atmospheric fluxes for the contaminants are available, they may also be directly compared with those obtained from the wetland cores. In any case, inventories of contaminants can be normalized to ^{210}Pb and compared with similar values in subtidal sediments. This approach

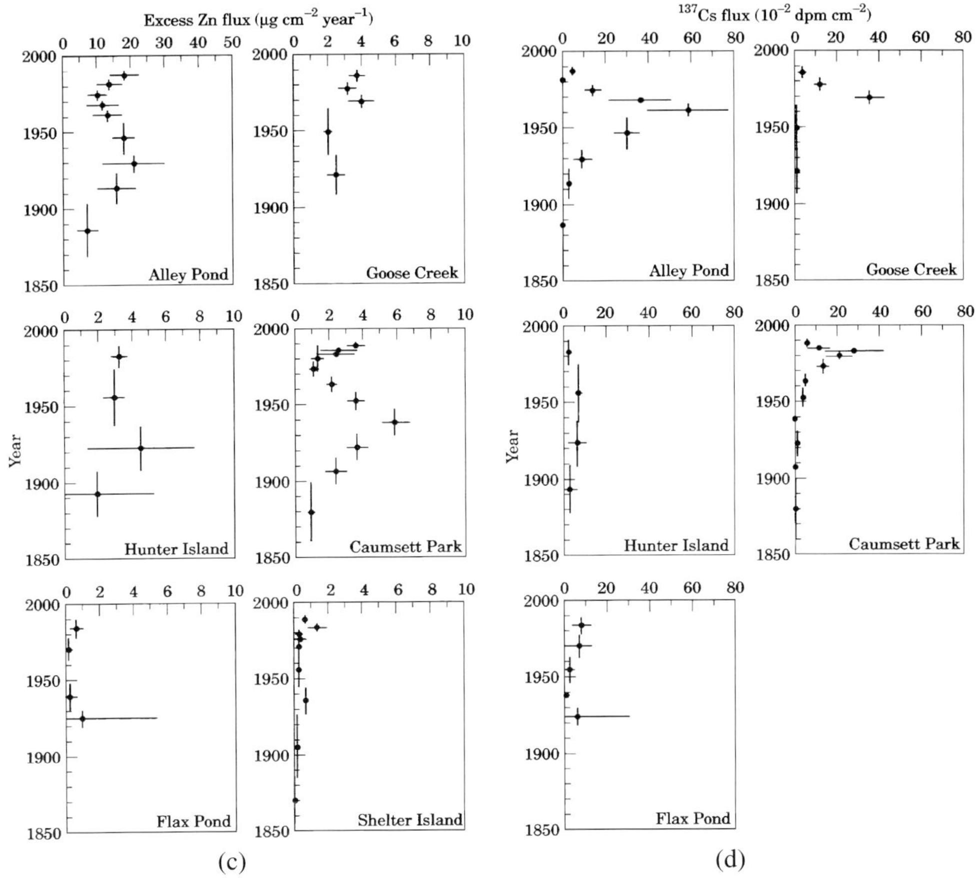

Fig. 6. (Continued.)

has been applied in Connecticut and New York, USA, and in the lagoon of Venice, Italy (McCaffrey & Thomson, 1980; Turekian et al., 1980; Cochran et al., 1998a, b). In these sites the atmospheric input of Pb was a significant, indeed dominant, pathway by which Pb was added to the coastal zone.

5. Eutrophication and blooms

5.1. *Submarine groundwater discharge as a potential source of nutrients to coastal areas*

As coastal areas develop, nutrient inputs to coastal waters have increased. This eutrophication has caused enhanced coastal production, including nuisance and harmful algal blooms (HABs). HABs can have serious consequences for human health if individuals consume shellfish that are contaminated with toxins from the algae. Economic consequences also may result from declines in fisheries following harmful algal blooms, either related to contamination of shellfish which cannot then be sold for consumption or negative impacts on fisheries linked to decreases in water quality (e.g. low dissolved oxygen caused by decomposition of algae).

Nutrient loading to coastal waters occurs from the watershed via rivers as well as from point source wastewater inputs. Increasingly, inputs from submarine groundwater discharge (SGD) and from the atmosphere are being perceived as important in the nutrient balance of coastal areas. For example, groundwater inputs of nitrate have been shown to be important in the coastal waters of New York, Massachusetts, Australia, Jamaica and other locations (IAEA, 2001). These pathways may also add other contaminants, including metals, organic compounds and anthropogenic radionuclides to the coastal zone. Natural radionuclides can help determine the magnitude of groundwater fluxes to the coastal zone and thus evaluate fluxes of nutrients and contaminants associated with these inputs. Radon-222 and the Ra isotopes (226, 228, 224, 223) are especially good tracers of SGD because they are mobilized in sediments or aquifers and are not strongly particle reactive once in the water column.

The input of radon (^{222}Rn) and radium isotopes (e.g. ^{226}Ra, ^{228}Ra, ^{224}Ra) with SGD has been recognized for about 20 years, usually as anomalously high activities of these radionuclides in coastal waters (Fanning et al., 1981). Radon and the radium isotopes are produced from parent radionuclides in the mineral structures of sediments, enter sediment pore waters through recoil and can then migrate out of the sediments. Mass balance calculations in many settings show that diffusion alone is insufficient to cause the elevated activities in the water column, implicating flow through the sediments and input via SGD. Moore (1996) noted the significant enrichments of ^{226}Ra in coastal waters of the South Atlantic Bight, USA, and ascribed them to SGD. Indeed Moore (1999) subsequently coined the term 'subterranean estuary' to connote the increases in dissolved Ra isotopes as a consequence of mixing of salt water and fresh water in coastal aquifers. Similar enrichments of ^{222}Rn have been linked to SGD (Fanning et al., 1981; Cable et al., 1996).

Several approaches are possible in determining rates of SGD using radon and radium. Evaluating radionuclide mass balances is one method that has been applied to ^{222}Rn (e.g. Corbett et al., 2000). Radon-222 inventories in coastal water columns can be determined and compared with the losses to the atmosphere, open ocean and radioactive decay to determine the input from the sediments. This can be transformed into a rate of SGD if the radon activity of the pore water is known.

Application of the suite of Ra isotopes to estimate SGD relies on the establishment of gradients as the isotopes enter the overlying water and are mixed away from the coast. The multiple half-lives permit eddy diffusive mixing coefficients to be determined (for example, using ^{223}Ra and ^{224}Ra; Moore, 2000) and then applied to the long-lived ^{226}Ra to estimate the offshore flux of this isotope. This flux must be balanced by input of Ra from SGD, and knowing the ^{226}Ra activity in groundwater permits estimation of the SGD flux.

5.2. *Reconstructing prior history of blooms*

Nuisance and harmful algal blooms (HABs) are increasingly common in coastal waters. Such blooms have impacted human health and caused economic losses in fisheries around the world, including the USA, Europe, Korea, Japan, Chile and elsewhere. In many cases it is

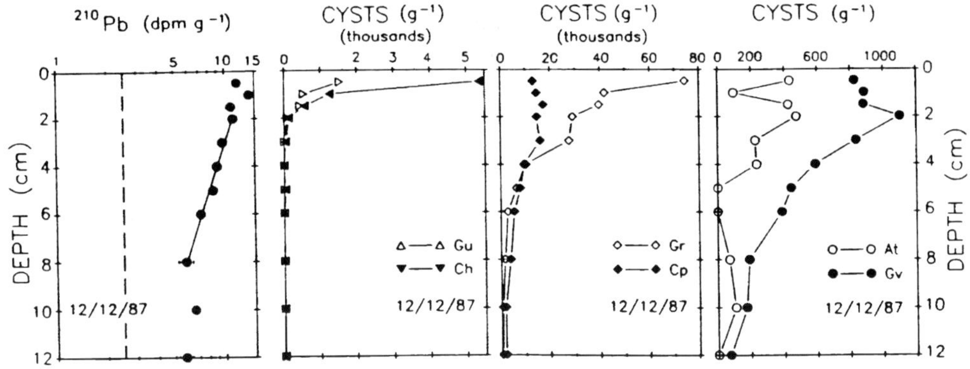

Fig. 7. Lead-210 and cyst profiles for Perch Pond (Cape Cod). For ^{210}Pb, the dashed and the solid lines indicate supported ^{210}Pb levels and regression analysis for the determination of the sedimentation rate (0.29 cm y^{-1}), respectively. The surface mixed layer is confined to the upper ~2 cm. Gu: *Gyrondinium uncatenum*, Ch: *Cochlodinium heterolobatum*, Gr: *Gonyaulax rugosum*, Cp: unknown *Scripsiella* sp., At: *Alexandrium tamarense*, Gv: *Gonyaulax verior*. (From Keafer et al., 1992, with permission from Elsevier.)

unclear whether bloom organisms have been present historically at low concentrations in the environment and their growth has been stimulated by changing nutrient balances and ecosystem dynamics, or whether they have recently been introduced into the system. Moreover, many species that are important in HABs have a dormant stage involving cyst formation. Cysts can settle out of the water column and be deposited at the sediment–water interface. In sediments, the cysts can be buried by the processes of sediment accumulation and mixing by the benthic macrofauna (Keafer et al., 1992). Their subsequent fate and importance in seeding new blooms depends on their ability to remain viable in an environment that is commonly devoid of oxygen. Thus there is interest in characterizing the down-core distribution of cysts with respect to the sediment chronology.

Reconstruction of the history of HABs involves collection of sediment cores and the determination of sediment chronologies and cyst abundance with depth in the sediment. However, as noted above, the presence of particle mixing by infauna often compromises the ability to extract a deposition chronology from a core. Nevertheless, an understanding of the dynamics of mixing and accumulation vis-à-vis cyst abundance with depth in the sediments provides valuable information of how cysts are maintained in the sediments and on their ability to seed new blooms. In a sheltered salt pond environment on Cape Cod, for example, Keafer et al. (1992) found that the mixing was confined to the upper 2 cm, a zone from which cysts could trigger blooms in the overlying water column. Once deposited deeper than 2 cm, however, the cysts were isolated from the overlying water (Fig. 7).

5.3. *Nutrient cycling and turnover*

The rate of nutrient cycling in the upper water column is an important parameter that is affected by the input of 'new' nutrients, primary and secondary production, and the decomposition and export of organic matter. The role of phosphorus in limiting the growth and distribution of marine phytoplankton is not well known. The cosmogenic P isotopes ^{32}P (half-life

14.3 d) and ^{33}P (half-life 25.3 d) are useful in determining turnover in the various P reservoirs. Lal et al. (1988) and Lal & Lee (1988) suggested that if the inputs of ^{32}P and ^{33}P to the surface ocean from the atmosphere and their activities in the various P reservoirs in the upper ocean are known, it is possible to calculate rates of P uptake and turnover, averaged over time scales of weeks. In particular, the $^{33}P/^{32}P$ ratio in the various P reservoirs compared with the input ratio from the atmosphere provides a means of aging the various P reservoirs, assuming that mass-dependent isotopic fractionation is negligible (Waser et al., 1996). Basic models to estimate P turnover times rely on the balance between the gain of ^{32}P or ^{33}P through ingestion and loss by decay, excretion and formation of large sinking particles. Such models can assume either that plankton ingest P throughout their life (continuous uptake model) or only early in their life (age model) (Waser et al., 1996).

Attempts to use the $^{33}P/^{32}P$ ratio as a chronometer for P turnover have begun by measuring the ratio in rain. Although the individual activities (per liter of rain) displayed over an order of magnitude variation (Lal et al., 1957, 1960; Goel et al., 1959; Lal & Peters, 1967; Waser et al., 1994; Waser & Bacon, 1995; Benitez-Nelson & Buesseler, 1999a), the $^{33}P/^{32}P$ ratio varies from $\sim$0.6–1.6, with an annual mean of $\sim$0.9. Benitez-Nelson & Buesseler (1999b) noted that the $^{33}P/^{32}P$ ratio in the dissolved inorganic P and small plankton (i.e. picoplankton) was similar to the ratio of rain in the Gulf of Maine, suggesting rapid turnover of the P. In contrast, values in plankton are generally greater than 1 (Lal & Lee, 1988; Waser et al., 1996; Benitez-Nelson & Buesseler, 1999b), with increases in higher trophic levels (i.e zooplankton; Waser et al., 1996). The results suggest longer P turnover times in higher trophic levels, ranging from several weeks for phytoplankton to several months for zooplankton. Benitez-Nelson & Buesseler (1999b) also showed that dissolved P recycling rates varied significantly from season to season. These authors stated that the rapid cycling of soluble reactive phosphorus indicated that low P concentrations could support high levels of primary production, while bacteria remineralize a small fraction of soluble non-reactive phosphorus. Indeed this process may provide a new source of inorganic nutrients to phytoplankton, and thus have implications for nutrient limitation.

6. Fisheries: Ages of size classes of fish

Effective fisheries management depends upon knowledge of the composition of fish stocks with respect to age, or in other words the relationship between fish size class and age. Indeed the age structure of a population often provides the first evidence of overfishing and may imply the need to impose quotas on harvests (Jennings et al., 2001). Although a number of approaches are used to determine the age of fish, a common method is through examination of the calcium carbonate otoliths. Typically otoliths are removed from a fish specimen and sectioned. Growth increments preserved in the otolith are then used to determine age. The relationship between number of growth increments and age may be known; often, however, the relationship is assumed, and this assumption can lead to error and, thereby, mismanagement. For accurate age assessments, age validation methods are required. Natural radionuclides provide a useful means of validating the relationship between fish age and otolith increment analysis.

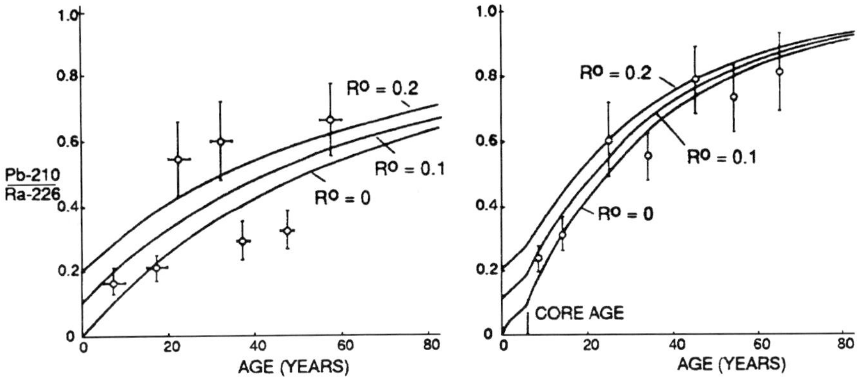

(a) (b)

Fig. 8. Ingrowth curves for the $^{210}Pb/^{226}Ra$ activity ratio as a function of fish age for linear growth models. Experimental data is for whole otoliths from different age classes or redfish (a) and sets of otolith cores from redfish in given age classes (b). Values of R^0 are initial $^{210}Pb/^{226}Ra$ activity ratios used to construct the growth curves. (From Smith et al., 1991, with permission from Kluwer Academic Publishers.)

As otoliths form they incorporate natural radionuclides from the seawater in which the fish grows. ^{210}Pb and the Ra isotopes (^{226}Ra and ^{228}Ra) are especially able to substitute for Ca in the $CaCO_3$ of which the otolith is composed. Several approaches have been taken to extract chronologies from the otoliths (Bennett et al., 1982; Fenton et al., 1990; Campana et al., 1990; Smith et al., 1991; Campana et al., 1993). Individual otoliths, if large enough, can be sectioned and clusters of growth increments analyzed, or multiple otoliths from a given size class can be analyzed together. In the latter case, a growth model for the otoliths must be assumed or determined. It is recommended, however, to restrict the assay to the otolith core, as it reflects elapsed time since core formation, which in turn is very similar to the age of the fish (Campana, 2001, and references therein; Fig. 8).

In both cases, the goal is to follow the decay of ^{210}Pb or ^{228}Ra as a function of time. The useful time range of these chronometers is about five half-lives, or 100 years for ^{210}Pb and 30 years for ^{228}Ra. Each radionuclide also produces through ingrowth a daughter that is in turn radioactive, and which is itself not incorporated into the otolith during growth. Thus the ingrowth of ^{210}Po and ^{228}Th from decay of ^{210}Pb and ^{228}Ra, respectively, can be followed (Fig. 8). Radioactive equilibrium is attained in these parent–daughter pairs in about 2 years and 10 years, respectively. With this suite of radionuclides, chronologies can be developed for fish growth over time scales of <2 years to ~ 100 years.

The accurate determination of fish age is a critical component for determining whether overfishing is occurring and thus leads to more effective management of a fishery. An example of a fishery in which accurate knowledge of growth could have been useful is the orange roughy fishery on the continental slope of New Zealand. This fish is long-lived (perhaps to 150 years), making it an appropriate target for the ^{210}Pb method described above. Lack of knowledge of the rate of growth and age at sexual maturity of the orange roughy lead to overfishing and a decline in the population in just a few years (GESAMP, 2001). The application of radionuclides to calibration of the relationship between growth increments and time in fish otoliths is perhaps best applied to deep-water fisheries, where such information is often lacking and direct observations of growth are difficult. However these fisheries are often considered coastal zone resources.

Acknowledgments

The authors' involvement in some of the research described in this paper has been supported by the U.S. National Science Foundation, the National Oceanic and Atmospheric Administration and its Sea Grant Program, the U.S. Dept. of Energy, the Hudson River Foundation, the Ministerio de Educación y Cutura of Spain and the European Union. Financial support to PM through a postdoctoral fellowship from the Government of Spain and the Fulbright Commission is gratefully acknowledged.

References

Allen, J. R. L. & Rae, J. E. (1988). Vertical salt-marsh accretion since the Roman period in the Severn Estuary, Southwest Britain. *Marine Geology*, *83*, 225–235.

Aller, R. C. & Cochran, J. K. (1976). Th-234/U-238 disequilibrium in near-shore sediment: Particle reworking and diagenetic time scales. *Earth and Planetary Science Letters*, *29*, 37–50.

Aller, R. C., Benninger, L. J. & Cochran, J. K. (1980). Tracking particle associated processes in nearshore environments by use of ^{234}Th/^{238}U disequilibrium. *Earth and Planetary Science Letters*, *47*, 161–175.

Appleby, P. G. & Oldfield, F. (1978). The calculation of Pb-210 dates assuming a constant rate of supply of unsupported Pb-210 to the sediment. *Catena*, *5*, 1–8.

Armentano, T. V. & Woodwell, G. M. (1975). Sedimentation rates in a Long Island salt marsh determined by Pb-210 dating. *Limnology and Oceanography*, *20*, 452–456.

Benitez-Nelson, C. R. & Buesseler, K. O. (1999a). Phosphorus 32, phosphorus 33, beryllium 7 and lead 210: Atmospheric fluxes and utility in tracing stratosphere/troposphere exchange. *Journal of Geophysical Research*, *104*, 11745–11754.

Benitez-Nelson, C. R. & Buesseler, K. O. (1999b). Variability of inorganic and organic phosphorus turnover rates in the coastal ocean. *Nature*, *398*, 502–505.

Bennett, J. T., Boehlert, G. W. & Turekian, K. K. (1982). Confirmation of longevity in *Sebastes dipoproa* from Pb-210/Ra-226 measurements in otoliths. *Marine Biology*, *71*, 209–215.

Benninger, L. K., Aller, R. C., Cochran, J. K. & Turekian, K. K. (1979). Effects of biological sediment mixing on the ^{210}Pb chronology and trace metal distribution in a Long Island Sound sediment core. *Earth and Planetary Science Letters*, *43*, 241–259.

Benoit, G. K., Turekian, K. K. & Benninger, L. K. (1979). Radiocarbon dating of a core from Long Island Sound. *Estuarine and Coastal Marine Science*, *9*, 171–180.

Cable, J. E., Burnett, W. C., Chanton, J. P. & Weatherly, G. L. (1996). Estimating groundwater discharge into the northeastern Gulf of Mexico using Radon-222. *Earth and Planetary Science Letters*, *144*, 691–604.

Campana, S. E. (2001). Accuracy, precision and quality control in age determination, including a review of the use and abuse of age validation methods. *Journal of Fish Biology*, *59*, 197–242.

Campana, S. E., Oxenford, H. A. & Smith, J. N. (1993). Radiochemical determination of longevity in flying fish *Hirundichthys affinis* using Th-228/Ra-228. *Marine Ecology Progress Series*, *100*, 211–219.

Campana, S. E., Zwanenburg, C. T. & Smith, J. N. (1990). Pb-210/Ra-226 determination of longevity in redfish. *Canadian Journal of Fisheries and Aquatic Sciences*, *47*, 163–165.

Canuel, E. A., Martens, C. S. & Benninger, L. K. (1990). Seasonal-variations in Be-7 activity in the sediments of Cape Lookout Bight, North Carolina. *Geochimica et Cosmochimica Acta*, *54*, 237–245.

Cochran, J. K., Frignani, M., Hirschberg, D. J. & Barnes, C. (1995). Thorium isotopes as indicators of scavenging rates in the Venice Lagoon. *Marine and Freshwater Research*, *46*, 215–221.

Cochran, J. K., Frignani, M., Salamanca, M., Bellucci, L. G. & Guerzoni, S. (1998a). Lead-210 as a tracer of atmospheric input of heavy metals in the northern Venice Lagoon. *Marine Chemistry*, *62*, 15–29.

Cochran, J. K., Hirschberg, D. J., Wang, J. & Dere, C. (1998b). Atmospheric deposition of metals to coastal waters (Long Island Sound, New York, USA): Evidence from salt marsh deposits. *Estuarine and Coastal Shelf Science*, *46*, 503–522.

Corbett, D. R., Dillon, K., Burnett, W. & Chanton, J. (2000). Estimating the groundwater contribution into Florida Bay via natural tracers, Rn-222 and CH_4. *Limnology and Oceanography*, *45*, 1546–1557.

Fanning, K., Byrne, R. H., Breland, J. A., Betzer, P. R., Moore, W. S. & Elsinger, R. J. (1981). Geothermal springs of the west Florida continental Shelf: Evidence of dolomitization and radionuclide enrichment. *Earth and Planetary Science Letters*, *52*, 343–354.

Feng, H., Cochran, J. K., Hirschberg, D. J. & Wilson, R. E. (1998). Small-scale spatial variations of natural radionuclide and trace metal distributions in sediments from the Hudson River Estuary. *Estuaries*, *21*, 263–280.

Feng, H., Cochran, J. K. & Hirschberg, D. J. (1999a). ^{234}Th and ^{7}Be as tracers for transport and sources of particle-associated contaminants in the Hudson River Estuary. *Science of the Total Environment*, *237/238*, 401–418.

Feng, H., Cochran, J. K. & Hirschberg, D. J. (1999b). ^{234}Th and ^{7}Be as tracers for the transport and dynamics of suspended particles in a partially mixed estuary. *Geochimica et Cosmochimica Acta*, *63*, 2487–2505.

Feng, H., Cochran, J. K. & Hirschberg, D. J. (1999c). ^{234}Th and ^{7}Be as tracers for the sources of particles to the turbidity maximum of the Hudson River Estuary. *Estuarine and Coastal Shelf Science*, *49*, 629–645.

Fenton, G. E., Ritz, D. A. & Short, S. A. (1990). Pb-210/Ra-226 disequilibria in otoliths of Blue Grenadier, *Macruronus novaezelandiae*: Problems associated with radiometric ageing. *Australian Journal of Marine and Freshwater Research*, *41*, 467–473.

GESAMP (2001). A sea of troubles (35 pp.). IMO/FAO/UNESCO–IOC/WMO/WHO/IAEA/UN/UNEP Joint Group of Experts on the Scientific Aspects Maine Environmental Protection. GESAMP Report No. 70.

Goel, P. S., Narasappaya, N., Prabhakara, C., Rama Thor & Zutshi, P. K. (1959). Study of cosmic ray produced short-lived ^{32}P, ^{33}P, ^{7}Be, and ^{35}S in tropical latitudes. *Tellus*, *11*, 91–100.

Goldberg, E. D., Gamble, E., Griffin, J. J. & Koide, M. (1977). Pollution history of Narragansett Bay as recorded in its sediments. *Estuarine and Coastal Shelf Science*, *5*, 549–556.

Goldberg, E. D., Hodge, V., Koide, M., Griffin, J., Gamble, E., Bricker, O. P., Matisoff, G., Holden, G. R. & Braun, R. (1978). A pollution history of Chesapeake Bay. *Geochimica et Cosmochimica Acta*, *42*, 1413–1425.

Gustafsson, Ö., Buesseler, K. O., Geyer, W. R., Moran, S. B. & Gschwend, P. M. (1998). On the relative significance of horizontal and vertical transport of chemicals in the coastal ocean: Application of a two-dimensional Th-234 cycling model. *Continental Shelf Research*, *18*, 805–829.

Hirschberg, D. J., Chin, P., Feng, H. & Cochran, J. K. (1996). Dynamics of sediment and contaminant transport in the Hudson River Estuary: Evidence from sediment distributions of naturally occurring radionuclides. *Estuaries*, *19*, 931–949.

Honeyman, B. D. & Santschi, P. H. (1989). The role of particles and colloids in the transport of radionuclides and trace metals in the oceans. In J. Buffle & H. P. van Leewen (Eds), *Environmental Particles* (pp. 370–423). Boca Raton: Lewis Publishers.

IAEA (2001). Report of the IAEA Advisory Group Meeting on "The application of isotope methods for water resources assessment and management of coastal zones and small islands" (43 pp.). Draft Report. Vienna: IAEA.

Jennings, S., Kaiser, M. J. & Reynolds, J. D. (2001). *Marine Fisheries Ecology*. London: Blackwell Science Ltd.

Kaufman, A., Li, Y.-H. & Turekian, K. K. (1981). The removal rates of ^{234}Th and ^{228}Th from waters of the New York Bight. *Earth and Planetary Sciences Letters*, *54*, 385–392.

Keafer, B. A., Buesseler, K. O. & Anderson, D. M. (1992). Burial of living dinoflagellate cysts in estuarine and nearshore sediments. *Marine Micropaleontology*, *20*, 147–161.

Kearney, M. S., Stevenson, J. C. & Ward, L. G. (1994). Spatial and temporal changes in marsh vertical accretion rates at Monie Bay: Implications for sea-level rise. *Journal of Coastal Research*, *10*, 1010–1020.

Koide, M., Bruland, K. W. & Goldberg, E. D. (1973). Th-228, Th-232 and Pb-210 geochronologies in marine and lake sediments. *Geochimica et Cosmochimica Acta*, *37*, 1171–1187.

Krishnaswami, S., Lal, D., Somayajulu, B. L. K., Weiss, R. & Craig, H. (1976). Large-volume in-situ filtration of deep Pacific waters: Mineralogical and radioisotope studies. *Earth and Planetary Sciences Letters*, *32*, 420–429.

Krishnaswami, S., Benninger, L. K., Aller, R. C. & Von Damm, K. L. (1980). Atmospherically-derived radionuclides as tracers of sediment mixing and accumulation in near-shore marine and lake-sediments – evidence from Be-7, Pb-210, and Pu-239, Pu-240. *Earth and Planetary Science Letters*, *47*, 307–318.

Lal, D. & Lee, T. (1988). Cosmogenic ^{32}P and ^{33}P used as tracers to study phosphorus recycling in the upper oceans. *Nature, 333*, 752–754.

Lal, D. & Peters, B. (1967). Cosmic ray produced radioactivity on the Earth. In K. Sitte (Ed.), *Handbuch der Physik*, Vol. 46/2 (pp. 551–612). New York: Springer-Verlag.

Lal, D., Narasappaya, N. & Zutshi, P. K. (1957). Phosphorus isotopes ^{32}P and ^{33}P in rain water. *Nuclear Physics, 3*, 69–75.

Lal, D., Rama, T. & Zutshi, P. K. (1960). Radioisotopes ^{32}P, ^{7}Be and ^{35}S in the atmosphere. *Journal of Geophysical Research, 65*, 669–673.

Lal, D., Chung, Y., Platt, T. & Lee, T. (1988). Twin cosmogenic radiotracer studies of phophorous recycling and chemical fluxes in the upper ocean. *Limnology and Oceanography, 33*, 1559–1567.

Matsumoto, E. (1975). Th-234-U-238 radioactive disequilibrium in the surface layer of the oceans. *Geochimica et Cosmochimica Acta, 39*, 205–212.

McCaffrey, R. J. & Thomson, J. (1980). A record of the accumulation of sediment and trace metals in a Connecticut salt marsh. In B. Saltzman (Ed.), *Estuarine Physics and Chemistry: Studies in Long Island Sound. Advances in Geophysics*, Vol. 22 (pp. 129–164). New York: Academic Press.

McKee, B. A., DeMaster, D. J. & Nittrouer, C. A. (1984). The use of Th-234/U-238 disequilibrium to examine the fate of particle-reactive species on the Yangtze continental shelf. *Earth and Planetary Sciences Letters, 68*, 431–442.

McKee, B. A., DeMaster, D. J. & Nittrouer, C. A. (1987). Uranium geochemistry on the Amazon shelf – evidence for uranium release from bottom sediments. *Geochimica et Cosmochimica Acta, 51*, 2779–2786.

Moore, W. S. (1992). Isotopes of the U and Th decay series in the estuarine environment. In M. Ivanovich & R. Harmon (Eds), *Uranium Series Disequilibrium: Applications to Earth, Marine and Environmental Sciences* (pp. 396–422). Oxford: Clarendon.

Moore, W. S. (1996). Large groundwater inputs to coastal waters revealed by ^{226}Ra enrichments. *Nature, 380*, 612–614.

Moore, W. S. (1999). The subterranean estuary: A reaction zone of groundwater and sea water. *Marine Chemistry, 65*, 111–125.

Moore, W. S. (2000). Determining coastal mixing rates using radium isotopes. *Continental Shelf Research, 20*, 1993–2007.

Moore, W. S. & Shaw, T. J. (1998). Chemical signals from submarine fluid advection onto the continental shelf. *Journal of Geophysical Research, 103*, 21543–21552.

Quigley, M. S., Santschi, P. H., Hung, C.-C., Guo, L. & Honeyman, B. D. (2002). Importance of acid polysaccharides for ^{234}Th complexation to marine organic matter. *Limnology and Oceanography, 47*, 367–377.

Santschi, P. H., Li, Y. H. & Bell, J. (1979). Natural radionuclides in the water of Narragansett Bay. *Earth and Planetary Science Letters, 45*, 201–213.

Sañudo-Wilhelmy, S. A. & Flegal, A. R. (1992). Anthropogenic silver in the Southern California Bight: A new tracer of sewage in coastal waters. *Environmental Science and Technology, 26*, 2147–2151.

Smith, J. N. (2001). Why should we believe ^{210}Pb sediment geochronologies? *Journal of Environmental Radioactivity, 55*, 121–123.

Smith, J. N., Nelson, R. & Campana, S. E. (1991). The use of Pb-210/Ra-226 and Th-228/Ra-228 disequilibria in the ageing of otoliths of marine fish. In P. J. Kershaw & D. S. Woohead (Eds), *Radionuclides in the Study of Marine Processes* (pp. 350–359). New York: Elsevier.

Sommerfield, C. K., Nittrouer, C. A. & Alexander, C. R. (1999). Be-7 as a tracer of flood sedimentation on the northern California continental margin. *Continental Shelf Research, 19*, 335–361.

Thomson, J., Turekian, K. K. & McCaffrey, R. J. (1975). The accumulation of metals in and release from sediments of Long Island Sound. In L. E. Cronin (Ed.), *Estuarine Research*, Vol. 1 (pp. 28–44). New York: Academic Press.

Toole, J., Baxter, M. S. & Thomson, J. (1987). The behavior of uranium isotopes with salinity change in 3 UK estuaries. *Estuarine, Coastal and Shelf Science, 25*, 283–297.

Turekian, K. K., Nozaki, Y. & Benninger, L. K. (1977). Geochemistry of atmospheric radon and radon products. *Annual Review Earth Planetary Sciences, 5*, 227–255.

Turekian, K. K., Cochran, J. K., Benninger, L. K. & Aller, R. C. (1980). The sources and sinks of nuclides in Long Island Sound. In B. Saltzman (Ed.), *Estuarine Physics and Chemistry: Studies in Long Island Sound. Advances in Geophysics*, Vol. 22 (pp. 129–164). New York: Academic Press.

Valette-Silver, N. (1993). The use of sediment cores to reconstruct historical trends in contamination of estuarine and coastal sediments. *Estuaries, 16*, 577–588.

Varekamp, J. C., Thomas, E. & Vandeplassche, O. (1992). Relative sea-level rise and climate change over the last 1500 years. *Terra Nova, 4*, 293–304.

Waser, N. A. D. & Bacon, M. P. (1995). Wet deposition fluxes of cosmogenic ^{32}P and ^{33}P and variations in the $^{33}P/^{32}P$ ratios at Bermuda. *Earth and Planetary Science Letters, 133*, 71–80.

Waser, N. A. D., Fleer, A. P., Hammar, T. R., Buesseler, K. O. & Bacon, M. P. (1994). Determination of natural ^{32}P and ^{33}P in rainwater, marine particles and plankton by low-level beta counting. *Nuclear Instruments and Methods in Physics Research A, 338*, 560–567.

Waser, N. A. D., Bacon, M. P. & Michaels, A. F. (1996). Natural activities of ^{32}P and ^{33}P and the $^{33}P/^{32}P$ ratio in suspended particulate matter and plankton in the Sargasso Sea. *Deep-Sea Research II, 43*, 421–436.

Woodruff, J. D., Geyer, W. R., Sommerfield, C. K. & Driscoll, N. W. (2001). Seasonal variation of sediment deposition in the Hudson River Estuary. *Marine Geology, 179*, 105–119.

MARINE RADIOACTIVITY
Hugh D. Livingston (Editor)

Chapter 2

Linking legacies of the Cold War to arrival of anthropogenic radionuclides in the oceans through the 20th century

T. F. Hamilton

Environmental Science Division, Lawrence Livermore National Laboratory, Livermore, CA 94551-0808, USA

List of acronyms

AMAP	Arctic Monitoring and Assessment Program
AEC	Atomic Energy Commission
CRESP	Coordinated Research and Environmental Surveillance Programme
DOE	Department of Energy
DU	Depleted Uranium
EARP	Enhanced Actinide Removal Plant
ERDA	Energy Research and Development Administration
EURT	East Urals Radioactive Trace
GEOSECS	Geochemical Ocean Section
HEU	Highly Enriched Uranium
HLW	High-Level Waste
IAEA	International Atomic Energy Agency
IASAP	International Arctic Sea Assessment Project
LDC	London Dumping Convention
LEU	Low Enriched Uranium
MED	Manhattan Engineer District or 'Manhattan Project'
MPA	MAYAK Production Association
MRI	Meteorological Research Institute
NAS	National Academy of Sciences
NDRC	National Defense Research Committee
NEA	Nuclear Energy Agency
NTS	Nevada Test Site
OECD	United Nations Organization for Economic Co-operation and Development

OSRD	Office of Scientific Research and Development
PNE	Peaceful Nuclear Explosions
RCU	Radiocarbon Unit
PUREX	Plutonium Uranium Extraction
REDOX	Phosphate Precipitation, Reduction, Oxidation
SIXEP	Site Ion Exchange Effluent Plant
TR	Tritium Ratio
TRU	Transuranic
TU	Tritium Units
WIPP	Waste Isolation Pilot Plant

1. Introduction

The nuclear age has presented social, moral, political and environmental challenges that have impacted on the security and prosperity of mankind and his planet. Significant quantities of man-made radionuclides have been released into the environment as a result of atmospheric weapons testing, nuclear weapons production activities and nuclear power fuel-cycle operations. Other releases have resulted from accidents involving nuclear materials, intentional disposals, and the general use of radioactive materials in medicine, industry, research and space exploration. Societal fear of radiation, growing concerns about disposal of nuclear waste, and public opposition to nuclear power have virtually ended the development of the nuclear energy industry as a viable alternative to the burning of fossil fuels. Yet, the uncontaminated natural environment, including the marine environment, is inherently radioactive.

Knowledge about specific sources and exposure pathways of environmental radioactivity provide a scientific basis for estimating health risks and developing appropriate safety standards to regulate releases and minimize exposures. For most individuals, exposure to ionizing radiation from natural sources far exceeds that delivered by the man-made sources in the environment. The same can be said of the global marine ecosystem. The oceans cover an area of about 3.6×10^8 km^2, or about three-quarters of the surface area of the Earth, and constitute an enormous natural reservoir of radioactivity. Approximately 12,600 Bq m^3 of naturally occurring radioactivity in seawater is due largely to the presence of long-lived radionuclides of primordial origin: ^{40}K, ^{87}Rb and radionuclides arising from the decay chains of ^{238}U, ^{235}U and ^{232}Th. The total activity of the natural marine radiation environment, including the upper few meters of deep-sea sediment, exceeds 5×10^7 PBq.[1] It is also widely acknowledged that naturally occurring ^{210}Po and ^{210}Pb provide the major dose to marine organisms and, in turn, the dietary intake of ^{210}Po is a major contributor to the natural background dose for those population groups consuming large quantities of fish and shellfish.

Studies of the natural radiation environment provide us with a useful perspective for assessing the impacts of intentional or accidental releases of radionuclides into the environment. In addition to answering key questions about the total amount and potential dose contribution from man-made radionuclides entering the environment, researchers are able to predict the fate and transport of individual radionuclides from knowing the transport properties of naturally occurring stable and radioactive analogs.

[1] 1 PBq = 1×10^{15} Bq.

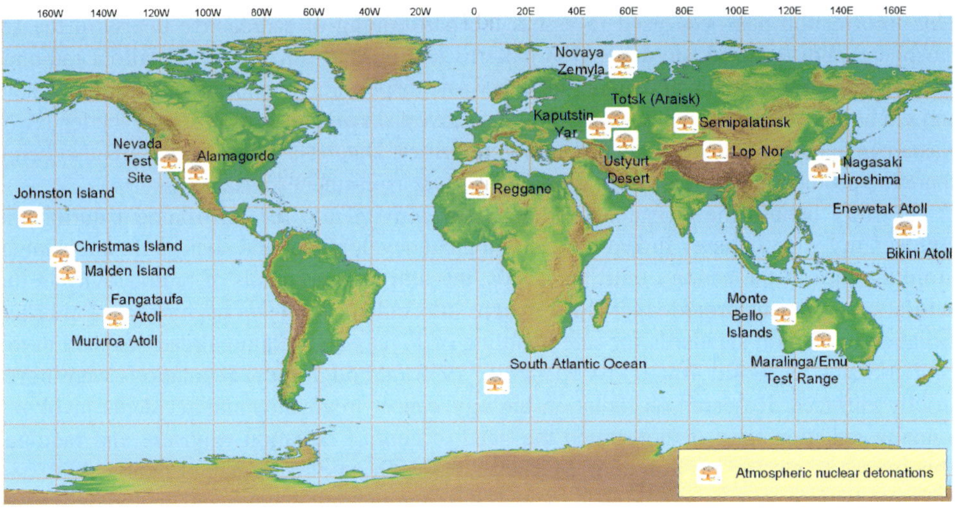

Fig. 1. Sites where atmospheric nuclear detonations have occurred.

The global dispersion and deposition of debris from atmospheric nuclear weapons is by far the largest source of artificial radioactivity released into the global environment. The location of declared sites where atmospheric nuclear detonations have taken place is shown in Figure 1. The levels and distributions of fallout radionuclides in the oceans are reasonably well described, although new information and data about accidental and deliberate disposals at sea have raised concerns about the environmental consequence of such actions on a local and regional scale. Public concern about the potential sources, levels and effects of radionuclide releases into the environment continue to grow as new information becomes available. Moreover, the guidance on radiation protection and standards associated with potential, low-level exposures compared to those delivered by the natural radiation environment are cause for scientific debate and controversy. The fact that a substantial quantity of artificial radioactivity has entered the environment over the past five decades without obvious detriment to the environment, or significantly increasing the measured collective dose to the world's population, is contrary to public perception.

International organizations such as the United Nations Committee on the Effects of Atomic Radiation continue to evaluate global radiation sources and the effects of radiation exposure as a scientific basis for estimating radiation risk (UNSCEAR, 2000). However, the behavior and transport of radionuclides through the biosphere may be influenced by the type and nature of the source, the release conditions, radionuclide speciation, aging effects and/or other environmental factors. There is also a need to develop a better understanding of the impacts of radioactive contamination on biota to test the assumptions that existing radiation protection standards and controls for humans provide adequate protection for all living organisms. Consequently, efforts are ongoing to develop a better understanding of the behavior, fate and transport of man-made radionuclides in the environment, especially in relation to source-specific assessments and localized releases. For example, new data on the total number and yields of individual atmospheric nuclear weapons tests have only recently become available.

This information has been used to assess the partitioning of radioactive debris in the atmosphere, and to reevaluate deposition patterns and doses from individual fallout radionuclides. New information and data on nuclear waste disposal and handling operations on land and sea have also served to help quantify the relative risks and hazards posed by the legacy of waste management and weapons production activities. Database compilations are also being shared to promote the development of more accurate assessment tools.

The story behind the arrival of artificial radioactivity into the environment during the 20th century is an odyssey linked by technological development and socio-political change. From the turn of the century until the 1940s, the first measurements of radioactivity in the natural environment provided an early insight into radiation effects on humans. Published radiation protection standards for safe handling of X-rays and radium already existed in the United States by World War II (NCRP, 1936, 1938) although these regulations were never strictly enforced. Research on radiation biology, cancer treatments and accidents involving ionizing radiation have demonstrated the need to control radiation exposure and radionuclide releases into the environment. Moreover, research on environmental radioactivity and radiation biology accelerated interest in studies on the potential health and ecological consequences of other pollutants in the environment. Today, studies on environmental radioactivity form an integral part of interdisciplinary research on climate change and carbon sequestration; atmospheric sciences; ocean dynamics and particle transport; archeology; and studies on the origin, fate and transport of nonradioactive pollutants in the oceans.

During the period between 1945 and 1963, environmental radioactive contamination studies were concerned primarily with the behavior of radioactive fallout from atmospheric nuclear weapons tests. The principal nuclear powers agreed to a test-ban treaty on atmospheric nuclear weapons testing in 1963, after which the annual deposition of radionuclides produced in atmospheric nuclear testing rapidly diminished. Secrecy surrounding nuclear weapons production processes and overriding public concerns about the rapid buildup of nuclear weapons arsenals tended to hide the emerging problems in nuclear waste management and environmental contamination. The production and stockpiling of nuclear materials and weapons required an extensive reprocessing effort that generated large volumes of radioactive waste. In the aftermath of the Cold War, the nuclear weapons states have begun the process of identifying legacy waste generated from nuclear weapons production activities, providing environmental restoration, and nuclear materials and facilities stabilization.

Public anxiety about nuclear technology has historically been linked to the consequences of nuclear weapons testing and the threat of nuclear war. The nuclear weapons testing era of the 1960s was followed by a period of great optimism about the potential peaceful uses of atomic energy. The civilian nuclear power industry had established a good safety record since its inception in 1956, and nuclear energy production was growing at an average annual rate of over 20% per year (1970–1986). Nuclear power production peaked in the United States during the late-1970s because the industry was plagued by economic problems, lower demands for energy, and public and political controversy about reactor safety and radioactive emissions. Public opposition to the nuclear power industry was already on the rise in the aftermath of the 1979 reactor accident at Three Mile Island in Pennsylvania and came to the forefront of international attention following the catastrophic accident at the Chernobyl nuclear power plant in the Ukraine during May 1986. Chernobyl fallout was dispersed around the globe with measurable deposition occurring throughout the Northern Hemisphere. Worldwide public attention on environmental radioactivity shifted from the effects of nuclear weapons testing and the Cold War to the safety of nuclear power plants. Waste handling and disposal practices of the past were no longer tolerated by the public or regulatory agencies. A conservative policy in waste management added to the nuclear legacy with large quantities of nuclear waste – from both military and peaceful uses – being held in interim storage. Within the United States, growing concerns about nuclear energy including stockpiling of nuclear waste and associated environmental issues, caused parts of the weapons production complex to close and orders for new nuclear power plants to be canceled.

The Cold War ended with the United States and the Russian Federation agreeing on arms reductions but dismantling nuclear weapons, closing surplus production facilities, stabilizing waste, and environmental cleanup have only added to the nuclear waste stockpile. By the mid-1990s, the United States Department of Energy (DOE) managed over 3.6×10^7 m^3 of solid and liquid waste containing about 37,400 PBq of radioactivity (DOE/EM, 1997). Most of the high-level waste from chemical processing of spent nuclear fuel and irradiated target assemblies is being held in interim storage at the four sites where it was originally generated: Hanford, the Idaho National Engineering Laboratory, the Savannah River Site and the West Valley Demonstration Project (DOE/EM, 1997). The waste stockpile has been categorized into high-level waste ($\sim$35,500 PBq) composed of transuranic elements and fission products in concentrations that require permanent isolation, transuranic (TRU) waste ($\sim$140 PBq) resulting almost exclusively from weapons production processes, and low-level waste ($\sim$1850 PBq) characterized as all other radioactive waste not classified as high-level waste, TRU waste, spent fuel or natural uranium and thorium byproduct material. Prior to 1970, portions of all waste categories were disposed of as effluents from nuclear facilities in shallow burial trenches, by sea burial, or by deep underground injection. These practices have since been discontinued. TRU waste is segregated and placed in retrievable storage above- or below-ground, typically in metal drums on soil-covered storage pads. More than 300,000 barrels of such waste are either buried or in temporary storage throughout the United States. The barrels await permanent disposal at the Waste Isolation Pilot Plant (WIPP), a planned geologic repository near Carlsbad, New Mexico. The DOE is also responsible for managing materials that are either not required for immediate use or no longer meet the current mission of the Department (DOE/EM, 1997). Materials in inventory include natural uranium, highly enriched uranium (HEU), low enriched uranium (LEU), depleted uranium (DU), plutonium and other nuclear materials used for nuclear research and weapons production; spent nuclear fuel; and lithium and lithium compounds used in the manufacture of nuclear weapons and production of tritium. About half the total of 8.2×10^8 kg of legacy materials held in inventory up until the mid-1990s has resulted from nuclear weapons production activities. Depleted uranium comprises 71% of the mass of radioactive materials in inventory, the majority of which is maintained at the three gaseous diffusion plants in Paducah, Kentucky; Portsmouth, Ohio; and Oak Ridge, Tennessee (DOE/EM, 1997). As of 1996, the DOE had identified approximately 5000 of its 20,000 facilities as surplus and accumulated approximately 1.9×10^9 m^3 of contaminated environmental media as waste from site cleanup and facilities stabilization programs.

The United States has begun the process of addressing the environmental legacy of over five decades of nuclear weapons production. The end of the arms race appears to have become more problematic for the Russia Federation. Large quantities of nuclear waste built up from nuclear military operations of the Former Soviet Union have already been abandoned. Military shipyards in the Russian Arctic region are littered with idle nuclear-powered submarines and containers of spent nuclear fuel extracted during refueling operations. New information and data have recently confirmed that the Former Soviet Union dumped high-level nuclear waste, including reactor assemblies, some with spent nuclear fuel in the Russian Arctic Seas (1964–1992) (IAEA, 1999). Concerns about the potential impacts of nuclear waste dumping in the Arctic led to the development of the Arctic Monitoring and Assessment Program (AMAP) and renewed efforts by the International Atomic Energy Agency (IAEA) to gather, compile, and evaluate information and data on source terms and radionuclide inventories. Scientific expeditions to the region during the 1990s indicate that elevated levels of radioactive contamination are confined to the immediate vicinity of waste containers and dumped objects. General levels of radioactive contamination in regional waters are said to be lower than those observed in the 1970s. These findings suggest that previous waste disposal operations at sea have had a negligible radiological impact on the environment. Nonetheless, the continued practice of sea disposal of radioactive waste has been widely condemned by the international community.

This chapter is intended to summarize the sources and occurrences of man-made radioactivity in the oceans, as well as provide a technical basis for addressing public misunderstandings about the risks posed by environmental radioactivity. The main contribution of dispersed radioactivity entering the environment has come from testing of atmospheric nuclear weapons from 1945 to 1980. The measured total global deposit of long-lived fission products such as ^{90}Sr and ^{137}Cs is in agreement with the estimated fission yields and partitioning of radioactive debris in the atmosphere. Other unrestrained sources of radioactive contamination of the marine environment include direct discharges of radioactive effluents into rivers and coastal seas from reprocessing and fuel cycle operations, and runoff or leakage from other land-based sources of contamination. Human activities involving nuclear materials will always be subject to accidents and accidental releases of radioactivity. However, the greatest challenge in controlling the man-made radiation environment (i.e. limiting exposures to both humans and biota) and, to some extent, in maintaining international peace and nuclear security is seen as the management of legacy waste and nuclear materials in inventory. Today, the risk of illicit trafficking, terrorism, safety and facility vulnerabilities, and the temptation to use substandard waste containment or disposal options continue to grow with the increase in the global nuclear-waste burden. New challenges are also arising from the growing accumulation of stored plutonium from both civilian and military operations. By the end of 1997, the plutonium in storage included 170 t of separated plutonium from civilian reprocessing operations, and another 100 t of excess plutonium from dismantled warheads no longer required for defense purposes (Oi, 1998). Moreover, the cumulative amount of plutonium in spent fuel from power reactors worldwide is predicted to exceed 1700 t by 2010.

2. Discovery of elements beyond uranium

The early production history of man-made elements beyond uranium (element 92), and the discovery of their nuclear properties, is one of the most fascinating periods in science. Following the discovery of an uncharged primary particle – the neutron – by Sir Charles Chadwick in 1932, experimenters continued the study of the atom's nucleus by bombarding ordinary elements with neutrons, producing a range of radioactive isotopes one or two atomic numbers removed from that of the target nucleus. In 1934, Enrico Fermi and his co-workers at the University of Italy in Rome investigated the irradiation of uranium with slow neutrons and found that they produced radioactive species whose chemistry did not fit the properties of existing elements 90, 91 or 92. Fermi reasoned that he must have produced new elements 93 or 94. It was not until December of 1938 that two German scientists, Otto Hahn and Fritz Strassman, conclusively identified that one of the products of neutron capture by uranium was actually an isotope of a known element, element 56 (barium). They communicated their findings in the scientific journal *Nature*, and privately to former co-worker Lise Meitner, who had escaped from Nazi Germany. Meitner met up with her nephew Otto R. Frisch in Copenhagen and together they worked out a detailed explanation of what Hahn and Strassman had observed: the breakup of the nucleus of a heavy atom into two lighter isotopes or fission products with a force unparalleled on an atomic scale. Enrico Fermi, working at Columbia University, was joined by Neils Bohr, a Danish physicist at Princeton University, and Ernest O. Lawrence at the University of California, Berkeley, to formulate a strong contingent of nuclear physicists and chemists working in the United States on nuclear fission. Before the theory behind nuclear fission could be published, two separate groups of experimenters at Columbia University, headed by Fermi and Leo Szilard, demonstrated that two or more neutrons were released in the fissioning of a uranium nucleus, and postulated that under the right conditions a chain reaction was possible. The military significance of producing a chain reaction was realized immediately. Scientists in the United States agreed to withhold information related to the military use of atomic energy from publication but F. Joloit and co-workers in Paris reached the same conclusion and published their findings (Compton, 1956). By the time of the German invasion of Poland on September 1, 1939, more than 100 papers related to nuclear fission had appeared in the scientific literature along with evidence that uranium fission occurred by neutron capture on ^{235}U and not ^{238}U, the most abundant isotope of uranium. With this knowledge, nuclear physicists speculated that a chain reaction could potentially be sustained by using ^{235}U separated from natural uranium or by devising a technique to slow the passage of neutrons to enable their more selective absorption on ^{235}U.

The flow of information on nuclear physics research diminished with the outbreak of World War II. In the summer of 1939, Hungarian-born refugee physicists Leo Szilard, Eugene Wigner and Edward Teller convinced Albert Einstein of the need to alert President Roosevelt about the potential dangers of Nazi Germany developing and using atomic weapons in a world conquest. Einstein outlined details of recent advances and the imposing dangers of harnessing atomic energy in a personal letter to the President. President Roosevelt immediately decided to establish a joint Army–Navy uranium committee to give government financial assistance to those engaged in uranium fission research. The work of the uranium committee got off to a slow start reporting that the military applications of atomic energy ... must be regarded only as possibilities (Compton, 1956). At the same time, British and German scientists received substantial support aimed directly at atomic bomb production and were making significant advances in developing techniques for separation of fissionable ^{235}U. As events in Europe intensified, it was those scientists working outside of government at American universities with private funding that helped lay the true foundation for advancing the early wartime effort in nuclear fission. Links between American universities, the National Academy of Sciences (NAS) and military services were further strengthened by the formation of a National Defense Research Committee (NDRC) in June of 1940 to oversee the work of the Government's Uranium Committee as well as other government projects. By the time the United States entered into World War II there was a broad representation of American science devoted to national service.

The first artificially produced elements beyond uranium (elements 93 and 94) were discovered at the University of California at Berkeley. Plutonium was synthesized by bombarding uranium with charged particles (deuterons) to produce ^{238}Np, which decayed by beta emission to ^{238}Pu. The isotope ^{239}Pu was isolated in the spring of 1941 and found to undergo slow neutron-induced fission (Kennedy et al., 1946). The discovery of an artificially produced fissile element brought new life to the military objective of nuclear energy. In January of 1942, Present Roosevelt gave approval for the development of the atomic bomb. The exploratory stage came under the direction of what was known as the S-1 Committee organized under the Office of Scientific Research and Development (OSRD) and headed by Dr. Vannevar Bush at the Carnegie Institute in Washington. Research on various methods of ^{235}U separation and plutonium production proceeded on parallel fronts. Tracer experiments on the chemical and nuclear properties of plutonium continued through the early 1940s using cyclotron production facilities at the University of California's 'Old Radiation Laboratory' at Berkeley and the wartime Metallurgical Laboratory at the University of Chicago. The atomic bomb development program was re-assigned to a new district within the U.S. Army Corps of Engineers in June 1942 in what was known as the Manhattan Engineer District (MED) or 'Manhattan Project' under the direction of General Leslie Groves. A detailed history of the Manhattan Project and the events leading up to the development of the first atomic bomb can be found elsewhere (Gosling, 1994; Rhodes, 1986; Jones, 1985; Hewlett & Anderson Jr., 1962). Under the Atomic Energy Act of 1946, management of the United States nuclear programs was re-assigned to the Atomic Energy Commission (AEC, 1946–1975) and, over the next 20 years, consolidated into a large, government-owned research and development complex. The AEC was abolished in 1975. Management of the United States nuclear weapons complex was transferred to the Energy Research and Development Administration (ERDA) and, later, to the United States Department of Energy (DOE) (1977-present).

In general, waste management practices used in the United States over the past two decades have come under increasing scrutiny. The United States has largely suspended nuclear weapons production activities since the early 1980s and begun to reduce the size of the weapons complex under a stockpile stewardship program. Growing public awareness and stringent agency controls related to the treatment, processing, storage, transport and disposal of radioactive waste have increased the burden of nuclear waste management and environmental stewardship. Negligent nuclear waste management practices used by the Former Soviet Union are also presenting serious technical, political and economic problems within the Russian Federation. However, the collapse of the Soviet Union and end of the Cold War have ushered in a new era of cooperation between the United States and the Russian Federation in linking the legacies of nuclear weapons production processes and their environmental consequences. Considerable information has become available on the nuclear waste management and extent of environmental contamination within the Former Soviet Union (Bradley, 1997; OTA, 1995).

Table 1
Current inventory of radioactive releases to the environment in the Western Siberian Basin and adjoining territories of the Russian Federation

Site	Measured or estimated releases to the environment (PBq, 1×10^{15} Bq)
Production reactor and reprocessing waste:	
Tomsk-7 waste injection	37,000
Krasnoyarsk-26 waste injection	16,700
Mayak reservoirs, lakes, Techa River	4500
Tomsk-7 reservoirs	4800
Krasnoyarsk-26 reservoirs	>0.7
Mayak production reactor coolant water discharges	4.9
Tomsk-7 production reactor coolant water discharges	1.4
Krasnoyarsk-26 production reactors coolant water discharges	3.9
Mayak-1957 high-level waste storage tank explosion	1.6
Mayak-1967 air borne releases from Lake Karachai	0.02
Subtotal	63,000
Other radioactive releases to the environment:	
Chernobyl accident	58
Uranium mill tailings	220
Dimitrovgard waste injection	3.3
To Irtysh River from weapons testing	3.3
Other nuclear power plant operations	1.3
Subtotal	290
Grand total	63,290

Responsibility for the Russian nuclear weapons complex and associated waste management operations falls largely under the Ministry of Atomic Energy (Minatom). This organization is said to be arguably the owner of the world's largest nuclear waste stockpile (Bradley, 1997). Estimates of releases to the environment through the end of 1996 exceed 6.3×10^4 PBq compared with less than 100 PBq in the United States (Table 1). Approximately 97% of the radioactive waste entering the environment has been disposed of by underground injection or discharged into surface waters. A major problem facing the Russian nuclear complex today is the lack of adequate facilities for the safe handling, treatment and storage of nuclear waste. Large quantities of untreated waste and spent nuclear fuel assemblies are being placed in interim storage on sea and land. These circumstances are seen as dramatically increasing the risk of radiation accidents and uncontrolled releases to the environment.

3. Spent nuclear fuel reprocessing and nuclear weapons production

Enrico Fermi and his associates at the University of Chicago brought the first nuclear reactor (the Chicago Pile, CP-1) into operation on December 2, 1942. Reactor technology developed quickly during War World II and less than a year later a 3.8 thermal megawatt (MWa) research reactor (Clinton Pile, X-10) began operation at a site in eastern Tennessee, now called Oak Ridge. The X-10 was used to test reactor operations and plutonium separation technologies, and by February of 1944 was producing several grams of plutonium per month. A test pile (the Hanford 305 Test Pile) constructed at Hanford, near Richland, Washington also served as a platform to develop and research materials for use in full-scale reactors. Three full-scale, single-pass plutonium production reactors were initially constructed at Hanford during 1944 to produce plutonium for the Manhattan Project. The Atomic Energy Commission (AEC) built eleven additional production reactors between 1948 and 1955, including five single-pass reactors and a new generation N Reactor at Hanford, and five heavy water moderated reactors (R, P, L, K and C) at the Savannah River Site near Aiken, South Carolina. By the early 1960s, demand for weapons-grade plutonium was being adequately met by the Savannah River Site reactors and the original single-pass Hanford reactors were closed down (1964–1971). The N reactor at Hanford was shut down in 1987. The P, L, K and C reactors at the Savannah River Site were converted to use highly enriched uranium (HEU) fuel in 1968, and continued to produce tritium and other radioisotopes for nuclear weapons programs into the late 1980s.

Low-level and transuranic waste management operations at nuclear weapons production and fabrication plants in the United States have historically included land and sea disposals, ground injection, and temporary storage in tanks, seepage basins and ponds. The single-pass Hanford reactors used water from the Columbia River for cooling: the water passed through the reactors into retention ponds and, after a few hours, was released back into the river. The effluent stream contained induced radioactivity from neutron activation of naturally dissolved minerals, water treatment chemicals and other entrained corrosive products. Uranium fuel element failures within the reactor added fission products and other fuel products to the effluent stream. Transport of Hanford radioactivity down the Columbia River to the Pacific Ocean possibly represented the first significant occurrence of contamination of the marine environment by man-made radioactivity. Plutonium production at the Savannah River Site differed by the fact the reactors used a closed-loop cooling system so that discharges under routine operations were considerably lower than at Hanford. However, significant quantities of radioactivity escaped the Savannah reactors from nonroutine occurrences such as reactor purges, heat exchanger leaks, and fuel element failures. The reactor effluents included activation, fission and fuel element products, as well as significant quantities of tritium. During the first year of operation of the Savannah reactors, coolant water and disassembly basin effluents were released directly into local stream and creeks. Two artificial lakes, known as the PAR Pond and L Lake, later served as coolant water reservoirs to allow the reactor effluent to cool before final discharge. The lake sediments are now contaminated with ^{137}Cs and transuranic elements built up from historical discharges (DOE/EM, 1997). Other low-level liquid wastes generated from fuel storage and disassembly, cleaning and decontaminating reactor equipment, use of toxic water treatment chemicals such as hexavalent chromium, and oils and other fuel products were disposed of in Hanford-style cribs, ponds or seepage basins. The seepage basins at the Savannah River Site were closed and backfilled during the early 1960s, and then replaced with seepage basins with containment. Through 1991, an estimated 1.3×10^9 m^3 of waste water containing about 52 PBq of radioactivity was discharged into the ground at the Hanford Site alone (DOE/EM, 1997).

The bulk of the DOE high-level waste was generated when plutonium and/or uranium was chemically separated from spent nuclear fuel either by bismuth phosphate precipitation, reduction oxidation (REDOX) or plutonium uranium extraction (PUREX). Chemical process waste included cladding waste produced by removal of coatings from irradiated fuel assemblies. Miscellaneous low-level and transuranic wastes streams also were generated from plutonium concentration and finishing processes, waste volume-reduction operations, uranium solidification, laboratory analysis, spill cleanup and other operations. High-level radioactive wastes are stored at four main sites in the United States including Hanford, the Savannah River Site, Idaho National Engineering Laboratory and the West Valley Demonstration Project. Significant quantities of transuranic waste are also stored or buried at the Rocky Flats Environmental Technology Site in Colorado and at the Los Alamos National Laboratory in New Mexico. Most high-level radioactive and hazardous wastes generated from chemical processing at Hanford are stored in underground tanks. These tanks contain alkaline liquid, salt cake, and sludge with an average activity concentration around 30 TBq m^{-3}. Spills and leaks associated with the tank operations and other waste management practices have released large quantities of radioactivity into the environment causing a flow of contaminated ground water toward the Columbia River. As of 1995, the Hanford Site managed 240,000 m^3 of high-level radioactive waste containing about 11,800 PBq. About 40% of the high-level waste stored at Hanford is contained in highly radioactive capsules. The capsules contain cesium and strontium salts segregated from high-level waste generated from the REDOX and PUREX plants. This was done to allow additional low-level waste to be discharged to the environment and to conserve available tank space. The Savannah River Site manages about 130,000 m^3 of high-level waste containing in excess of 19,500 PBq. The primary radionuclides in storage at both sites include ^{137}Cs, ^{90}Sr, ^{90}Y, ^{137m}Ba and ^{241}Pu. In 1996, defense waste reprocessing plants at Savannah River Site and West Valley Demonstration Project began producing vitrified forms of high-level waste. During the period between 1954 and 1988, hazardous and low-level radioactive wastes from chemical reprocessing at the Savannah River Site were reportedly discharged into seepage basins, and after evaporation, some waste discharged into local streams (DOE/EM, 1997). Generation of high-level waste has decreased significantly since the early 1990s when the DOE stopped reprocessing of spent nuclear fuel. All high-level waste produced in acceptable forms for long-term storage will eventually be disposed of in a geologic repository.

The first plutonium production reactors in the Former Soviet Union began operation in June of 1948 at the MAYAK Production Association (MPA). Mayak is located about 70 km north of Chelyabinsk adjacent to the city of Ozersk (formerly Chelyabinsk-65) on the border of the West Siberian Basin (Fig. 2). The Mayak facility occupies an area of about 200 km^2 and borders an extensive system of rivers, lakes and marshes. The flood plain connects to the Techa River system providing an important transport vector for passage of contaminated surface waters and runoff down the Ob River into the Kara Sea. Today, the MPA continues to operate two production reactors to produce radionuclides for military and civilian use, an isotope production plant for commercial sale of isotopes and radiation sources, scientific research and manufacturing laboratories, as well as waste management and storage facilities. Other major Russian plutonium production, reprocessing, and waste storage facilities include Krasnoyarsk-26 and Tomsk-7, all three sites being located within the Siberian Basin (Fig. 2). Krasnoyarsk-26, now known as Zheleznogorsk or the 'Iron City', is situated underground about 50 km north of Krasnoyarsk on the eastern bank of the Yenisey River. Tomsk-7 and the

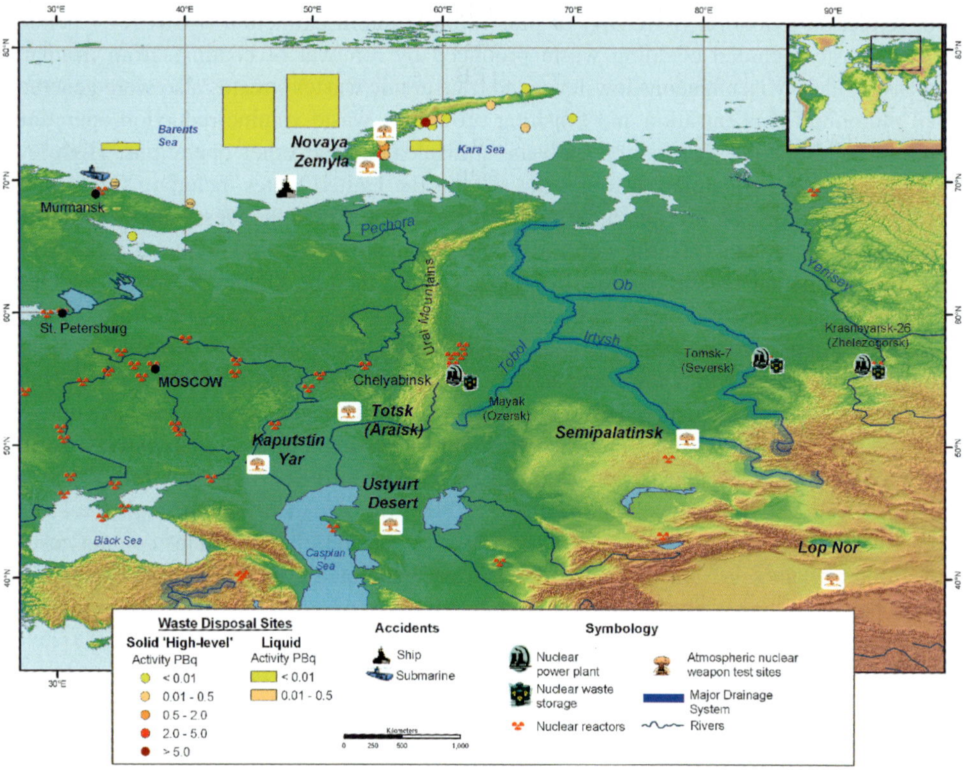

Fig. 2. Map of weapons production reactors and reprocessing facilities, nuclear test sites, and waste disposal sites within the western Siberian Basin and adjoining territories of the Russian Federation.

associated city of Seversk are situated on the Tom River about 25 km from the city of Tomsk.

The Former Soviet Union operated five graphite-modified, single-pass, water-cooled reactors at Mayak (1948–1990), two reactors at Krasnoyarsk-26 (1958–1992) and a single reactor at Tomsk-7 (1955–1990). Although no official data are available, cooling water passing through these 'Hanford style' reactors was most likely contaminated with neutron-activation products formed from naturally occurring minerals and chemical additives in the coolant water or from entrainment of corrosion products from fuel cladding materials. Nonroutine releases of fission products, and fuel element failures and leaks, were also common in this type of reactor design. The reactor coolant water was released directly into nearby reservoirs, lakes, and rivers. Five additional graphite-moderated, production reactors with closed circuit cooling were brought into operation during the early 1960s: four of these reactors were located at Tomsk-7 and the other reactor at Krasnoyarsk-26. The original eight single-pass production reactors and two of the Tomsk-7 closed-circuit reactors were shut down between 1987 and 1992. Russia has continued to reprocess spent nuclear fuel from two remaining plutonium production reactors at Tomsk-7 and the single remaining reactor at Krasnoyarsk-26. Much of this material has been placed in storage for potential use in future energy and/or weapons production programs. The estimated total inventory of radionuclide releases from the Mayak production reactors to Lake Kyzyl-Tash on the Upper Techa River is estimated to be around 4.9 PBq (Table 1; Bradley, 1997). Associated releases from production reactors at the Krasnoyarsk-26 and Tomsk-7 sites are 3.9 PBq and 1.4 PBq to the Yenisey and Tom River, respectively (Table 1).

Beginning in 1948, the MPA commissioned their first radiochemical plant for separation of plutonium from spent nuclear fuel and a plant for conversion of plutonium to high purity metallic components for atomic bomb production. High-level chemical process waste was initially routed to storage tanks but tank storage was soon overwhelmed. A decontamination process was introduced to conserve tank space and allow a portion of the waste stream to be diverted into the Techa River. The technological waste-management process failed with large quantities of radioactivity being released into the Techa River (Degteva et al., 2000; Vorobiova et al., 1999). About 76 million m^3 of liquid radioactive waste containing an estimated 100 PBq of radioactivity was reportedly discharged directly into the Techa River between 1949 and 1956. Ninety-five percent of the release occurred between March 1950 and November 1951 (Vorobiova et al., 1999; Degteva et al., 1994). The bulk of the radioactivity consisted of isotopes of ruthenium (^{103}R, ^{106}Ru) and the rare-earth elements along with an estimated 12 PBq of ^{90}Sr and 13 PBq of ^{137}Cs (Cochran et al., 1993). Approval was obtained to divert the bulk of the chemical process waste into Lake Karachai (reservoir 9) with lesser amounts ($\sim$4–7 TBq d^{-1}) entering the Techa River. Over the next 10–12 years a cascade of natural lakes, dams and by-pass canals were constructed for the management of low-level and intermediate-level wastes in an effort to contain radioactive contamination to the Upper Techa River catchment and reduce radiation exposures to residents living downstream. The reservoirs acted as sedimentation ponds for adsorbed radionuclides but the discharges were still sufficient to cause severe contamination of the entire floodplain. Systematic measurements of radioactive contamination of the Techa River began in 1951, including monitoring of river water, bottom sediments, flood-plain soils, vegetation, fish, milk and other foodstuffs, and external gamma-exposure rates. The practice of discharging chemical reprocessing waste into Lake Karachai was terminated towards the end of 1956. Construction on a second reprocessing plant (Complex BB) began in 1954 but did not commence operations until 1959. Complex BB was designed with improved radiation protection controls to reduce worker exposure and quickly reached its production goals enabling authorities to close down the original plant (Bradley, 1997). Reprocessing of defense reactor fuel was discontinued at Mayak in 1987. The Mayak production reactors remained in operation until 1990 with spent nuclear fuel shipped to Tomsk-7 for reprocessing. A total of 123,000–136,000 t of defense reactor fuel was reprocessed at Mayak between 1949 and 1987 (Bradley, 1997). Modernization and expansion of the site continued through the 1970s. Beginning in 1972, the original reprocessing plant (Complex B) was upgraded to receive, store and reprocess spent fuel from different types of reactors. An additional 3400 t of spent nuclear fuel has since been processed in this facility, known as the RT-1 plant. The fuel has come from Russian-designed VVER-440, BN-350 and BN-550 power reactors; breeder reactors; research reactors; and nuclear-powered submarines and icebreakers (Bradley, 1997). The total present-day inventory of liquid radioactive waste released into the environment at Mayak (1949–1995), consisting mostly of ^{137}Cs and ^{90}Sr, is estimated at 4500 PBq (Table 1). By comparison, the total combined inventory of radioactivity released into the near-surface environment at the Hanford (Washington) and Savannah River (Georgia) sites in the United States is around 60 PBq (Bradley et al., 1996).

Other secondary sources of radioactive contamination of the Upper Techa have included the explosion of a waste storage tank at Mayak in 1957 (the Kyshtym accident) and wind-borne releases from Lake Karachi. The Kyshtym explosion was caused by the failure of a cooling system allowing highly explosive nitrate salts inside a high-level waste storage tank to overheat (Nikipelov & Drozhko, 1990). An estimated 74 PBq of radioactivity was injected into the atmosphere and dispersed by wind to form the East Urals Radioactive Trace (EURT) (Botov, 1992). Deposition densities along the contamination track ranged from 5 PBq km^{-2} near the source to about 4 GBq at up to ~100 km distance (Nikipelov et al., 1990). At the time of the accident, the tank contained about 740 PBq of high-level liquid waste consisting mostly of short-lived fission products including ^{144}Ce (^{144}Pr), ^{95}Zr (^{95}Nb), ^{90}Sr (^{90}Y) and lesser amounts of ^{137}Cs (Drozhko et al., 1989). The present-day residual environmental inventory attributed to the Kyshtym accident is dominated by ^{90}Sr (^{90}Y) and is estimated to be ~1.6 PBq (Table 1).

Wind-borne releases from Lake Karachai were first detected in 1967 when a combination of meteorological conditions led to the resuspension of contaminated dust and silt from the exposed shoreline of the lake (Bol'shakov et al., 1991; Botov, 1992). Radioactive contamination was dispersed by wind up to distances of 50–75 km from the MPA site. Deposition from wind-blown sources of contamination have added about 15–370 GBq km^{-2} of radioactivity to regional soils, with some hot spots containing activity levels up to 1850 GBq km^{-2} (Cochran et al., 1995). An estimated 22 TBq of radioactivity was associated with this contamination event (Table 1). Countermeasures introduced during the late 1960s to help reduce the spread of secondary contamination from Lake Karachai included covering the exposed shoreline with sand and improving lake embankments. Large quantities of soil and rock as well as hollow cement blocks have since been added to reduce the water holding capacity of the lake and stabilize muddy bottom deposits. Today, effluent discharges from waste management operations are still discharged into Lake Karachai to help maintain the water level and prevent further wind erosion of the shoreline.

As will be shown, the 4900 PBq of radioactivity released to surface waters of Lake Karachai and the Upper Techa River represent only a small fraction of the total inventory released to the environment at Tomsk-7 and Krasnoyarsk-26 by deep underground injection (Table 1). Discharges of radioactive liquid effluents from the Mayak facility have decreased from a maximum rate of ~0.16 PBq per day during the early 1950s (Degteva et al., 1994) to about 10 PBq per year during the 1990s (Christensen et al., 1995). As a result, the average annual concentrations of ^{90}Sr and ^{137}Cs in Techa River water collected near the village of Muslyumovo, located about 80 km downstream from the MPA site, have reportedly decreased from 1900 and 1.5 kBq l^{-1}, in 1951, to 0.15 and 0.015 kBq l^{-1}, in 1988, respectively (Jachmenev, 1995).

The construction of reservoir dams and canals to divert the Techa River from flowing into the reservoirs has helped regulate the outflow of water, silt, and associated secondary contamination into the Lower Techa River system. The closure of original defense reactor reprocessing plants, waste minimization efforts, increased availability of tank storage, and use of improved reprocessing and waste management technologies have all helped reduce radioactive waste and associated discharges into the environment. Nonetheless, high levels of residual radioactive contamination within the Mayak region have made it necessary to establish a health protection zone covering an area of approximately 350 km^{-2} (Aarkrog et al., 2000). Both agricultural practices and permanent residences are forbidden within the region. A larger area has been designated as an observation zone where agriculture practices are permitted with environmental surveillance.

As of 1994, high-level waste derived from chemical reprocessing of spent nuclear fuel at Mayak has been vitrified into solid forms (Bradley, 1997), and steps have been taken to improve technologies for handling intermediate- and low-level liquid waste. Some releases to the environment are expected from storage (disposal) of legacy liquid and solid wastes in underground trenches and concrete repositories, but the inputs are probably insignificant when compared with previously declared releases to the environment. Through 1990, approximately 30,000 PBq of liquid and solid radioactive waste had been accumulated on the MPA site (Aarkrog et al., 2000) – about 75% of the radioactive waste inventory was classified as high-level waste (HLW). About one third of the HLW on site was vitrified into solid forms while the remainder was stored as nitric acid liquors or suspensions in stainless tanks or concrete tanks lined with stainless steel. By 1996, the MPA was generating about 16–20,000 m^3 of liquid intermediate-level waste (ILW) per year with a total radioactivity content of about 30 PBq. ILW and liquid low-level waste (LLW) from the facilities sewage treatment plant, and coolant water from the reactor and waste management operations, continue to be discharged into the industrial reservoirs and lakes.

Radioactive contamination of the Techa River basin system remains a significant human health and ecological concern. Past activities at Mayak have severely contaminated the Techa River and associated floodplains that feed successively into the Islet, Tobol and Ob Rivers out to the Kara Sea: surface water runoff, overflow from the Mayak reservoirs and canals, and drainage from contaminated marshes near the former village of Assanov continue to provide a source-term for secondary contamination of the Techa River. Moreover, Russian scientists believe that infiltration of contaminated groundwater will ultimately become a major source of secondary contamination of rivers flowing into the Arctic Ocean. The reasoning is twofold. The main source of water supplying the Techa River during the summer months is groundwater seepage (Vorobiova et al., 1999). Raising the height of the Mayak dams to control the overflow of contaminated water into the Techa River has been offset by an increase in the rate of groundwater recharge. Secondly, the concentration of radioactivity in seepage water is increasing because the Mayak reservoirs contain highly mineralized water from discharges of chemical reprocessing waste; the natural sorption capacity of bottom deposits is practically exhausted leading to more rapid mobilization and migration of radioactive contaminants. The front of a southward-migrating mound of contaminated groundwater formed under Lake Karachai has already advanced 25 km over the past 40 years and is approaching the Mishelyak River (Bradley et al., 1996).

Discharges of radioactive waste into Lake Karachai on the MPA site have captured international attention but higher quantities of spent-fuel reprocessing waste were discharged into reservoirs and open pits at the Tomsk-7 nuclear materials production and reprocessing site. Recent estimates indicate that the Tomsk-7 reservoirs contain about 4800 PBq of radioactivity (Moscow Interfax, 1994) Large quantities of liquid radioactive waste have regularly been discharged into Romashka and Tom Rivers including more than 17.5 PBq of radioactivity from a single-pass production reactor that operated from 1955 through 1990. The present inventory of radioactivity released from reactor operations (about 1.4 PBq) is dominated by activation products ^{63}Ni, ^{55}Fe and ^{60}Co, formed from corrosion of aluminum fuel cladding and process tubes (Bradley and Jenquin, 1995). The reservoirs and basins are now mostly covered over with sand, concrete or asphalt. Other waste management practices employed at Tomsk-7 have included storage of solid and liquid radioactive waste in tanks and special concrete repositories (buildings or underground tanks), but the predominant form of waste disposal is by far deep-well injection (Table 1). Well-injection technology has been in use at Tomsk-7 since the early 1960s, and the practice continues today. Radioactive waste injected into the ground at Tomsk-7 accounts for nearly 90% of the total waste released or about 37,000 PBq. The waste is reportedly injected into sandy Cretaceous strata at depths of up to 450 m under very high pressure (Bradley, 1997). Russian authorities believe that radioactive waste injected into underground formations at Tomsk-7 (and Krasnoyarsk-26) will remain isolated from the surrounding environment for the next 500–1000 years allowing sufficient time for the bulk of the radioactive material to decay. Other scientists are concerned about contamination of the local water supply, located at depths of only 20 m below ground, and possible long-term migration of radioactive contamination into the Tom and Ob Rivers.

Other sources of radioactive contamination at Tomsk-7 include atmospheric emissions (e.g. ^{85}Kr, ^{3}H, ^{131}I and alpha-emitting radionuclides) and incidents involving reactor or waste management operations. The most serious radiation accident occurred on April 6, 1993, when a waste storage tank exploded. The tank contained nearly 9 t of uranium in the form of partially processed uranyl nitrate solution, about 310 g of plutonium, tribute phosphate (TBP) and paraffin (Nuclear Fuel, 1993; Nuclear Waste News, 1993). The blast released about 0.03 PBq of radioactivity composed mainly of short-lived beta- and gamma-emitting radionuclides such as ^{95}Nb, ^{106}Ru and ^{95}Zr. Fortunately, wet snow began to fall soon after the blast limiting dispersion of radioactive aerosols to nearby forests and uninhabited regions up to a distance of about 8–15 km. Only the village of Georgievka and surrounding districts required some form of decontamination.

As with Tomsk-7, little quantitative information is available on waste management practices and environmental contamination at Krasnoyarsk-26. The Krasnoyarsk-26 complex known as the Mining and Chemical Combine was built underground to ensure survival from a nuclear strike. The first graphite-moderated plutonium production reactor at Krasnoyarsk-26 was commissioned in 1958. A second graphite reactor began operations in 1962 followed by the commissioning of a closed-circuit production reactor and spent-fuel reprocessing plant in 1964. The original single-pass defense reactors were decommissioned in 1992 (ITAR-Tass, 1992), while the closed circuit reactor and reprocessing plant have apparently remained in operation. Construction of a separate reprocessing plant, known as RT-2, began operation in 1983 to reprocess VVER-1000 spent reactor fuel but prospects for its completion are uncertain (Bradley, 1997). As of January 1995, the spent-fuel storage facility within the RT-2 plant contained 1100 t of VVER-1000 fuel and had a holding capacity of 3000 t. Solid and liquid high-level radioactive wastes are also stored on site in underground tanks.

Sources of radioactive contamination of the Yenisey River from Krasnoyarsk-26 include discharges of coolant water from the original single-pass defense reactors (Tass World Service, 1992) and water overflow from open reservoirs contaminated with chemical reprocessing waste. The present inventory of radioactive waste released to the Yenisey River from reactor operations is estimated to be around 3.9 PBq (Table 1). The four main reservoirs at Krasnoyarsk-26 contain about 0.7 PBq of radioactivity but, unlike the reservoirs at Mayak and Tomsk-7, the activity composition of the residual radioactivity is dominated by plutonium and other long-lived radionuclides. Radioactive contamination of soils within the floodplain of the Yenisey River have been traced over distances of 1500 km, ranging from a maximum of 1.5 TBq km^{-2} near the discharge point to less than 4 GBq km^{-2} at 500–1500 km distance (Bradley, 1992). The principal radionuclides contained in bottom sediments include ^{51}Cr, ^{54}Mn, ^{60}Co, ^{90}Sr, ^{137}Cs, ^{238}Pu and $^{239+240}Pu$. External exposure rates near the discharge point in the Yenisey River are known to have exceeded permissible standards for radiation protection. Local inhabitants continue to be exposed to elevated levels of ^{32}P, ^{24}Na, ^{65}Zn and ^{60}Co by consuming fish contaminated by discharges from the Krasnoyarsk-26 site.

Liquid wastes generated from spent fuel reprocessing have been injected into the ground at Krasnoyarsk-26 since 1963. The main injection site is located on a high terrace about 750 m from the Yenisey River. Radioactive (and industrial) wastes of different types are injected into a sandy clay deposit about 450 m thick – the deposit is further divided into beds of quartz-feldspar, sandstones and kaolin/mica clays. The current inventory of radioactive waste injected underground at Krasnoyarsk-26 is estimated at 16,700 PBq (Table 1). The residual inventory of radioactive waste injected underground at this site is much greater than that discharged into surface water bodies. Although the geologic formation is favorable to radioactive waste disposal, a number of concerns have been raised concerning reported leaks in a 15-km-long pipeline used to transport radioactive waste from the main site to the well-injection installation. Other Russian scientists are skeptical about the impermeability of the geologic formation and concerned about possible long-term infiltration of contaminated ground water into local water supplies and the Yenisey River. Deep-well injection has also been used to dispose of radioactive waste at the Scientific Research Institute for Nuclear Reactors in Dimitrovgrad (Bradley et al., 1996) (Table 1).

Other potential land-based sources of radioactive contamination of surface and subsurface waters in the Siberian Basin include nuclear weapons testing, peaceful nuclear explosions, power reactor operations, the 1986 Chernobyl accident in Ukraine, and disposal of ore tailings and liquid wastes from uranium mining and milling operations (Table 1). However, the total radioactivity released from all these events including Chernobyl is much less than 1% of the 63,000 PBq discharged directly into the environment from nuclear weapons production activities. The Western Siberian Artesian Basin is one of largest shallow water basins in the world. Surface water hydrology within the basin is dominated by an extensive system of rivers, lakes and marshes that act as sinks for runoff and contaminant flow to the Arctic. The region is also believed to be a single groundwater basin with pervasive artesian character (Bradley, 1997). As a result, discharges of radioactive waste from the Mayak, Tomsk-7 and Krasnoyarsk-26 sites are potential long-term radioactive source terms to both the local and regional surface and subsurface hydrologic systems. Much has been done to improve waste management practices used by nuclear weapons complexes over the past five decades. The Russian Federation and the United States are also working together to evaluate the impacts of past (and present) releases including assessments on the long-term impacts on the Arctic region. Research is also continuing on quantifying the regional hydrology of the Western Siberian Basin and establishing boundary conditions to model contaminate flow and evaluate options for cleanup and remediation of contaminated sites.

4. Nuclear weapons testing

The full military potential of nuclear fission was first realized with the successful detonation of an 'atomic bomb' in July 1945 over a New Mexico desert near the town of Alamagordo in the United States. The atomic bombings of Hiroshima and Nagasaki followed a few weeks later where the immediate loss of human life led to an end of World War II. Early nuclear weapons were pure fission devices fueled by either ^{239}Pu or ^{235}U. During the fission of a heavy element, the compound nucleus consisting of the original nuclei plus a neutron splits into two lighter elements, simultaneously emitting gamma rays, 200–220 MeV of fission energy, and two or three neutrons that propagate the fission of other target nuclei. The probability that fission will occur is defined by the effective critical mass of the fissionable material, the energy of the neutrons inducing fission, and the relative number of protons and neutrons of the target nucleus. To achieve rapid supercriticality of fissionable materials, a conventional explosive device was used either to bring two or more subcritical masses together to exceed the critical mass, or to compress a subcritical mass to become supercritical. Uranium-235 and ^{239}Pu contain an even number of protons and odd number of neutrons and are capable of undergoing fission with neutrons of virtually any energy. Nuclei having an even number of protons and neutrons (e.g. ^{240}Pu, ^{238}U and ^{232}Th) can only be fissioned by high-energy (fast) neutrons above a threshold value of about 1 MeV – the higher the energy the greater the probability of fission.

The first thermonuclear or hydrogen bomb was detonated by the United States on Enewetak Atoll in the Marshall Islands on 1 November 1952. A significant fraction of the energy released by thermonuclear devices comes from nuclear fusion where the high temperature generated by the primary fission reaction is used as a trigger to fuse deuterium and tritium with the concomitant release of neutrons and vast amounts of energy.

The explosive energy of a nuclear explosion is conventionally expressed in units of energy released by a ton (907 kg) of the explosive TNT (1 kiloton (kt) $= 4.18 \times 10^{12}$ J). The complete fission of one kilogram of ^{239}Pu containing about 2.5×10^{24} atoms will produce an explosive equivalent of 17.5 kt. The terms 'dirty' and 'clean' bombs have sometimes been used to describe the relative amounts of radioactivity produced by nuclear tests (Eisenbud & Gesell, 1997). Pure fission devices typically generate more radioactivity than weapons whose energy is primarily derived from maximizing the fusion yield.

A nuclear explosion produces a cloud of incandescent gas and vapor in excess of 10–100 million degrees called a fireball (Fig. 3). The size and height of stabilization of the fireball are a function of the explosive yield of the device, the altitude of denotation and the meteorological conditions at the time of the blast. The fireball of a 1-Mt explosion reaches a diameter of around 2000 after about 10 seconds, rising rapidly at initial speeds of several hundred kilometers per hour, and then slowing as it cools (Kathryn, 1984). As the fireball dissipates it assumes a toroidal shape where strong convective forces uplift cool air and surface debris into the cloud. The fireball from an atmospheric nuclear detonation spreads out into a typical mushroom shape as it reaches the top of the troposphere (15–26 km) and may continue to rise well into the stratosphere. The cloud from a 1-Mt nuclear explosion reaches a maximum height about 10 minutes after denotation and may exceed 40 km in altitude (Kathryn, 1984).

Fig. 3. Detonation of a high-energy nuclear device showing the formation of a fireball.

4.1. *Modes of production of radioactive debris*

Nuclear fission produces hundreds of short-lived, neutron-rich, nuclides of about 35 different elements with half-lives ranging from fractions of a second to 17 million years. Fission products normally decay by beta-emission through isobaric chains to longer-lived and finally stable nuclides. The detailed mass distribution of fission products depends on the target nucleus and the energy of the fissioning neutron. The production probabilities or mass yields for ^{239}Pu and ^{235}U target nuclides are shown in Fig. 4. The highest fission yields are associated with those isotopes with closed nuclear shells centering on mass numbers from 85 to 104 and from 130 to 139. The principal radionuclides of environmental significance produced from fission by slow and fast neutrons on ^{235}U, ^{239}Pu or ^{238}U include ^{79}Se, ^{85}Kr, ^{90}Sr (^{90}Y), ^{93}Zr, ^{95}Zr, ^{99}Tc, ^{103}Ru, ^{106}Ru (^{106}Rh), ^{106}Rb, ^{107}Pd, ^{113m}Cd, ^{121m}Sn, ^{126}Sn, ^{125}Sb (^{125m}Te), ^{129}I, ^{131}I, ^{134}Cs, ^{135}Cs, ^{137}Cs, ^{140}Ba, ^{141}Ce, ^{144}Ce, ^{147}Pm and ^{151}Sm. The ^{137}Cs/^{90}Sr ratio in fallout can be used to provide a measure of the relative fission yields from target nuclides within a mix of fuel components (IAEA, 1998).

The chemical and physical properties of particles formed in nuclear explosions (e.g. the size, distribution, shape, composition and color) are known to vary according to the altitude of the denotation and composition of materials incorporated into the fireball (Crocker et al., 1966). The energy released from a large, near-surface denotation is sufficient to lift and immediately vaporize several hundred thousand tons of soil and associated material. Vapors formed within the fireball begin to condense within seconds of denotation and under certain conditions

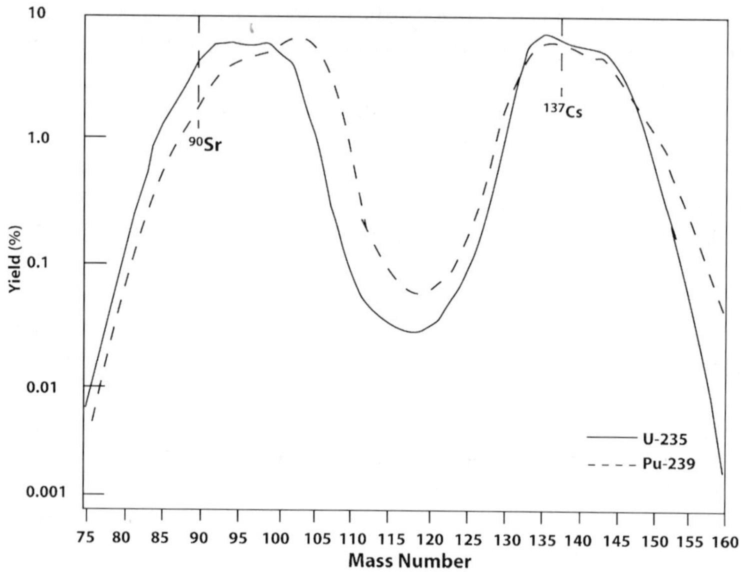

Fig. 4. Yield curves for fission of ^{235}U and ^{239}Pu.

fractionation between volatile and refractory materials may occur. Decay chain dynamics also play an important role in early condensation and fractionation of radioactive debris, especially for nuclides formed from short-lived noble gas precursors such as ^{137}Cs and ^{90}Sr. Soil particles entering the fireball may also serve as nuclei on which radioactive debris or other condensation products attach. Refractory nuclides tend to be incorporated into larger particles (0.4–4 mm) formed from condensation of iron, aluminum and other refractory materials. Larger particles settle to earth quickly and produce what is known as local or close-in fallout deposition. The volatile elements tend to be associated with, or deposited onto, the surfaces of smaller-sized particles (<0.3 mm in diameter) and are more likely to enter the global environment. It is assumed, on average, about 50% of the volatile fission products produced in near-surface detonations are deposited in the local or regional environment, while the remainder of the radioactive debris is widely dispersed into the global atmosphere (Peterson, 1970).

In addition to fission products, radioactive species are produced from neutron activation of non-fuel bomb and mounting materials, the atmosphere, and in soil and water within the immediate vicinity of the exploding device. The main activation products produced in atmospheric weapons tests of environmental significance include ^{3}H, ^{14}C, ^{36}Cl, ^{41}Ca, ^{45}Ca, ^{55}Fe, ^{59}Ni, ^{60}Co, ^{59}Ni and ^{63}Ni. The residual gamma spectra of typical environmental materials collected near former atmospheric nuclear test sites are usually dominated by ^{137}Cs, ^{60}Co and the europium isotopes 152 and 155. Europium-152 and ^{154}Eu are considered products of neutron capture. Tritium is both a fuel residue and a fuel product of thermonuclear explosions. Radiocarbon (^{14}C) is produced by the interaction of neutrons on atmospheric nitrogen based on the ^{14}N (n, p) ^{14}C reaction. The total residual tritium and radiocarbon released from atmospheric nuclear weapons testing has been estimated at 186,000 PBq and 213 PBq, respectively (UNSCEAR, 2000). As will be shown, these reactions have produced a marked increase in the natural background concentration of tritium and radiocarbon in the surface ocean.

The main components of a nuclear weapon may include tritium, lithium deuteride, uranium or plutonium. Tritium is primarily used for boosting a fission device but may also be used in the secondary stage of thermonuclear weapons. Tritium is both an intermediate product and fuel for thermonuclear denotations where it is produced through neutron reactions on lithium. Weapons grade uranium contains greater than 90% ^{235}U. Weapons grade plutonium contains greater than 93% ^{239}Pu and less than 7% ^{240}Pu. Some ^{241}Pu is normally present in a typical isotope mix of weapons grade plutonium. Plutonium-241 has a half-life of 14.35 years and decays to ^{241}Am; therefore, the amount of ^{241}Am present in weapons grade plutonium will depend on the age of the nuclear fuel.

Reactions of a nuclear explosion would ideally consume all the available nuclear fuel. Early nuclear weapons used a solid sphere or hollow shell (known as a pit) of ^{239}Pu or ^{235}U and consumed a relatively small fraction of the fuel. It was soon realized that introduction of deuterium–tritium gas could produce additional neutrons. These neutrons caused new fission chains and helped boost the overall yield of the device. Much higher yields were obtained by separating the fission trigger from the thermonuclear package in a Teller–Ulam 'H bomb' configuration. The secondary stage of thermonuclear weapons produce very high neutron fluencies with sufficient energy to fission and consecutive neutron capture reactions on uranium, in staged weapons. Natural uranium will capture a single neutron to form ^{239}Pu from decay of ^{239}U. The high-mass uranium isotopes are short-lived, and they decay by beta emission to longer-lived plutonium isotopes (e.g. ^{239}Pu, ^{240}Pu, ^{241}Pu and ^{242}Pu) over the course of a few days. The relative abundance of mass chains 239, 240, 241 and 242 in global fallout is 1 to 0.18 to 0.013 to 0.0043 (UNSCEAR, 2000; Krey et al., 1976). ^{237}Np may also be formed in thermonuclear detonations by (n, 2n) reactions on ^{238}U. Plutonium isotopes in the primary capture single neutrons to form ^{240}Pu, ^{241}Pu and ^{242}Pu. A (n, 2n) reaction takes place when one neutron is absorbed and two neutrons are ejected; in this case, to form ^{237}U which then decays to ^{237}Np by beta emission. High-energy neutrons also produce ^{238}Pu and ^{234}U. Early weapons containing large quantities of uranium were known as 'dirty bombs' and produced a significant fraction of the total radioactive debris in global fallout.

Other radioactive species associated with atmospheric nuclear testing programs included activation products of neutron fluence monitors. A number of stable elements incorporated into critical components of nuclear devices were commonly used to determine the neutron energy spectrum and flux (and for other diagnostic purposes) by measuring single and multiple (n, 2n) products. Residual quantities of ^{235}U and ^{238}U remaining in the environment from nuclear weapons testing are often overlooked because of the high natural uranium content of soils, seawater, and other environmental materials.

4.2. *Atmospheric nuclear weapons testing*

The main contribution to the global man-made radiation environment has come from the testing of nuclear weapons in the atmosphere. Updated listings of the date, type, total number and explosive yields of individual nuclear tests as reported by country have recently been compiled by UNSCEAR (2000). There were reportedly 543 atmospheric nuclear denotations carried out

Table 2
Atmospheric nuclear tests (1954–1980) by number, country, site, year and fission yield (data source: UNSCEAR, 2000)

Region	Test site	Country	Years	Number of test	Yield (Mt) Total	Yield (Mt) Fission
Equatorial Pacific	Bikini Atoll	United States	1946–1858	23	76.8	42.2
	Christmas Island	United Kingdom	1958	6	6.65	3.35
	Christmas Island	United States	1962	24	23.3	12.1
	Enewetak Atoll	United States	1947–1958	42	31.7	15.5
	Johnson Atoll	United States	1958–1962	12	20.8	10.5
	Pacific Ocean	United States	1955–1962	4	0.102	0.102
			Total	111	159	84
Northern temperate latitudes	Algeria	France	1960–1961	4	0.073	0.073
	Japan[a]	United States	1945	2	0.036	0.036
	Kapustin Yar	Former Soviet Union	1957–1962	10	0.98	0.68
	Lop Nor	China	1964–1980	22	20.72	12.2
	New Mexico	United States	1945	1	0.021	0.021
	Nevada Test Site (NTS)	United States	1951–1962	86	1.05	1.05
	Semipalatinsk	Former Soviet Union	1949–1962	116	6.59	3.74
	Totsk, Aralsk	Former Soviet Union	1954–1956	2	0.04	0.04
			Total	243	29.5	17.8
Polar-north	Novaya Zemlya	Former Soviet Union	1955–1962	91	239.6	80.8
			Total	91	239.6	80.8
Southern Hemisphere	Atlantic	United States	1958	3	0.0045	0.0045
	Fangataufa Atoll	France	1966–1970	4	3.74	1.97
	Malden Island	United Kingdom	1957	3	1.2	0.69
	Maralinga/Emu Test Ranges	United Kingdom	1953–1957	9	0.080	0.08
	Monte Bello Islands	United Kingdom	1952–1956	3	0.1	0.1
	Mururoa Atoll	France	1966–1974	37	6.38	4.13
			Total	59	11.5	7.0
Total	*All sites*	*All countries*		504[#]	440	189

[#]Includes 39 safety tests: 22 by the USA, 12 by the United Kingdom, and 5 by France.
[a]Two cases of military combat use.

between 1945 and 1980 with an estimated yield of 440 Mt. The annual number and total fission yields of atmospheric nuclear tests performed by all countries are summarized in Table 2, and the location of test sites shown in Fig. 1. The number of reported detonations includes 39 safety trails used to assess the impacts of handling or operations accidents. Atmospheric nuclear tests include detonations performed as airdrops, suspensions from balloons, launchings by rockets, mountings on towers, placement on barges, and from anchorage points under water. Through the end of 1998, an additional 1876 underground nuclear detonations have been reportedly carried out with a combined yield of 90 Mt (UNSCEAR, 2000). Cratering test are known to have released radioactive debris into the atmosphere but, in general, releases to the near-surface environment from underground explosions are only occasionally observed.

The first atomic bomb – code named Trinity – was detonated by the United States in July 1945 near the town of Alamagordo in New Mexico. Less than a month later, nuclear weapons were detonated under wartime conditions over the Japanese cities of Hiroshima and Nagasaki. The United States established a nuclear weapons testing program shortly after the end of World War II on Bikini Atoll in the northern Marshall Islands. Operation Crossroads commenced in June of 1946 and involved two nuclear tests: the ABLE shot (21 kt) detonated at an altitude of about 150 m and the BAKER shot (21 kt) detonated 30 m underwater inside the lagoon. Three more nuclear devices were detonated in 1948 on Enewetak Atoll before the Former Soviet Union tested their first nuclear device on 29 August of 1949. The United Kingdom followed with their first nuclear test during October of 1952. Atmospheric nuclear weapons testing continued at an accelerated rate from 1954 to 1958. A nuclear weapons-test moratorium was declared in the fall of 1958; by which time a total of 261 nuclear detonations had taken place having a cumulative explosive yield of 152 Mt. Virtually all these tests were carried out in the atmosphere with unrestrained release of radioactivity. Nine detonations had estimated explosive yields equal to or greater than 4 Mt representing more than 55% of the estimated total cumulative fission yield for all tests to 1958. These high-energy tests ($\geqslant$4 Mt) were all detonated in the near-surface environment of Bikini and Enewetak Atolls in the northern Marshall Islands.

A nuclear weapons testing moratorium was observed through 1959 before France detonated its first nuclear weapon on 13 February of 1960 in Algeria. The Former Soviet Union resumed nuclear testing in September of 1961 prompting the United States to take a similar course of action in April of 1962. This second phase of atmospheric nuclear testing (1960–1962) involved 180 detonations having a total estimated explosive yield of 257 Mt. There were 118 atmospheric nuclear tests conducted in 1962 alone. There were also 15 atmospheric nuclear tests with total yields equal to or greater than 4 Mt and in contrast to high-energy tests detonated between 1954 and 1958, the majority of these tests were located in the polar atmosphere. The two exceptions were airdrops detonated over Johnson Atoll and Christmas Island in the equatorial mid-Pacific. China conducted the only additional high-energy nuclear test on 17 November of 1976 at the Lop Nor test site in Sinkiang Province.

The Former Soviet Union, the United States and the United Kingdom agreed to an atmospheric test-ban treaty in 1963. France and China were not signatories to the treaty. China detonated its first nuclear weapon in October of 1964, and through 1980 conducted a total of 22 nuclear tests at the Lop Nor test site. In 1966, France moved its atmospheric nuclear testing program to the Tuamotu Islands, in French Polynesia, where they detonated 4 tests on Fangataufa Atoll and 37 tests on Mururoa Atoll.

As will be shown, the injection and partitioning of radioactive debris in the atmosphere can be estimated from the location and yield of each test. There were 111 atmospheric nuclear tests conducted in the equatorial-north Pacific Ocean, 243 nuclear tests at northern temperate latitudes, 91 nuclear tests in the polar-north and 59 tests conducted in the Southern Hemisphere (Table 2). Therefore, the total fission energy released for partitioning of radioactive debris in the atmosphere is largely divided between the equatorial Pacific (43%) and the polar-north (44%). Approximately 64% of the total fission energy released in the polar north was delivered by 13 high-energy detonations ($\geqslant$4 Mt) between 1961 and 1962, and 64% of the total fission energy released in the equatorial Pacific was delivered by 11 high-energy ($\geqslant$4 Mt) denotations, most of which took place between 1954 and 1958. Taken together, high-energy denotations account for 66% of the total yield and 56% of the fission yield of all atmospheric nuclear tests carried out between 1945 and 1980.

4.3. *Dispersion and deposition of radioactive debris*

The nature and partitioning of radioactive fallout between the local environment, the troposphere and the stratosphere are determined by

(1) the type, location and altitude of the test;
(2) the energy yield; and
(3) the quantity and type of materials interacting with the device (Hamilton et al., 1996).

Radioactive debris deposited at or near test sites within a few hours from the time of detonation is described as local or close-in fallout deposition. The extent to which a given nuclear explosion will produce local or regional fallout depends on the explosive yield of the test, its height above ground and the physical characteristics of the particles formed. Airbursts are defined as those tests occurring in the atmosphere at or above a height of $55Y^{0.4}$ m (where Y is the total yield in kiloton) (Petersen, 1970). Below this height the fireball is expected to interact with the Earth's surface and produce a significant amount of local and regional fallout contamination. The apportionment of radioactive debris in the atmosphere is coupled to the stabilization height of cloud formation following a nuclear explosion (Petersen, 1970) and can be estimated empirically from partitioning yield estimates. It is assumed that on average about 50% of the volatile radionuclides (e.g. ^{90}Sr, ^{137}Cs and ^{131}I) produced in near-surface denotations entered the local or regional environment. The remainder of the debris from near-surface denotations and all the debris from airbursts entered the global environment (UNSCEAR, 2000) producing a worldwide pattern of global fallout deposition.

Updated annual partitioning yields for injection of radioactive debris into the various atmospheric regions were recently published by UNSCEAR (2000) and are shown graphically in Fig. 5. The total partitioning yield contributing to worldwide dispersion of radioactive debris is estimated to be around 160.5 Mt compared with about 29 Mt deposited locally or regionally (UNSCEAR, 2000). The latter estimate is somewhat uncertain because of varying conditions between tests and the seasonality of atmospheric transport processes, but the uncertainty is small compared with the total fraction injected into the global atmosphere. Partitioning yields into the troposphere, stratosphere and high equatorial atmosphere were 15.6 Mt, 139 Mt and 6.4 Mt, respectively. About 17 Mt was released into the Southern Hemisphere compared with 144 Mt in the Northern Hemisphere. Injection of nuclear debris into the northern equatorial stratosphere was most pronounced in 1954 followed by a significant pulse into the polar-north stratosphere during 1961–1962. About 44.6 Mt and 87.4 Mt of radioactive debris was injected into the northern equatorial and polar stratospheres, respectively, and together these two regions account for more than 82% of the total radioactive debris dispersed globally by atmospheric nuclear weapons testing.

Concern over the fate and transport of radioactive debris from nuclear weapons tests conducted in the late 1950s led to the establishment of a series of global monitoring networks.

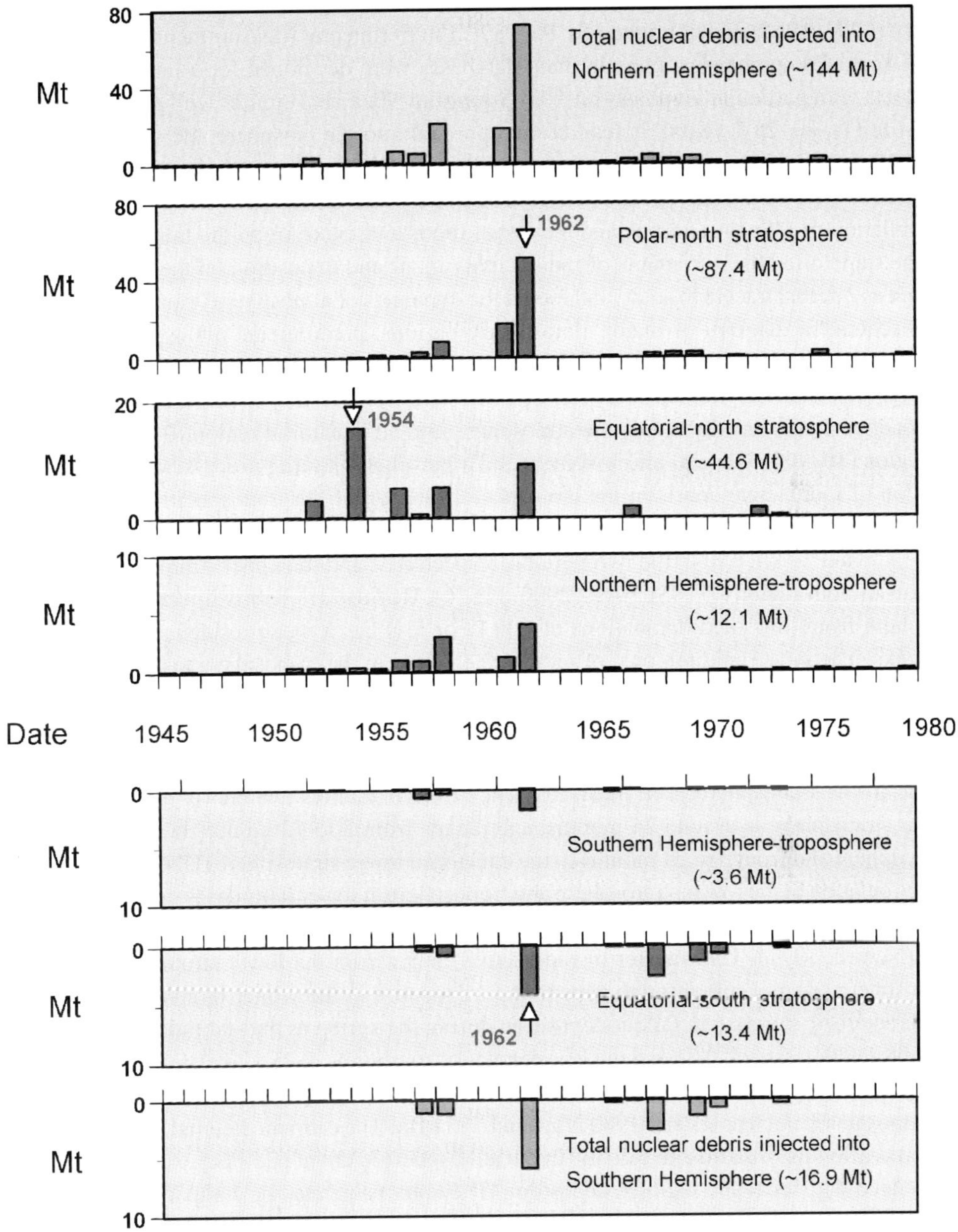

Fig. 5. Annual partitioning yields from nuclear detonations and apportionment of debris in the atmosphere. Partitioning from equatorial sites such as Christmas Island and high-altitude tests on Johnson Island were assumed equally divided between the Northern and Southern Hemispheres. For tests conducted at temperate sites (30–60°) releases were essentially averaged between the equatorial and polar atmospheres depending on the month of the year the nuclear test was conducted (data source: UNSCEAR, 2000).

A surface-air monitoring program was initially established by the United States Naval Laboratory (1957–1962) and continued by the DOE Environmental Measurements Laboratory in New York. Comprehensive measurement databases were developed for a number of fission products with particular emphasis on ^{90}Sr. Strontium-90, a beta emitter with a relatively long half-life ($T_{1/2} = 28.7$ years), is readily incorporated into the biosphere and shares chemical properties with calcium, an essential element for most organisms including humans. A primary concern over the fate of radioactivity released by nuclear weapons testing programs was assimilation of ^{90}Sr into marine and terrestrial foods and exposure to the human population. At the same time, measurements of radioactivity in air and deposition on the Earth's surface served as internal tracers to study and model the dynamics of atmospheric transport processes.

A schematic diagram of major atmospheric regions, predominant atmospheric transport processes, and the measured ^{90}Sr deposit averaged over 10-degree latitude bands is shown in Fig. 6. Empirical models used to describe atmospheric dispersion and deposition of radioactive debris usually divide the atmosphere into an equatorial region (0–30°) and a polarregion (30–90°). General air movement and atmospheric mixing processes control the dispersion of radioactive debris in the atmosphere. The top of the troposphere averages about 9 km in the polar region and 17 km in the equatorial region. The lower stratosphere extends to 17 km and 24 km within the two regions, respectively, and the upper stratosphere to about 50 km in both regions (UNSCEAR, 2000; Fig. 6). Radioactive debris injected into the troposphere from low-yield detonations of 100 kt or less has a mean residence time of about 3 weeks. The most rapid removal of radioactive debris in the troposphere takes place during rainout events occurring locally or regionally within a few thousand kilometers from the test site. Radioactive particles injected into the stratosphere behave as aerosols and descend more slowly by gravitational settlement in the upper stratosphere and by eddy diffusion processes in the lower stratosphere. The mean residence time of radioactive debris injected into the upper stratosphere is around 24 months, and ranges from 3 to 12 months in the polar lower stratosphere and from 8 to 24 months in the equatorial lower stratosphere (UNSCEAR, 2000). Air circulation in the lower stratosphere and troposphere at lower latitudes is driven by Hadley cell formation. Hadley cells tend to increase or decrease in size and shift in latitude with season (Newell, 1971). The transfer of radioactive debris from the lower stratosphere into the troposphere often occurs through gaps in the tropopause in the winter months and produces a characteristic increase in fallout deposition during the spring at mid-latitudes. The normalized production rate of ^{90}Sr in a nuclear denotation, assuming 1.45×10^{26} fissions per Mt, is around 3.88 PBq Mt^{-1} (UNSCEAR, 2000). The corresponding amount of ^{90}Sr produced in all atmospheric nuclear tests to 1980 is around 733 PBq. The estimated global release inferred from partitioning of radioactive debris in the atmosphere is about 623 PBq excluding releases associated with local and regional deposition. The latter value may be compared with a hemispheric ^{90}Sr deposit of 612 PBq using global fallout measurements (after UNSCEAR, 2000; Monetti, 1996). The partitioning yield estimate of 160.5 Mt provides better agreement with the measured deposition than previous fission yield estimates. Moreover, the measured results indicate that about 2% of the ^{90}Sr injected into the global atmosphere decayed before deposition and infers that the average residence time of radioactive debris in the atmosphere was $\sim$1 year.

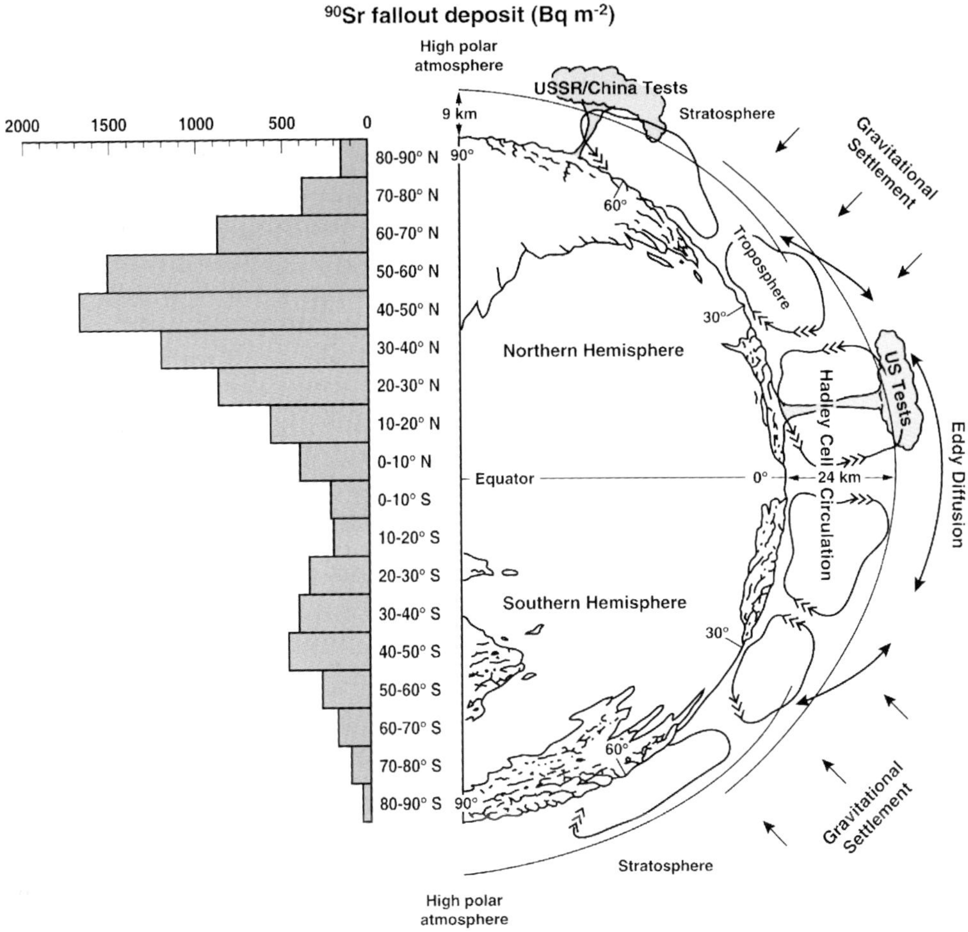

Fig. 6. Schematic diagram of transfer mechanisms between atmospheric compartments (to right) and the measured latitudial fallout deposit of ^{90}Sr (through 1990, to left) (Monetti, 1996; modified after Kathren, 1984).

The measured global cumulative deposit reached a maximum of around 460 PBq in 1967–1972 (UNSCEAR, 2000); since this time the cumulative deposit has decreased because radioactive decay of the global ^{90}Sr burden has been more rapid than inputs from the atmosphere (Fig. 7). The calculated annual deposition of ^{90}Sr derived from model calculations shows close agreement with measured values up until the early 1980s but becomes more uncertain with time (Fig. 7). Deviations between the measured deposition and that calculated using empirical models can be attributed to the uncertainty of the measurements, resuspension of previously deposited material, and the influence of ^{90}Sr deposited during 1986 after the Chernobyl accident in the Ukraine. About 1.4 PBq of the ^{90}Sr deposit was associated with the Chernobyl accident. The measured global deposit of ^{90}Sr in 1990 was around 311 PBq. By 2000, the ^{90}Sr deposit would have decayed to about 245 PBq.

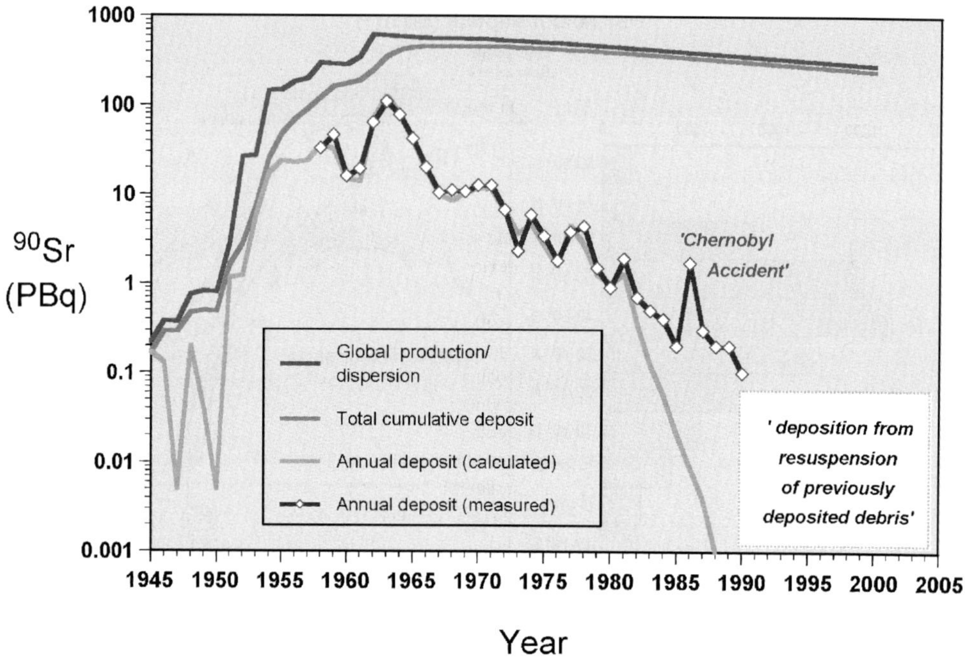

Fig. 7. Estimated global production, and the cumulative and annual deposition of ^{90}Sr calculated from individual fission yields of tests and atmospheric model predictions (after UNSCEAR, 2000) compared with the measured fallout deposit of ^{90}Sr through 1990 (Monetti, 1996).

5. Sources of anthropogenic radionuclides in the oceans

Key radionuclides produced and dispersed globally by weapons testing, and their respective half-lives, fission yields, production modes, and global releases into the atmosphere are shown in Table 3 (updated after UNSCEAR, 2000).

Partitioning of fission yield estimates between the local, regional and global environment vary from test to test, and on the fractionation of volatile and refractory elements. The global production and dispersion estimates shown in Table 3 do not include debris injected into local and regional environments. As previously formulated, the measured cumulative deposit of volatile radionuclides in integrated fallout is consistent with global dispersion and deposition of all the debris from airbursts and 50% of the debris from near-surface detonations. For the volatile radionuclides, about 29 Mt of fission energy is assumed to be deposited locally and regionally; of which, about 28 Mt comes from injections from near-surface detonations at test sites in the Pacific. Partitioning estimates for refractory radionuclides in near-surface detonations assume 50% of the debris is deposited within the immediate vicinity of the test site while an additional 25% is deposited regionally (Beck & Krey, 1983; Hicks, 1982). In general, the radionuclide composition of fallout debris produced in near-surface detonations can be adequately described as having all the volatile radionuclides and half of the refractory radionuclides present in unfractionated debris (Hicks, 1982). The corresponding partitioning

Table 3
Production and global release of key radionuclides in atmospheric nuclear tests

Nuclide	Half-life (years)	Fission yield (%)	Production mode	Global release into the atmosphere[a] (excluding local fallout)	
				PBq	kg
^{3}H	12.33	–	Fuel residue and fuel product	186,000	518
^{14}C	5730	–	(n, p) in device and environment; ^{14}N (n, p); ^{13}C (n, γ)	213	1290
^{55}Fe	2.73	–	(n, γ), (2, 2n) and (n, α) in device, and (n, γ) in the environment	1530	17
^{90}Sr	28.78	3.5	Fission product	623	122
^{99}Tc	211,100	5.8	Fission product	0.14	222
^{125}Sb	2.76	0.4	Fission product	741	20
^{129}I	15,700,000	1.7	Fission product; ^{129}Xe (n, p)	0.0006	87
^{137}Cs	30.07	5.6	Fission product; ^{137}Ba (n, p)	948	295
^{237}Np[b]	2,200,000	–	^{238}U (n, 2n) ^{237}U, $\tilde{\beta}$-decay	0.03	1270
^{238}Pu[b]	87.7	–	Fuel residue and fuel product; ^{239}Pu (n, 2n)	0.28	0.18
^{239}Pu[b]	24,100	–	Fuel residue and fuel product; ^{238}U (n, β^{-}) ^{239}Np, $\tilde{\beta}$-decay	6.52	2835
^{240}Pu[b]	6500	–	Fuel residue and fuel product; ^{238}U (2n, 2β^{-})	4.35	512
^{241}Pu[b]	14.4	–	Fuel residue and fuel product; ^{238}U (3n, 2β^{-})	142	37
^{242}Pu[b]	375,000	–	Fuel residue and fuel product; ^{238}U (4n, 2β^{-})	0.002	11
^{241}Am[c]	433	–	Fuel residue and fuel product; β-decay of ^{241}Pu	(4.8)	(38)

[a]Total global dispersion of radioactive debris resulting from 160.5 Mt of fission energy and 250.6 Mt of fusion energy (after UNSCEAR, 2000).
[b]Estimated from isotopic ratios in integrated global fallout.
[c]Maximal deposit from decay of the global ^{241}Pu deposit (expected in 2032 at a level equivalent to ~44% of the present global $^{239+240}Pu$ activity deposit).

yield for dispersion of refractory radionuclides into the global environment is estimated at 130 Mt. The remainder of the debris (or equivalence of about 59 Mt of fission energy) was presumably deposited as local or regional fallout. The most significant localized releases were derived from high-energy, near-surface nuclear detonations on Bikini and Enewetak Atolls in the northern Marshall Islands (Fig. 8). Much of the local and regional deposition from these tests entered the marine environment forming a significant source-term, especially in the Northwest Pacific Ocean.

Sixty-six nuclear devices were detonated on Bikini and Enewetak Atolls. Near-surface blasts on towers, barges or underwater produced large quantities of partially or completed vaporized CaO, $Ca(OH)_2$ and $CaCO_3$ (Joseph et al., 1971; Adams et al., 1960). The physical/chemical characterization of local and regional fallout from the Pacific Proving Grounds

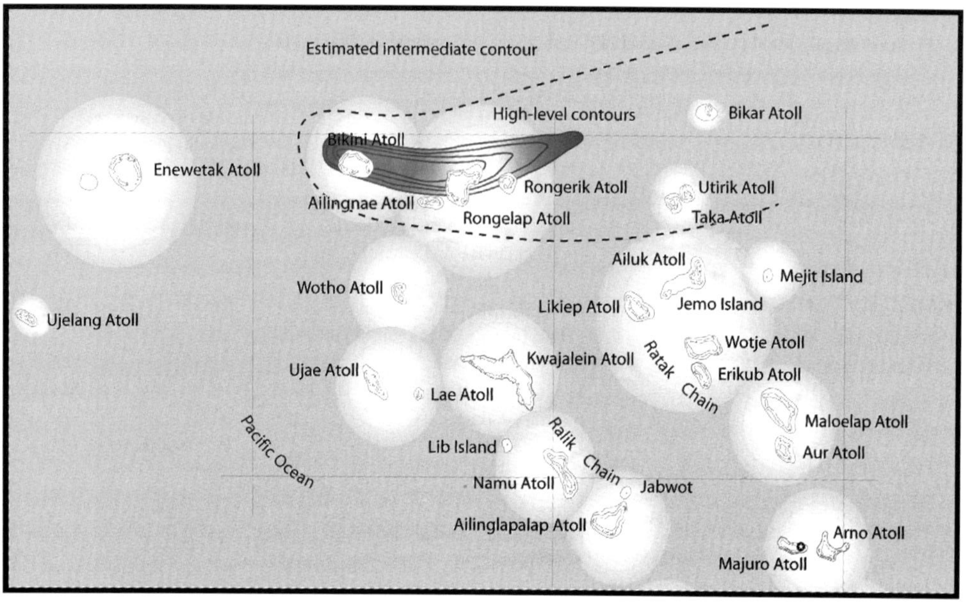

Fig. 8. Local and intermediate fallout following the detonation of a 15 Mt thermonuclear test 'Bravo' on Bikini Atoll on 1 March of 1954 in the northern Marshall Islands.

is different to that found in globally dispersed debris. High-temperature vaporization and condensation processes produced different types and sizes of high-specific activity particles (Schell et al., 1980), which interacted with seawater (Adams et al., 1960). On hydration the particles swell forming a 'crumbly or fluffy structure' (Adams et al., 1960); this is accompanied by release of hydroxyl ions and interaction with magnesium ions in seawater to form an inert shell of magnesium hydroxide on the particles (Buesseler, 1997). These relatively insoluble particles deposited in the lagoon and continental shelf or slope sediments formed a reservoir and secondary source term to the equatorial Pacific. As of 1972, the $^{239+240}$Pu inventory in lagoon slope sediments of Bikini and Enewetak Atoll was 54.4 TBq and 44.4 TBq, respectively (Noshkin & Wong, 1979). The corresponding annual export of $^{239+240}$Pu to the open ocean was estimated at 0.12 TBq and 0.10 TBq, respectively. Similar studies at test sites in the South Pacific indicate that the annual export of $^{239+240}$Pu from Mururoa Atoll Lagoon to the open ocean is around 0.02 TBq (Bourlat et al., 1995). Radionuclide remobilization processes at these test sites are thought to be responsible for addition of up to 2 PBq of ^{137}Cs and 8–9 TBq of $^{239+240}$Pu to the Pacific Basin (Hamilton et al., 1996).

The total oceanic inventories of selected fission products and transuranic elements produced in atmospheric nuclear weapons tests are shown in Table 4. Radionuclide inventories were calculated by multiplying the measured latitudinal deposit in 10-degree bands with the corresponding fractional sea surface area across the world's oceans (taken from Baumgartner & Reichel, 1975). Marine contributions from nuclear weapons testing include 189 PBq of ^{90}Sr and 300 PBq of ^{137}Cs (Table 4). This compares with about 90 PBq of ^{90}Sr and 142 PBq of ^{137}Cs deposited globally in the terrestrial environment. Neglecting transfers between oceans, the Pacific contains about 60% of the total oceanic pool of volatile radionuclides and up to 70% of the refractory radionuclides (with the exception of ^{238}Pu at ~50%). Readers are reminded that 90% of the debris injected into the global atmosphere was deposited prior to the 1970s so present-day distributions of fallout radionuclides in the oceans will largely be controlled physical mixing and biogeochemical processes rather than a function of deposition patterns through the 1950s and 1960s (Hamilton et al., 1996, and references therein). The inventory estimates shown here for ^{90}Sr and ^{137}Cs are consistent with the results of the GEOSECS (Geochemical Ocean Section) expeditions with the exception of the Arctic Ocean. Discrepancies between the measured and predicted radionuclide inventories in the Arctic Ocean will be discussed in detail in the following sections. It is estimated that the Arctic, Atlantic, Indian and Pacific Oceans contain about 2.0 PBq, 51 PBq, 22 PBq and 114 PBq of ^{90}Sr, and about 3.2 PBq, 81 PBq, 31 PBq and 182 PBq of ^{137}Cs, respectively. Other important long-lived fallout radionuclides of radiological or geochemical significance include the transuranic elements along with ^{99}Tc and ^{129}I (Table 4).

The radionuclide inventories shown in Table 4 include estimated contributions from local and regional fallout as well as global dispersion of ^{238}Pu from the abortive reentry of a navigational satellite over the Indian Ocean. The SNAP-9A satellite was launched on April 21, 1964, but failed to reach orbital velocity and burned up in the high stratosphere. The SNAP-9A power unit contained about 0.63 PBq of ^{238}Pu and almost tripled the global deposit of ^{238}Pu from nuclear weapons testing (Hardy et al., 1973; Krey, 1967). Moreover, the SNAP-9A ^{238}Pu deposition distribution is an entirely different pattern compared with fallout from nuclear weapons tests. Approximately 80% of the SNAP-9A ^{238}Pu deposit was found in the Southern Hemisphere compared with about 20% from weapons fallout. In addition, about 1.7 PBq of ^{238}Pu and 1.2 TBq of ^{239}Pu entered the Tonga Trench in the North Pacific when an Apollo lunar probe was aborted in flight (Dobry, 1980). It is thought that the plutonium inside the lunar probe has remained intact in its containers at the bottom of the sea (Aarkrog, 1988). Localized inputs of plutonium into the marine environment also occurred in January of 1968 when an American B-52 aircraft carrying four nuclear weapons crashed on ice in Bylot Sound near Thule, Greenland. Plutonium contained in the weapons was released into the environment by a conventional chemical explosion and further distributed in the fuel fire following the crash. It has been estimated that about 1 TBq (or about 0.5 kg) of plutonium remained in bottom sediments of Bylot Sound after the initial cleanup operation (Smith et al., 1994; Aarkrog et al., 1984, 1987).

For comparison, about 3.6 t of plutonium was dispersed into the oceans from nuclear weapons tests (Table 4). The Thule plutonium was initially present in the form of discrete oxide particles of low solubility and is not thought to constitute a significant source of plutonium contamination in the surrounding environment (AMAP, 1998). In 1966, a similar incident involving the crash of a U.S. aircraft carrying nuclear weapons occurred near Palomores, Spain. In this instance, plutonium contamination was largely confined to agricultural lands surrounding the crash site. Impacts on the marine environment were apparently limited by the recovery of an intact nuclear weapon that fell into the Mediterranean Sea (unofficial source).

The world's oceans contain about 12 PBq of $^{239+240}Pu$ from global and local fallout deposition (Table 4). The Arctic, Atlantic, Indian and Pacific Oceans contain about 0.1 PBq, 2.3 PBq,

Table 4

Oceanic inventory of fission products and transuranium elements originating from globally dispersed debris, and local and regional deposits from atmospheric nuclear weapons test

Globally dispersed radioactive debris deposited in the oceans (including estimates of regional fallout from Pacific Ocean tests sites and deposition from SNAP-9A)

Radio-nuclide	Arctic Ocean		Atlantic Ocean		Indian Ocean		Pacific Ocean		Total oceanic inventory	
	PBq	kg	PBq	kg	PBq	kg	PBq	kg	PBq	kg
^{90}Sr	2.0	0.40	51	10	22	4	114[a]	23	189	37
^{99}Tc	0.001	1.8	0.03	46	0.01	20	0.07[b]	110	0.11	178
^{129}I	0.000005	0.72	0.00012	18	0.000050	7.7	0.0003[c]	43	0.0005	69
^{137}Cs	3.2	1.0	81	25	34	11	182[d]	57	300	93
^{237}Np	0.0003	11.3	0.007	263	0.002	94	0.02[e]	888	0.03	1256
^{238}Pu	0.002[m]	0.0032	0.13[n]	0.2	0.11[o]	0.17	0.50[f,p]	0.78	0.73	1.16
^{239}Pu	0.054	23	1.4	591	0.58	250	4.5[g]	2960	6.5	2820
^{240}Pu	0.036	4.2	0.90	106	0.38	45	4.0[h]	477	5.4	632
$^{239+240}Pu$	0.090	28	2.3	697	0.96	295	8.6[i]	2436	12	3456
^{241}Pu	0.17	0.046	4.3	1.1	1.8	0.48	24[j]	6.2	30	7.9
^{242}Pu	0.00001	0.091	0.0003	2.3	0.0002	1.0	0.003[k]	22	0.004	26
^{241}Am	0.04	0.29	0.92	7.2	0.39	3.1	3.7[l]	29	5.1	40

[#]Decay date 1 January, 2000.

[a]Includes an estimated 36 PBq of ^{90}Sr in local and regional fallout.

[b]Includes an estimated 0.025 PBq of ^{99}Tc in local and regional fallout.

[c]Includes an estimated 0.0001 PBq of ^{129}I in local and regional fallout.

[d]Includes an estimated 58 PBq of ^{137}Cs in local and regional fallout.

[e]Includes an estimated 0.013 PBq of ^{237}Np in local and regional fallout calculated from measured isotope ratios in global fallout deposition.

[f]Includes an estimated 0.11 PBq of ^{238}Pu in local and regional fallout calculated from measured isotope ratios in global fallout deposition.

[g]Includes an estimated 2.4 PBq of ^{239}Pu in local and regional fallout calculated from apportionment of radioactive debris in the atmosphere.

[h]Includes an estimated 2.7 PBq of ^{240}Pu in local and regional fallout using isotopic ratios observed in fallout from the 'Mike' detonation on Eneweatk Atoll.

[i]Includes an estimated 5.1 PBq of $^{239+240}Pu$ in local and regional fallout using isotopic ratios observed in fallout from the 'Mike' detonation on Eneweatk Atoll.

[j]Includes an estimated 17 PBq of ^{241}Pu in local and regional fallout using isotopic ratios observed in fallout from the 'Mike' detonation on Eneweatk Atoll.

[k]Includes an estimated 0.003 PBq of ^{242}Pu in local and regional fallout using isotopic ratios observed in fallout from the 'Mike' detonation on Eneweatk Atoll.

[l]Includes a maximal deposit of 2.3 PBq of ^{241}Am in local and regional fallout from decay of ^{241}Pu.

[m]Includes 0.0003 PBq of ^{238}Pu to the Arctic Ocean from SNAP-9A burnup (the ^{238}Pu deposit was distributed globally after Hardy et al., 1973).

[n]Includes 0.083 PBq of ^{238}Pu to the Atlantic Ocean from SNAP-9A burnup (the ^{238}Pu deposit was distributed globally after Hardy et al., 1973).

[o]Includes 0.090 PBq of ^{238}Pu to the Indian Ocean from SNAP-9A burnup (the ^{238}Pu deposit was distributed globally after Hardy et al., 1973).

[p]Includes 0.15 PBq of ^{238}Pu to the Pacific Ocean from SNAP-9A burnup (the ^{238}Pu deposit was distributed globally after Hardy et al., 1973).

1.0 PBq and 8.6 PBq of $^{239+240}$Pu, respectively. These data predict that a significant fraction of the plutonium present in the Pacific Ocean comes from local and regional injections rather than global fallout deposition. Water column inventories in the North Pacific do reflect an excess of plutonium above what is expected from global fallout deposition (Bowen et al., 1980). For example, the $^{239+240}$Pu inventory in the North Pacific water column during the 1973–1974 GEOSECS expeditions was estimated to be around 6.4 PBq or about 50% more than that expected from global fallout alone. Recent observations of anomalously high ^{240}Pu/^{239}Pu atom ratios in the water column and dated coral records (Buesseler, 1997) provide strong evidence the main source of this excess plutonium is local fallout from the Pacific Proving Grounds in the Marshall Islands. The average ^{240}Pu/^{239}Pu atom ratio in integrated global fallout is around 0.18 (Kiode et al., 1985; Krey et al., 1976). Much higher ratios have been observed in debris from high-energy nuclear tests conducted in the Marshall Islands; for example, debris from the 'Mike' thermonuclear test conducted on Enewetak Atoll in 1952 contained a ^{240}Pu/^{239}Pu atom ratio of ~0.36 (DOE, 1982). Plutonium isotopic abundances for the 'Mike' test were used to predict the local and regional inputs to the Pacific shown in Table 4. A mixture of 4.6 PBq of global fallout with a ^{240}Pu/^{239}Pu atom ratio of ~0.18 combined with 5.1 PBq of plutonium distributed as 'Mike' debris over the North Pacific would give rise to an elevated ^{240}Pu/^{239}Pu atom ratio in the water column of around 0.24. Observed ^{240}Pu/^{239}Pu atom ratios reported in North Pacific seawater and sediment samples over a wide geographical area of the North Pacific range from ~0.19 to ~0.34 (Buesseler, 1997); the average ratio is ~0.23 in close agreement with the predicted value given above. The elevated levels of plutonium in the water column combined with available source-specific plutonium isotopic information appear to confirm earlier speculation that local fallout from nuclear weapons tests conducted in the Marshall Islands, typically characterized by an elevated ^{240}Pu/^{239}Pu atom ratios (>0.2), may form a major plutonium source-term in the North Pacific. Moreover, the trend toward finding higher ^{240}Pu/^{239}Pu atom ratios in deep water samples and underlying sediments suggests plutonium geochemistry differs between sources of plutonium. Plutonium contained in local fallout is preferentially removed from the water column whereas global fallout plutonium behaves as a more soluble tracer. Source specific behaviors of Pu-bearing fallout particles can also be used to describe the relatively low ^{240}Pu/^{239}Pu atom ratios observed in some coastal sediments of the United States (Buesseler, 1986; Scott et al., 1983) because of preferential removal of Nevada fallout containing an average ^{240}Pu/^{239}Pu atom ratio of 0.035 (Hicks & Barr, 1984).

Other important sources of artificial radioactivity in the marine environment include the dumping of nuclear waste; controlled effluent discharges associated with the nuclear fuel cycle and nuclear weapons production activities; accidental releases from land-based nuclear installations; and other accidents and/or losses at sea involving nuclear materials. Accurate estimates of the potential incremental or 'pulse-like' releases of radionuclides contained in dumped nuclear waste or in nuclear materials lost at sea (e.g. nuclear reactors or intact nuclear weapons aboard sunken vessels; and other accidents involving nuclear weapons and/or sealed radioactive sources) are difficult to obtain and will only be reviewed here on a comparative basis (Fig. 9).

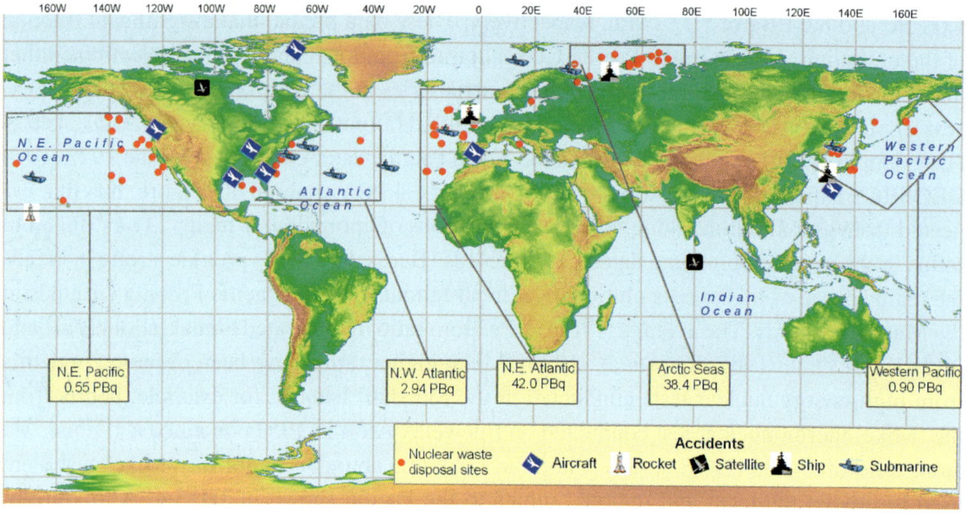

Fig. 9. Inventory of radioactive waste disposals at sea.

5.1. *Ocean dumping of nuclear waste*

Approximately 85 PBq of radioactive waste has been deliberately dumped into the oceans at more than 80 different locations worldwide (Fig. 9). The first sea disposal operation took place in 1946 off the coast of California in the Northeast Pacific. Pursuant to principals adopted by an inter-government conference held in London in 1972, a convention was adopted for the Prevention of Marine Pollution by Dumping of Wastes and Other Matter (IMO, 1972). The Contracting Parties to the London Convention 1972 agreed to promote the effective control of all sources of pollution of the marine environment. The International Atomic Energy Agency (IAEA) further pledged to take practicable steps to prevent pollution of the sea by dumping of waste and other matter liable to

(1) create hazards to human health,
(2) harm living resources and marine life,
(3) damage amenities, or
(4) interfere with other legitimate uses of the sea (IAEA, 1999).

The convention was ratified on 30 August of 1975 and became known as the London Dumping Convention (LDC). The contracting parties to the convention designated the IAEA as the international organization responsible for matters related to sea disposal of radioactive waste. The IAEA was mandated to establish criteria for sea disposal of radioactive waste in order to minimize impacts on man and the environment, and provide recommendations for environmental assessment methodologies. Through 1977, sea disposal operations took place under national authority using consultative mechanisms developed through the IAEA and United Nations Organization for Economic Co-operation and Development/Nuclear Energy Agency (OECD/NEA).

Between 1946 and 1982, five countries including Belgium, Japan, the Former Soviet Union, the United Kingdom and the United States regularly used sea disposal as a nuclear waste management option. During this period, about 46 PBq of low-level nuclear waste was dumped at sea, mostly in the Northeast Atlantic dumpsite (Fig. 9). The OECD/NEA maintained records of the waste disposal operations carried out by its Member States. Some indication of the composition and origin of the waste was usually given with beta–gamma emitters making up about 98% of the radionuclide inventory. The tritium inventory of radioactive waste disposal at sea to 1982 was around 15.6 PBq (IAEA, 1991) but represents less than 0.02% of the oceanic tritium deposit from atmospheric nuclear weapons testing (GESAMP, 1990). Other important radionuclides contained in radioactive waste dumped at sea include ^{90}Sr, ^{137}Cs, ^{55}Fe, ^{59}Co, ^{60}Co and ^{14}C, and lesser quantities ($<2\%$) of alpha-emitting radionuclides such as plutonium and americium. Radioactive waste dumped at sea consisted largely of low-level packaged waste from research, medical, industry and military activities. The waste packages contained mostly protective clothing, glass, and contaminated concrete, piping and other building materials encased in concrete or bitumen and/or placed in metal drums.

The OECD Council decided to develop a Coordinated Research and Environmental Surveillance Programme (CRESP) in 1977 as a key provision within a multilateral consultation and surveillance mechanism to keep sea disposal operations under review. The programme has conducted regular surveys of the Northeast Atlantic dumpsite since this time and the results have been published in reports (NEA/OECD, 1996). The IAEA Monaco Laboratory in cooperation with the Bundesforschungsanstalt fur Fischerrei (BFA), Germany, and the Fisheries Laboratory of the Ministry of Agriculture, Fisheries and Food, in the United Kingdom, conducted a major assessment of the Northeast Atlantic dumpsite in 1992. Analyses of seawater, sediments and biota indicate a local source of radioactivity entering the environment but negligible radiological impact (Baxter et al., 1995). The U.S. Environmental Protection Agency and the U.S. National Oceanic and Atmospheric Administration have also carried out radiological surveys of the Pacific and Northwest Atlantic dumpsites. Again, the results of these surveys show that migration of radioactivity is limited to transfers to water and sediment in close proximity to waste packages residing on the sea floor.

A voluntary moratorium on disposal of low-level radioactive waste at sea was introduced in 1985 by the Contracting Parties to the London Convention–Resolution LDC.21(9); a new resolution was adopted in 1993 prohibiting sea disposal of radioactive waste. Then in May 1993, the Russian Federation provided IAEA with information on previously undisclosed sea disposal operations adjacent to the territories of the Russian Federation in the Arctic and Far Eastern Seas. The Russian Federation published a report of historical waste disposal practices of the Former Soviet Union and the Federation through 1993 in what became commonly known as the 'White Book' (White Book, 1993). Radioactive materials dumped in the Arctic Ocean have included liquid waste, solidified packaged and unpackaged solid waste from nuclear installations, and, perhaps most alarmingly, high-level radioactive waste including reactor and reactor compartments with and without spent nuclear fuel. Information gathering about the dumping operations, extensive surveys of the dumpsites and surrounding environment, and radiological impact assessments are continuing (AMAP, 1998; IAEA, 1997; NRPA, 1996). According to information published by the Russian Federation, the total amount of radioactivity dumped in the Arctic Ocean was approximately 90 PBq at the time of disposal (White Book, 1993). The dumped objects included a total of 16 reactor assemblies or compartments, six nuclear reactor objects from submarines, and a reactor shielding assembly from the icebreaker *Lenin*, all containing spent nuclear fuel. Five of the reactor objects, including the reactor assembly from the icebreaker *Lenin*, were dumped in shallow fjords off the island of Novaya Zemlya between water depths of 20–50 m. The remaining reactor object with spent nuclear fuel was dumped in the Novaya Zemlya Trough at a depth of 300 m (Fig. 2). All the dumped objects containing spent nuclear fuel were filled with a special polymer mixture, identified as Furfurol™, prior to dumping. Deterioration of this protective barrier will eventually lead to the corrosion of the materials housing the reactor and the nuclear fuel, releasing residual radioactivity into the environment. A detailed assessment of the source terms indicate that the radionuclide inventory in the dumped reactors is about 40% of the value given in the White Book (IAEA, 1997; Sivintsev, 1994a, b; Yefimov, 1994). The Northeast Atlantic and Arctic Ocean dumpsites combined received about 95% of the radionuclide inventory of waste dumped in the world's oceans. The remaining sea disposal operations occurred in the Northeast Pacific, Western Pacific and Northwest Atlantic (Fig. 9); a small quantity of radioactive waste (about 1 TBq) was also dumped off the east coast of New Zealand. Sea disposal of radioactive waste has been prohibited under a resolution adopted by the Contacting Parties to the London Convention since 1993 (IMO, 1993).

The last documented waste disposal operation at sea occurred in 1993 when the Russian Federation released low-level liquid radioactive waste into the Sea of Japan (Danilov-Danilyan, 1993). A series of Japanese–Korean–Russian expeditions to the region during 1994–1995 indicated that the concentration of ^{90}Sr, ^{137}Cs, and plutonium isotopes in seawater and underlying sediments were within the range expected from global fallout from atmospheric nuclear weapons tests (Hirose et al., 1999; Pettersson et al., 1999). Approximately 98.6% of all radioactive waste disposal operations at sea were dumped in the form of packaged or unpackaged solid waste. The feasibility of using deep-ocean dumping was based on the premise that the integrity of the waste packages would remain intact, limiting the infiltration of seawater and subsequent remobilization of radionuclides and allowing the bulk of the radioactive materials to decay before entering the environment. Radioactive decay of the waste has already reduced the oceanic inventory of radioactive waste dumped at sea to about 10 PBq or a factor of 10 times less compared with the time of disposal. Low-level liquid effluents make up only a small fraction of the waste stream ($<1.4\%$) dumped in the oceans (IAEA, 1999) and are not expected to impact significantly on the marine radiation environment.

The International Arctic Sea Assessment Project (IASAP) was established in 1993, partly at the request of the Convention, but also to address deepening public concern about the potential health and ecological consequences of radioactive waste disposals in the Arctic region. Exploratory cruises to the Kara and Barents Seas were organized by a number of multidisciplinary teams of scientists, many under the auspices of the Arctic Monitoring and Assessment Program (AMAP). The activity concentrations of artificial radionuclides in the region are generally very low and, for the most part, can be explained by the deposition of global fallout or transfers of previously deposited debris. Consequently, the nuclear waste dumped in the Arctic Ocean appears to represent a negligible risk to human health and the environment. Nevertheless, elevated levels of selected radionuclides in the immediate vicinity of dumped objects were indicative of a gradual loss in waste containment (AMAP, 1998).

5.2. *Other sources of artificial radioactivity in regional seas*

Discharges of radioactive effluents from nuclear fuel reprocessing facilities into coastal waters have resulted in a significant increase in the inventories of a number of radionuclides in the North Atlantic (Livingston et al., 1982). Large quantities of radioactive waste have been discharged from the Sellafield nuclear complex (formally known as Windscale) on the west coast of Cumbria, in the United Kingdom. Reprocessing wastes have also been discharged into the English Channel from the La Hague reprocessing facility located on the northwest coast (Goury, Cherbourg Harbor) of France. The principal radionuclides and isotope ratios of radiological or oceanographic interest include ^{3}H, ^{90}Sr, ^{95}Zr, ^{95}Nb, ^{99}Tc, ^{106}Ru, ^{125}Sb, ^{129}I, ^{134}Cs, ^{137}Cs, ^{238}Pu, $^{239,240}Pu$, ^{241}Am, ^{241}Pu, $^{137}Cs/^{134}Cs$, $^{129}I/^{137}Cs$, $^{129}I/^{99}Tc$, $^{238}Pu/^{239+240}Pu$, $^{241}Am/^{239+240}Pu$, $^{237}Np/^{239}Pu$ and $^{241}Pu/^{239}Pu$.

Two air-cooled plutonium reactors, known as the Windscale Piles, were first brought into operation during 1951. Irradiated fuel from reactors was stored in open-water filled ponds and subsequently processed in a primary separation plant. The plutonium separation plant and two associated purifications plants (B202, B203) were transferred to an integrated facility in 1964; the second purification plant (B203) remained in operation until February 1987. The Windscale Piles were closed down following a fire in Pile 1 during October of 1957. The unit was replaced with spent nuclear fuel shipped in from the Calder Hall and other Magnox reactor facilities.

The two main sources of liquid radioactive effluents at Sellafield are chemical process liquors and purge waters from spent fuel storage ponds. Fuel storage pond water was discharged into the Irish Sea without treatment through until the late 1970s. Radioactive waste disposals increased significantly over this period because of increased fuel storage time and corrosion of the Magnox fuel (Fig. 10a). Historically, salt bearing liquors were an important source of ^{106}Ru that made up a significant fraction of the radionuclide inventory of the effluent stream (Gray et al., 1995) (Fig. 10a). Discharges of the transuranium elements also increased through the mid-1970s because of increased production levels and the addition of residues to the process stream (Fig. 10b). Major developments and advances in waste management technologies have proved very successful in reducing Sellafield discharges. Beginning in the early 1970s, discharges of radionuclides were progressively reduced by introducing primary treatments systems (e.g. sand filtration, ion-exchange and flocculation precipitation) and increasing the storage time of intermediate-level waste liquors prior to discharge. The disposal of intermediate-level waste liquors into the Irish Sea was terminated in 1980 (Gray et al., 1995). The significant reduction in radioactive waste disposals at sea after 1985 (Figs 10a and b) coincide with commissioning of the Site Ion Exchange Effluent Plant (SIXEP) and the Salt Evaporator. The SIXEP was used for treatment of storage pond waters. The Salt Evaporator allowed concentration and storage of chemical process liquors and other reprocessing effluents, reducing the discharge of short-lived fission products such as ^{106}Ru, ^{95}Zr and ^{95}Nb (Gray et al., 1995). The Enhanced Actinide Removal Plant (EARP) was commissioned in 1994 to treat the backlog of stored concentrates. Many by-products arising from the treatment are now being converted into vitrified forms for long-term storage. Contemporary discharges from the La Hague plant in France have also been reduced since commencement of operations at the site in 1966 (Betis, 1993).

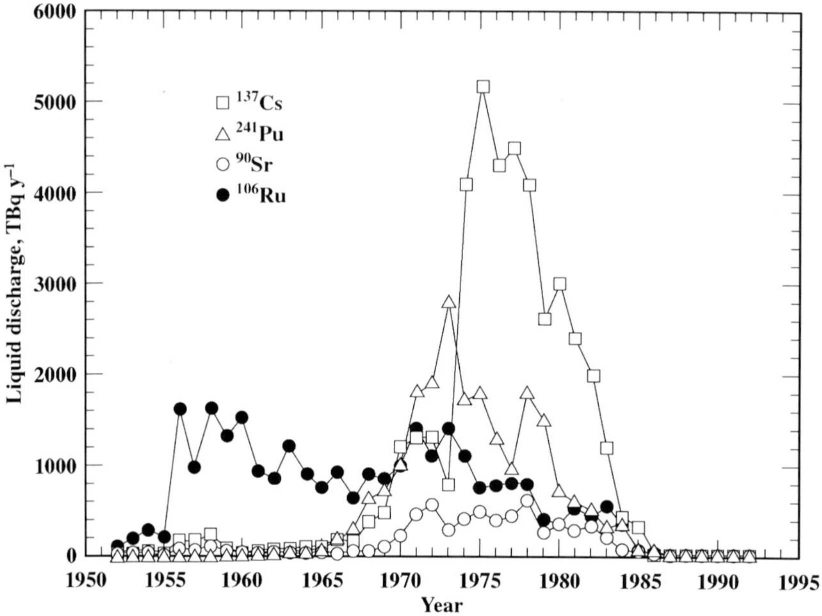

(a)

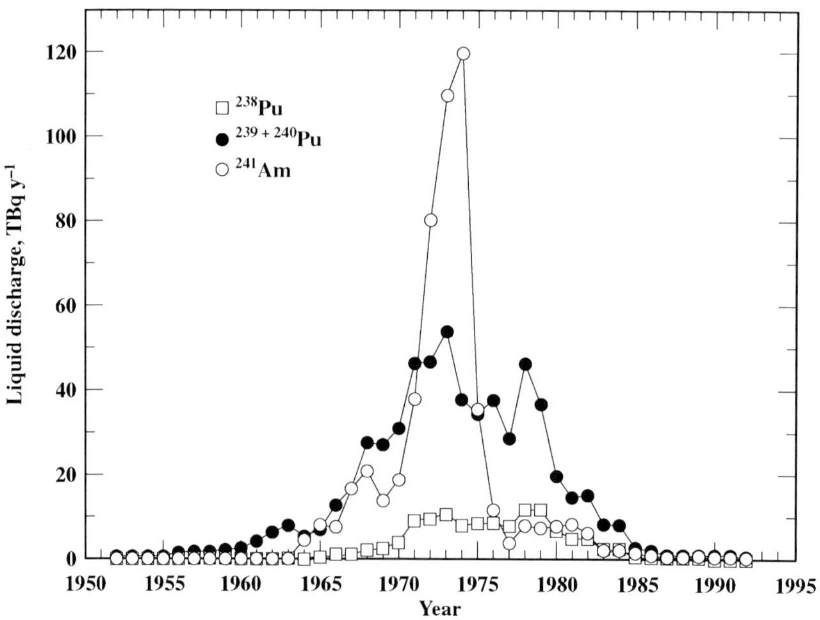

(b)

Fig. 10. Effluent discharges from the Sellafield reprocessing facility in the United Kingdom (1952–1992) to the Irish Sea: (a) beta/gamma emitters; (b) alpha emitters (data source: Gray et al., 1995).

In general, radionuclide discharges from western European reprocessing facilities were likely to have had a pronounced affect on radionuclide inventories in the Arctic Ocean, especially during periods of peak discharge in the mid-1970s (Kershaw & Baxter, 1995; Aarkrog et al., 1983). Discharges from La Hague and export of La Hague labeled seawater into the Norwegian and Barents Seas have largely been masked by Sellafield inputs, with the possible exception of ^{129}I and ^{125}Sb (Kershaw & Baxter, 1995; Raisbeck et al., 1993). The total contribution of La Hague to the marine inventory of artificial radionuclides discharged into the North Atlantic, expressed as a percentage of the Sellafield releases to 1992, is around 12.2%, 12.6%, 2.3% and 0.4% for ^{90}Sr, ^{99}Tc, ^{137}Cs and $^{238+239+240}$Pu, respectively (Kershaw & Baxter, 1995). Cumulative effluent discharges to sea from the Dounreay civilian reprocessing facility in Scotland are at least half as much again. Through 1992, fallout from nuclear weapons testing contributed about 79 PBq of ^{137}Cs and 1.8 PBq of $^{239+240}$Pu to the North Atlantic. The comparative contributions from Sellafield are 41 PBq and 0.6 PBq, respectively. The Sellafield contribution inside the latitude band where the discharges occurred is obviously even more pronounced. For example, the inventories of ^{90}Sr, ^{99}Tc, ^{137}Cs and $^{238+239+240}$Pu discharged through 1992 exceed the estimated global fallout deposit in the 50–60°N latitude band by factors of 0.8, >2000, 3.4 and 2.1, respectively.

The spatial and temporal distributions of ^{90}Sr, ^{99}Tc and ^{137}Cs in the North Atlantic over the past few decades have been described in many studies (Kershaw et al., 1992, 1999; Gueguenia et al., 1997; Kershaw & Baxter, 1995; Dahlgaard, 1993; Dahlgaard et al., 1986, 1988, 1991; Smith et al., 1990; Kautsky, 1989; Pentreath et al., 1985; Livingston et al., 1982, 1985; Aarkrog et al., 1983; Holm et al., 1983; Baxter et al., 1979). The initial dispersion of radionuclides in effluent streams is controlled by a number of factors including the physical–chemical form of the element, local hydrographic conditions, and the composition of bottom sediments. Strontium-90, ^{99}Tc, ^{129}I and the radiocesiums are relatively soluble in seawater and serve as conservative tracers for the passage of waters from western European reprocessing facilities to the Arctic (Gueguenia et al., 1997). Of interest, discharges of ^{99}Tc and ^{129}I from Sellafield have increased significantly through the mid-1990s reversing the progressive reductions seen for many other radionuclides (Kershaw et al., 1999). The pulsed release of ^{99}Tc provides a tracer to study and more accurately determine the transit times of North Atlantic water through the Norwegian Sea to the Arctic shelf seas and deep basins.

In contrast, reactive radionuclides released in particulate or hydrolyzed forms (e.g. plutonium and americium) are quickly removed to the underlying sediments by direct precipitation reactions or scavenging onto suspended particulate matter (Hetherington, 1975). About 0.6 PBq of $^{239+240}$Pu was discharged into the Irish Sea from Sellafield through 1992 (Kershaw et al., 1995). Most of the plutonium was retained in a relatively defined coastal zone of the eastern Irish Sea bounded by muddy subtidal and intertidal sediments. It has only been in recent years, as discharges from Sellafield have declined, that the Irish Sea has been reconciled as a potential source of plutonium to the North Atlantic rather than a sink (Kershaw et al., 1995). Based on plutonium isotope ratio measurements, resolubilized or resuspended plutonium leaving the Irish Sea can be distinguished from that contained in contemporary discharges and more closely reflects the cumulative Sellafield deposit found in sediments (Mitchell et al., 1999). The solubilized plutonium from Sellafield is exported into the North Sea and possibly as far as the Barents and Greenland Seas (Kershaw & Baxter, 1995; Holm et al., 1986).

5.3. *Sources of anthropogenic radionuclides in the arctic*

Over the past decade, a great deal of scientific and public attention has been given to investigations of radioactive contamination of the Arctic Ocean (Strand & Jolle, 1999; Strand & Cooke, 1995). It has been reported that the "Russian Arctic is filled with nuclear perils on sea and land, atomic waste litters the Murmansk Region" (International Herald Tribune, 1996). The Norwegian environmental organization, Bellona, has focused attention on Arctic contamination issues, especially the lack of adequate facilities for handling of liquid and solid radioactive wastes generated by naval and civilian operations on the Kola Peninsula. Tens of thousands of spent nuclear fuel assemblies from nuclear-powered submarines litter harbors and naval shipyards awaiting transfer to reprocessing facilities. In addition, a large quantity of liquid nuclear waste on the Kola Peninsula is held in temporary storage in tanks on sea and land, and aboard service ships and tenders (Nilsen et al., 1996). The Russian Federation face a difficult task to secure, store, and treat existing legacy waste while continuing the process of decommissioning nuclear submarines and dismantling nuclear weapons under strategic arms reduction treaties with the United States. According to Bradley (1997), the Russian Federation decommissioned a total of 147 submarines and service ships through the beginning of 1995, including 76 submarines in the Northern Fleet. About half of the decommissioned submarines remain in floating storage with the fuel on-board, and an additional 130 nuclear submarines remain to be decommissioned.

Extensive monitoring of radionuclides in seawater (and sediments) of the Barents, Kara and other regional seas suggests that current levels of radioactive contamination in the Arctic region remain relatively low and pose no immediate radiological concern. Concentrations of ^{137}Cs and $^{239+240}$Pu from surface activities in the Barents and Kara Seas between 1985 and 1996 are shown in Figs 11 and 12, respectively. Radionuclide distributions in the region can be adequately described by global fallout from nuclear weapons testing and the export of radionuclides from European reprocessing facilities in northwest Europe and the North Sea. Water flows northward into the Norwegian Sea via the Norwegian Coastal Current then splits off to the east into the Barents Sea or continues with the West Spitbergen Current through the Fran Strait into the Nansen Basin. About 1 Bq m^{-3} of ^{137}Cs in surface water of the Barents and Kara Seas is derived from Chernobyl fallout Chernobyl fallout (Dahlgaard et al., 1995) or as much as 10–20% of the total ^{137}Cs activity concentration. In all, about 100 PBq of ^{137}Cs was dispersed into the atmosphere from the 1986 Chernobyl nuclear reactor accident and deposited in the Northern Hemisphere, especially over the Ukraine, Belarus and western Russia. The plume crossed central Norway and Sweden with ^{137}Cs deposition exceeding 200 kBq m^{-2} (AMAP, 1998). Chernobyl fallout within the vicinity of the Arctic Circle was considerably less. For example, northern Finland received around 1–2 kBq m^{-2} but impacts of Chernobyl on the burden of artificial radioactivity in the Arctic Basin is difficult to quantify because of the heterogeneous or patchy nature of the deposition. Furthermore, the ^{137}Cs content of the Baltic Sea increased about 10-fold after the Chernobyl accident from immediate deposition and surface-water runoff. Export of Chernobyl labeled waters from the Baltic Sea and North Sea through the Norwegian Sea constitutes another important pathway for transport of artificial radioactivity into the Arctic Ocean. The influence of marine transport processes on the export of global fallout in the North Atlantic from low latitudes to the Arctic has often been overlooked. The maximal global fallout deposit from atmospheric nuclear weapons

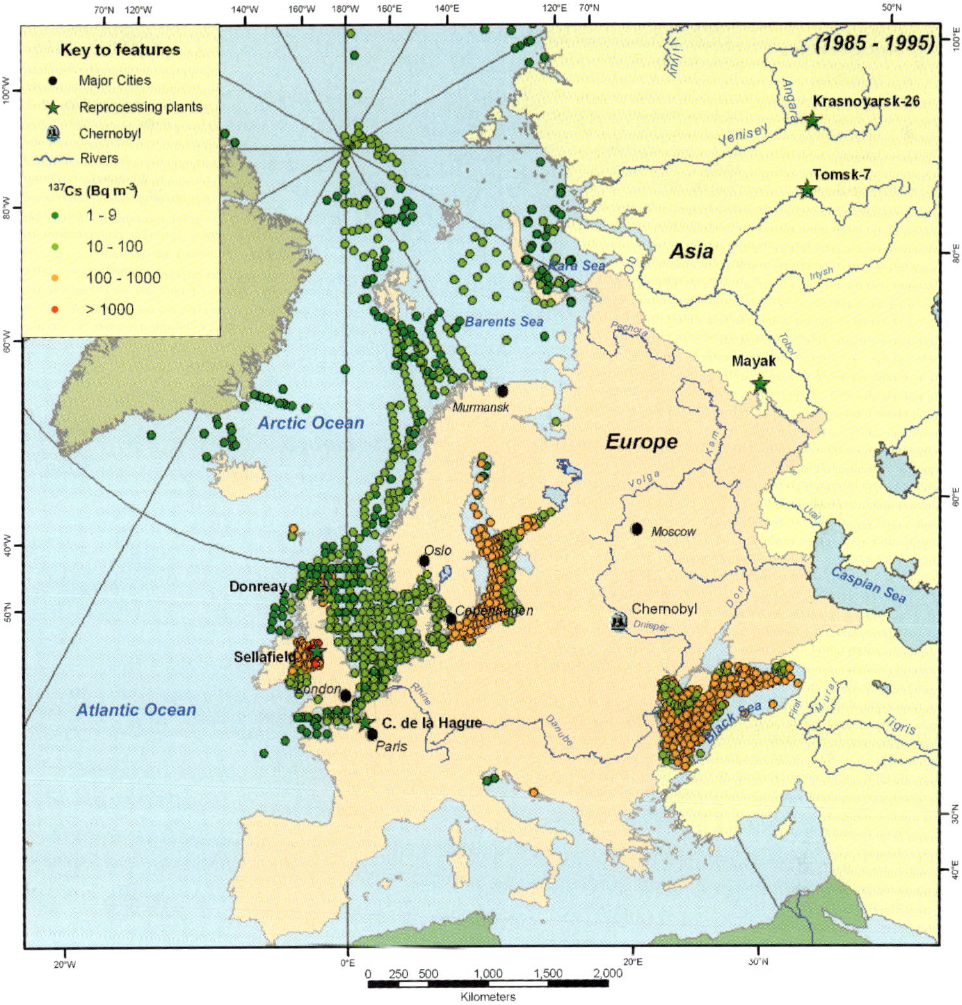

Fig. 11. Activity concentration of ^{137}Cs in surface waters of the Northeast Atlantic, Black Sea and Arctic Ocean (1985–1995) in relation to reprocessing facilities (data source: Dr. Kathy Crane, Office of Naval Research, USA).

testing occurred at mid-latitudes and decreased substantially toward the North Pole (Fig. 6). Redistribution of just 5% of the North Atlantic global fallout deposit essentially doubles the contribution in the Arctic Ocean. This scenario seems entirely plausible based on our knowledge of marine transport processes developed from studies of Sellafield-labeled waters. Sea ice rafted sediments may also play a role in the long-range transport and biogeochemical cycling of artificial radionuclides in the region (Cooper et al., 1998).

About 22% of the ^{137}Cs Sellafield discharge passes into the Barents Sea en route to the Nansen Basin with another 13% passing through the Fram Strait (Kershaw & Baxter, 1995). Quantifying the fluxes of other radionuclides can be more problematic (Kershaw & Baxter,

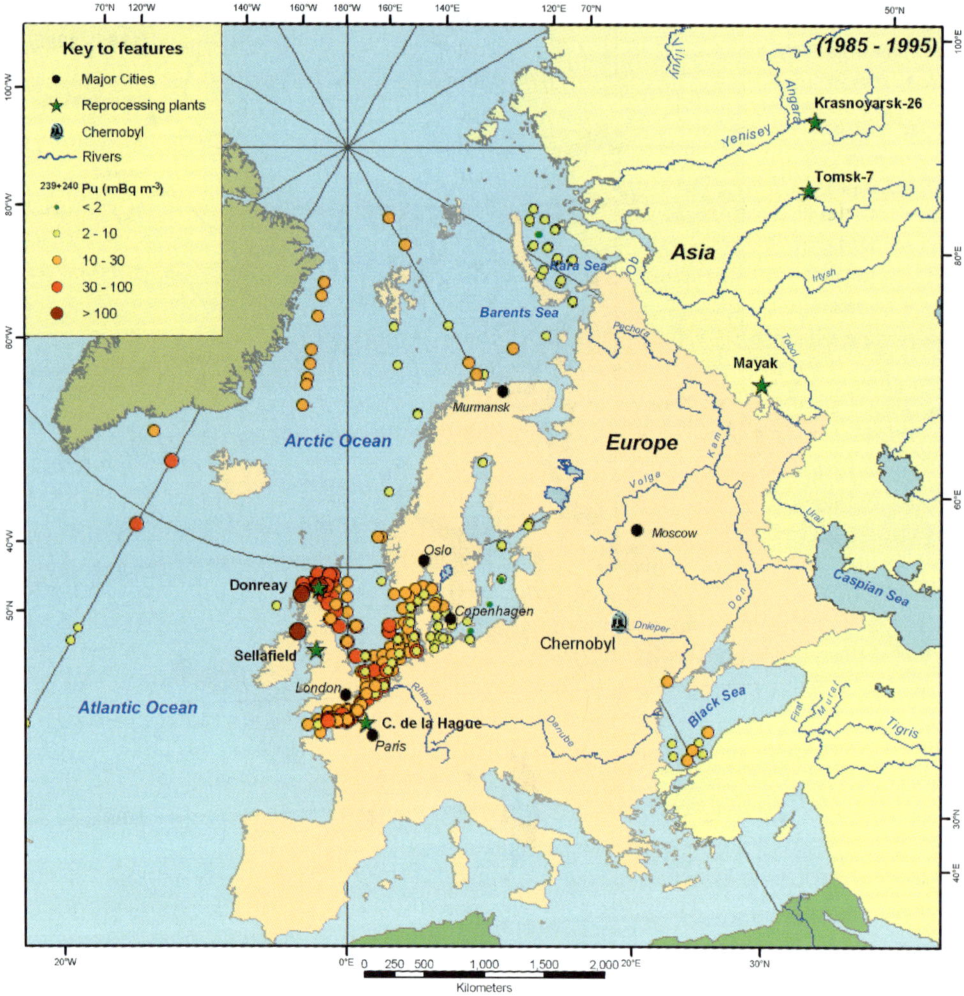

Fig. 12. Activity concentrations of $^{239+240}$Pu in surface waters of the Northeast Atlantic, Black Sea and Arctic Ocean (1985–1995) in relation to reprocessing facilities (data source: Dr. Kathy Crane, Office of Naval Research, USA).

1995) because inflowing North Atlantic water entering and mixing with Arctic waters is diluting radioactive contamination previously dominated by discharges from Sellafield during the 1950s and 1960s. Conversely, an increase in activity concentration of ^{137}Cs in western waters east of Greenland is attributed to old '^{137}Cs rich' Sellafield-labeled waters entering the Arctic circulation and being transported back to the North Atlantic through the East Greenland Current (AMAP, 1998). Transit times for export of ^{137}Cs to the Barents Sea are estimated to be around 5–6 years for Sellafield, and 17 to 36 months for La Hague-labeled waters (Guegueniat et al., 1997; Kershaw & Baxter, 1995).

Another major mechanism for water transport of radioactive contamination in the Arctic region relates to the unique surface-water hydrology and complex system of rivers, lakes and marshes that make up the Russian Arctic territories. Several large, northward flowing rivers serve as sinks for terrestrial contaminant transport into the Russian Arctic, including ground water contamination from what is described as the largest shallow water artesian basin in the world. The main sources of artificial radioactivity transported in river systems are previously deposited global fallout from atmospheric nuclear weapons testing and environmental releases from Russian nuclear production facilities. The Former Soviet Union also conducted a number of Peaceful Nuclear Explosions (PNE) for civilian use including cratering tests, denotations designed to stimulate gas and oil production or increase mineral recovery. The total yield from all PNEs conducted by the Former Soviet Union was around 1.6 Mt (Bradley, 1997). Many of these nuclear tests caused significant contamination adjacent to the Ground Zero (GZ) but it is not clear what impact these detonations may contribute to Arctic contamination. It is expected that releases to the Ob and Yenisey were negligible. It has been reported that about 3.2 PBq of radioactivity was released into the Irtysh River (Bradley, 1997), presumably in runoff from the Semipalatinsk test site (Table 1).

Other sources of radioactive contamination in the Arctic region include offshore transport and redistribution of radioactivity from nuclear sites on the southern coastline of Novaya Zemlya and other accidents involving nuclear ships, submarines and/or nuclear weapons. The Former Soviet Union submarine SSN *Komsomolets* sank off Bear Island (Fig. 2) on 7 April of 1989 after a fire broke out in the stern section of the vessel. The reactor was shut down prior to the vessel sinking but concerns have been raised about leakage of radioactivity from the reactor and two nuclear torpedoes aboard, one of which was fractured (Bergman & Baklanov, 1998). The reactor core and warheads reportedly contain a total of about 22 TBq of ^{239}Pu (10 kg), 2.4 PBq of ^{90}Sr and 2.7 PBq of ^{137}Cs (decay corrected to 1 January, 1995) (Høibråten & Thoresen, 1995). Scientific expeditions to the site indicate very little loss of radioactivity from the submarine (Kolstad, 1995).

The Former Soviet Union conducted three underwater nuclear detonations near Chernaya Bay, a 15-km fjordic inlet on the southern coastline of Novaya Zemlya. It is estimated that approximately 11 TBq of $^{239+240}$Pu from the tests has been retained in the local sediments (Smith et al., 2000). Chernaya Bay is among the world's most contaminated marine environments. Plutonium-239 + 240 levels in the central region of Chernaya Bay exceed 8000 Bq kg^{-1} and are characterized by low ^{240}Pu/^{239}Pu atom ratios ($\sim$0.03) associated with the detonation of a low-yield nuclear device (Smith et al., 1995). By exploiting the large difference in ^{240}Pu/^{239}Pu atom ratios between the Chernaya Bay fallout ($\sim$0.03) and atmospheric fallout ($\sim$0.18) end-members, Smith et al. (2000) estimated that an additional 2 TBq of Chernaya Bay labeled plutonium resides in the eastern Barents Sea from offshore transport from the embayment. A plume of low ^{240}Pu/^{239}Pu atom ratio plutonium is observed in a northwestern direction along the southern coastline of Novaya Zemlya indicating an additional pathway for transfer of previously deposited fallout debris into the Arctic Ocean.

The Mayak production association complex is located along the Techa River on the upper reaches of the Ob–Irtysh–Tobol–Iset River system. Approximately 37,000 PBq of radioactive waste has been generated since commencement of operations at Mayak in 1948. Today, authorities acknowledge that at least 4800 PBq of radioactivity was released directly into surface waters adjacent to the site (Hamilton, this report). Most of the radioactive contamination is held back in lakes and reservoirs on the Upper Techa; however, historical releases are known to have caused significant contamination of the riverbed and associated floodplain. A large number of residents living on the Techa River during the early 1950s received significant radiation exposures. Health studies on the population continue to this day. Overflow and seepage water from the dams, as well as migration of radionuclides deposited in the floodplain and especially runoff from the Asanov swamps, provide a possible source-term for transport of radioactive contaminants from Mayak to the Kara Sea. The Tomsk-7 and Krasnoyarsk-26 reprocessing facilities are located on tributaries to the Ob and Yenisey Rivers, respectively (Fig. 2). Despite the large releases of radioactivity from Russian reprocessing facilities and potential for riverine transport of the contamination to the Kara Sea, the major fraction of ^{137}Cs and plutonium deposited in delta sediments of the Ob and Yenisey Rivers is said to be derived from fallout from atmospheric nuclear weapons tests (Sayles et al., 1997; Baskaran et al., 1995). The latter findings are supported by analyses of Pu isotopes, and ^{238}Pu/$^{239+240}$Pu and $^{239+240}$Pu/^{137}Cs activity ratios, all of which were indistinguishable from global fallout. Using high quality mass-spectrometric measurements, Cooper et al. (2000) has shown that plutonium isotopic signatures in Ob and Yenisey Rivers sediments are distinctly different from those of northern-hemisphere stratospheric fallout, arguing that the ratios are more indicative of the presence of weapons grade plutonium from Russian reprocessing facilities located thousands of kilometers upstream. Observed ratios in sediments from the Eurasian Arctic Ocean were also shown to be inconsistent with significant contributions of plutonium to arctic sediments from western European facilities, namely Sellafield (Cooper et al., 2000). This work has re-addressed the need for a more thorough and accurate assessment of the fate and transport of radioactive effluents discharged into surface water from Russian reprocessing facilities (Hamilton, this report).

The USSR Hydrometeorological Service has maintained detailed records of the amount of ^{90}Sr entering the Kara Sea from the Ob and Yenisey Rivers. It is estimated about 1.1 PBq of ^{90}Sr was transported to the Kara Sea between 1961 and 1989 (SCRF, 1995; Vakulovsky et al., 1993). A concurrent set of data does not exist for ^{137}Cs. However, using the average ^{90}Sr/^{137}Cs activity ratio observed in river water over the same period ($\sim$0.1), the associated ^{137}Cs discharge to the Kara Sea is estimated at around 0.11 PBq. According to Aarkrog (1993), radionuclide contributions to the Arctic Ocean from atmospheric nuclear weapons fallout, discharges from Sellafield, and runoff from global fallout over land are 2.6 PBq, 1–2 PBq and 1.5 PBq for ^{90}Sr, and 4.1 PBq, 10–15 PBq and 0.5 PBq for ^{137}Cs, respectively. About 1–5 PBq of Chernobyl fallout was also delivered to the Arctic Ocean by direct deposition from the atmosphere and transport of Chernobyl labeled waters from the Baltic and North Seas.

In summary, large-scale contamination of the A rctic Ocean with artificial radionuclides (past and present) is controlled by four primary sources: global fallout from atmospheric nuclear weapons testing, discharges from European reprocessing facilities, fallout from the Chernobyl accident, and runoff from Siberian rivers. The relative contributions from different sources are both radionuclide and transport pathway specific. Through 1993, the inventory of ^{90}Sr in the Arctic increased about twofold over what was expected from global fallout deposition alone. The associated ^{137}Cs inventory increased about 4–6 times. The main source of the additional ^{90}Sr and ^{137}Cs in the Arctic has been traced to discharges from European reprocessing entering the Arctic via the Norwegian Coastal Current. Sellafield contributes as much as

90% of the excess ^{137}Cs inventory and 60% of the excess ^{90}Sr inventory. Chernobyl fallout is a significant contributor to the ^{137}Cs inventory in the Arctic Ocean and supplies anywhere from 5–40% of the additional ^{137}Cs inventory above the expected global fallout deposit. Chernobyl does not contribute significantly to the total ^{90}Sr inventory in the Arctic. The Dvina, Pechora, Ob and Yenisey Rivers in northwestern Russia and Siberia all drain into the Kara Sea and constitute an important pathway for export of previously deposited global fallout debris over land to the Arctic. Based on several studies conducted during the 1990s, about 40–60% of the additional ^{90}Sr and 3–5% of the additional ^{137}Cs in the Arctic Ocean may be from export of previously deposited fallout debris over land. This conjecture is mainly based on the lack of convincing evidence for significant riverine inputs to the Kara Sea of artificial radionuclides from Russian reprocessing facilities located upstream with the exception of weapons grade plutonium and, possibly, ^{129}I. Discharges of ^{99}Tc and ^{129}I from Sellafield and La Hague dominate the total oceanic inventory, especially those from La Hague. The estimated combined discharge of ^{129}I from reprocessing through 1994 is approximately 1440 kg (8.7 TBq) (Yiou, 1995) or about 16 times the global release from atmospheric nuclear weapons testing (Table 3). Other sources of radioactive contamination in the Arctic include indiscriminant dumping of nuclear waste, accidents involving nuclear ships and submarines, and offshore transport of plutonium from tests sites on Novaya Zemlya. Gaps in current understanding will require a continuing level of radiological surveillance of known and potential source-terms, source-term related assessments, and studies relating ocean dynamics and contaminant transport processes. There remain a number of critical challenges in developing appropriate long-term strategies to protect the Arctic and the wider regional environment from past, present, and potential future sources of artificial radioactivity. These are seen as negligent nuclear waste management practices in the handling of spent nuclear fuel, the management of legacy waste and nuclear materials in inventory, and risk of a catastrophic release from waste containment systems holding back high-level radioactive waste from Mayak and other Russian reprocessing facilities on the Ob and Yenisey Rivers. There is also a need to improve the safety of nuclear power plants at Kola and Bilibino. Some agreements involving Russian and bi- and trilateral cooperations with Norway, the United States and the European Community have already taken steps to improve the management of waste and spent nuclear fuel on the Kola Peninsula.

5.4. *The ocean water column*

Early measurements of artificial radionuclides in the oceans demonstrated that global fallout from atmospheric nuclear weapons tests penetrated the deep ocean. Interpretation of these data was complicated because of the heterogeneous nature of the global fallout deposition and questions concerning the quality of the measurements. No consensus could be reached about the significance of ^{90}Sr penetration into the deep ocean or on the total oceanic inventory (Volchok et al., 1971). These early studies highlighted the need for more systematic and long-term studies, and raised expectations about the value of using global fallout radionuclides as tracers of oceanographic processes. Modern day 'radionuclide oceanography' really only began with the commencement of the GEOSECS (Geochemical Ocean Section) program in 1973–1974 where very carefully collected, large volume hydrographic stations were occupied and fractions of these samples made available for analysis of ^{3}H, ^{14}C, ^{90}Sr, ^{137}Cs and the transuranic elements (Broecker et al., 1985; Livingston et al., 1985; Bowen et al., 1980). In general, ^{90}Sr and ^{137}Cs concentrations in the surface ocean have decreased steadily since the early 1960s to present-day levels of around <1 to 4 Bq m^{-3} (Bourlat et al., 1996). Early measurements showed clear latitudinal concentration gradients as expected from global fallout patterns but these trends have since become more obscure (Hamilton et al., 1996). Regional anomalies were also identified in association with discharges from reprocessing plants; upwelling and other oceanographic processes; and localized inputs from nuclear tests conducted in the Marshall Islands. As expected, the observed activity concentration of ^{90}Sr (and ^{137}Cs) in surface waters of the Southern Hemisphere was much lower than those reported for northern latitudes.

Depth distributions of ^{137}Cs and ^{90}Sr in the water column were characterized by sharp concentration gradients extending down to depths of 1000–2000 m (Nagaya & Nakamura, 1984, 1987). The depth distribution patterns were similar to those observed for fallout ^{3}H (Roether, 1974) and attributed to the conservative nature of ^{90}Sr and ^{137}Cs fallout. However, a significant fraction of ^{137}Cs and ^{90}Sr penetrated the deep ocean, and there was no clear understanding of the processes involved. The most probable explanation came from Martin (1970), Honjo (1980), Iseki (1981) and others who postulated that global fallout radionuclides are actively transported to the deep ocean on biogenic particles. Elevated levels of ^{137}Cs were also observed in waters immediately above the seafloor. This was attributed to preferential scavenging of ^{137}Cs into bottom sediments and remobilization processes, and/or advection of nuclide-rich water from other regions (Nakamura & Nagaya, 1975, 1985; Noshkin & Bowen, 1973).

The general depth distribution profiles of key radionuclides such as ^{90}Sr, ^{137}Cs and $^{239+240}$Pu do not appear to have changed significantly over the past 25 years (Hamilton et al., 1996) (Fig. 13). One of the more interesting features of plutonium behavior in the oceans is the widespread occurrence of a subsurface maximum ranging in depth from 250 to 1000 m (Nakanishi et al., 1995; Bowen et al., 1980; Noshkin & Wong, 1979). Although there is some evidence to suggest the depth of the subsurface maximum and proportion of plutonium residing in the deep ocean has increased over time (Nakanishi et al., 1984; Bowen et al., 1980), early transport models based on irreversible particle scavenging predicted a more rapid transfer of plutonium to the deep ocean. A reversible process of biologically mediated scavenging that varies in space and time best explains the depth distribution and transport of long-lived radionuclides such as plutonium in the deep oceans. Highly productive surface waters enhance the vertical transport of particle reactive species and provide a possible mechanism where radionuclides can be preferentially transported to the deep ocean and onto the seafloor. There is also strong evidence to suggest that the main vector for transport of particle reactive radionuclides to the deep ocean are large, rapidly sinking zooplankton faecal pellets (Fowler et al., 1983; Fisher & Fowler, 1983). Smaller particles and micropellets sink slowly and, hence, are more subject to biogeochemical cycling. Therefore, the appearance of a subsurface maximum for plutonium may be a function of both the particle reactivity of plutonium and dynamics of these transformation processes. A short residence time and higher affinity for removal of plutonium on sinking particles is supported by the general findings that particulate $^{239+240}$Pu concentrations range up to ~20% of the total activity concentrations in surface waters compared with less than 1% for ^{137}Cs (Hirose et al., 1992; Fowler et al., 1983; Noshkin & Wong, 1980). Although there is a small but significant flux of artificial radionuclides depositing on

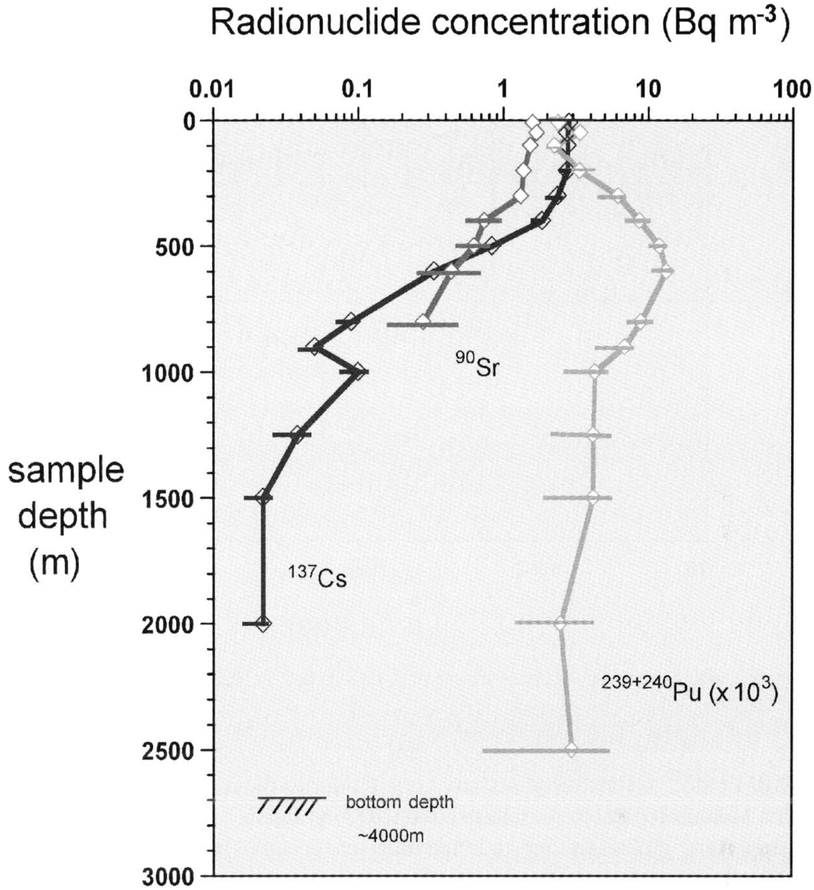

Fig. 13. Depth distribution of ^{90}Sr, ^{137}Cs and $^{239+240}$Pu in the ocean, Tuamoto Archipelago, French Polynesia (data source: Hamilton et al., 1996).

the sea floor, about 80% of the total $^{239+240}$Pu inventory and 95% of the total ^{137}Cs inventory in the open ocean remains in the water column. Long-term monitoring studies in French Polynesian waters indicate ^{137}Cs concentrations in surface water decrease with an apparent half-life of about 14 years (Fig. 14; Bourlat & Martin, 1992). Using a simple box model, Hirose et al. (1992) estimated that the residence times of $^{239+240}$Pu and ^{137}Cs in the surface-mixed layer of the western North Pacific were around 4 and 9.1 years, respectively.

Today, the global dispersion and deposition of fallout radionuclides on the Earth's surface is largely controlled by tropospheric resuspension of previously deposited debris on land (Monetti, 1996; Nakanishi et al., 1995). For example, seasonal patterns in the deposition of ^{137}Cs and $^{239+240}$Pu over Japan and on the Korean Peninsula are directly linked to 'yellow dust' events (Lee, 1994). According to Hamilton et al. (1996), the annual deposition rates of ^{137}Cs and $^{239+240}$Pu into the Japan Sea from resuspension of dust are about 0.8 MBq km^{-2}

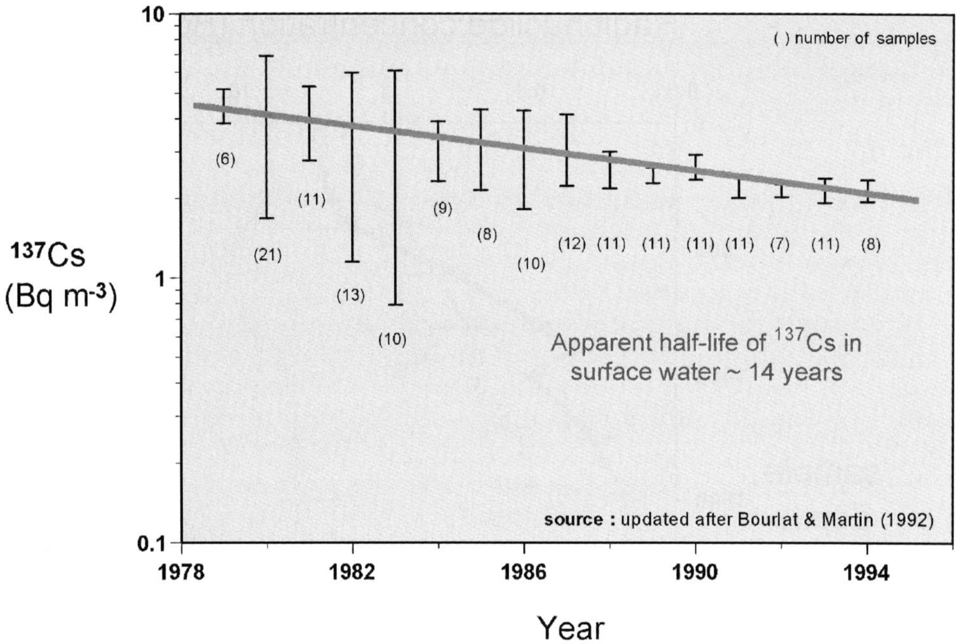

Fig. 14. Temporal pattern in the activity concentration of ^{137}Cs in French Polynesian surface waters (modified after Hamilton et al., 1996).

and 0.03 MBq km^{-2}, respectively; and are very similar to the leveling-off aerosol flux measured by the Meteorological Research Institute (MRI) in Japan.

Other important global fallout radionuclides of oceanic significance include tritium and ^{14}C. Very large quantities of tritium ($T_{1/2}$ = 12.3 years) and ^{14}C ($T_{1/2}$ = 5730 years) were produced in atmospheric testing of high-yield nuclear weapons during the 1950s and 1960s (Table 3). Bomb tritium generated in nuclear detonations was quickly incorporated into stratospheric water molecules and upon transfer to the troposphere; the tritiated water was rapidly scavenged onto raindrops and deposited by rainout over sea or land (Broecker et al., 1995). Bomb tritium preferentially resides in the high-latitude Northern Hemispheric oceans (Weiss & Roether, 1980) and is only present as tritiated water. Much of the bomb tritium deposited over land was transferred to the ocean in runoff or via evaporation. The delivery of tritium from land to sea is a major pathway but the input is poorly constrained. With the exception of small losses by atmospheric exchange processes there is no sink for oceanic tritium other than loss by radioactive decay. The concentration of tritium in the ocean is normally reported in TU (tritium units) used to denote the number of tritium atoms per 10^{18} atoms of hydrogen or synonymously, by the tritium ratio (TR). One TU is equivalent to 0.12 Bq, or in terms of atoms, 6.7×10^7 atoms per liter of water. The tritium oceanic inventory in 1973 was estimated to be around 60,000 PBq. The activity concentration of tritium in near-surface ocean waters is about 0.3 Bq kg^{-1} of water, but its isotopic concentration (i.e. the molar ratio of tritium to stable hydrogen) is extremely low (on the order of 10^{-18}). In contrast, ^{14}C was incorporated into CO_2 and transferred to the sea by invasion from the atmosphere. Differences

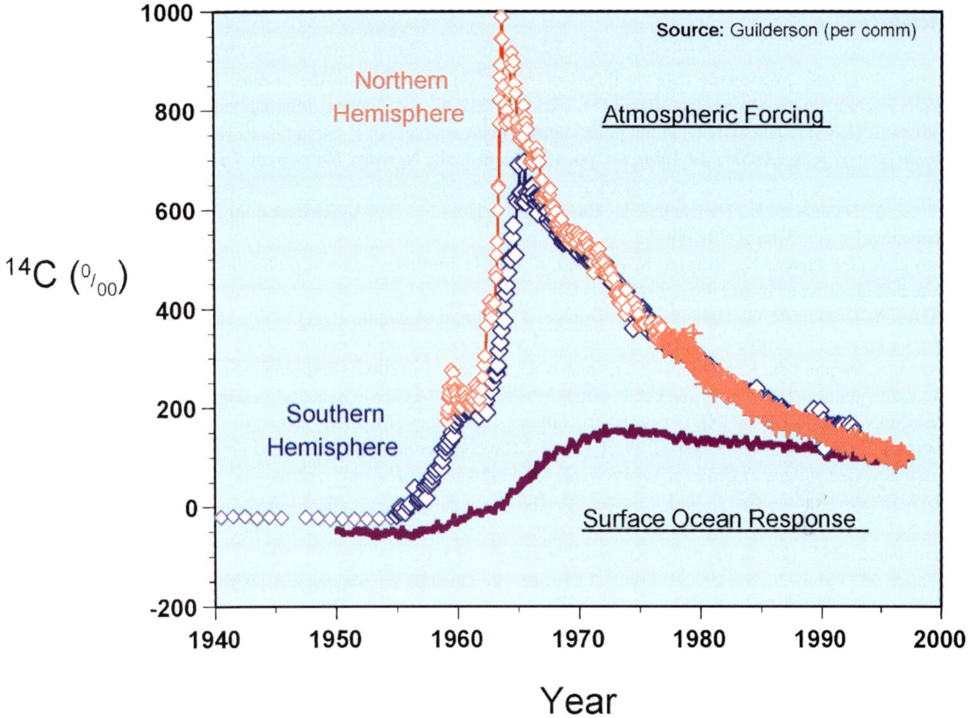

Fig. 15. Surface ocean response to global dispersion and deposition of ^{14}C in the ocean: a subtropical South Pacific high-resolution coral record from Rarotonga (21° S, 159° W) (data source: Guilderson, personal communication).

in environmental pathways between tritium and ^{14}C led to a major geographic separation between the two isotopes in the oceans. Naturally occurring ^{14}C is produced in the atmosphere by the action of neutrons from cosmic rays on nitrogen. Atmospheric nuclear testing more than doubled the natural cosmogenic atmospheric inventory. Bomb ^{14}C was spread uniformly throughout the atmosphere and into the oceans, and taken up into the biosphere by photosynthesis. Penetration of bomb ^{14}C in the surface ocean is illustrated in Fig. 15, showing a rapid increase in the amount of ^{14}C incorporated into a coral core extending back to the pre-nuclear age. The global ocean bomb ^{14}C inventory as of 1972 was about 305×10^{26} atoms (Radiocarbon Units-RCU or 116 PBq) (Broecker et al., 1985) increasing to about 435×10^{26} atoms (RCU or 167 PBq) by 1985 from post-1972 transfer of ^{14}C to the ocean (Lassey et al., 1988).

Acknowledgments

I thank Dr. Kathy Crane, Office of Naval Research (ONR), and Dr. Tom Guildersen, Lawrence Livermore National Laboratory (LLNL), for providing data for graphics. Ms. Lynn Wilder assisted with the preparation of maps and figures. Work performed under the auspices of the United States Department of Energy at Lawrence Livermore National Laboratory under contract W-7405-Eng-48.

References

Aarkrog, A. (1988). Worldwide data on fluxes of ^{239,240}Pu and ^{238}Pu to the oceans. In *Inventories of Selected Radionuclides in the Oceans* (pp. 103–138). IAEA-TECDOC-481. Vienna: International Atomic Energy Agency.

Aarkrog, A. (1993). Radioactivity in the polar regions-main sources. In P. Strand & E. Holm (Eds), *Environmental Radioactivity in the Arctic and Antarctic* (pp. 15–34). Østerås, Norway: Norwegian Radiation Protection Authority.

Aarkrog, A., Dahlgaard, H., Hallstadius, L., Hansen, H. & Holm, E. (1983). Radiocesium from Sellafield effluents in Greenland water. *Nature*, *304*, 49–51.

Aarkrog, A., Dahlgaard, H., Nilsson, K. & Holm, E. (1984). Further studies of plutonium and americium at Thule, Greenland. *Health Physics*, *46*, 29–44.

Aarkrog, A., Boellskifte, S., Dahlgaard, H., Duniec, S., Holm, E. & Smith, J. N. (1987). Studies of transuranics in an Arctic marine environment. *Journal of Radioanalytical and Nuclear Chemistry*, *1115*, 39–50.

Aarkrog, A., Simmonds, J., Strand, P., Christensen, G. & Salbu, B. (2000). Radiological assessment of past, present and potential sources to environmental contamination in the Southern Urals and strategies for remedial measures (SUCON, Risø-R-1243 (EN)) (71 pp.). RisøNational Laboratory (Roskilde, December 2000).

Adams, C. E., Farlow, N. H. & Schell, W. R. (1960). The compositions, structure and origins of radioactive fall-out particles. *Geochimica et Cosmoschimica Acta*, *18*, 42–56.

AMAP (1998). Radioactivity. In P. Strand, M. Balonoc, A. Aarkrog, M. J. Bewers, B. Howard, A. Salo & Y. S. Tsaturov (Eds), *AMAP Assessment Program Arctic Pollution Issues, Arctic Monitoring and Assessment Programme (AMAP)* (Ch. 8, pp. 525–610). Oslo, Norway.

Baskaran, M., Asbill, S., Santchi, P., Davis, T., Brooks, J., Champ, M., Makeyev, V. & Knlebovich, V. (1995). Distribution of $^{239+240}$Pu and ^{238}Pu concentrations in sediments from the Ob and Yenisey Rivers and the Kara Sea. *Applied Radiation and Isotopes*, *46*, 1109–1119.

Baumgartner, A. & Reichel, E. (1975). *World Water Balance: Mean, Annual Global, Continental and Maritime Precipitation, Evaporation and Run-off* (179 pp.). New York: Elsevier.

Baxter, M. S., McKinley, I. G., Mackenzie, A. B. & Jack, W. (1979). Windscale radiocesium in the Clyde Sea area. *Marine Pollution Bulletin*, *10*, 116–120.

Baxter, M. S., Ballestra, S., Gastaud, J., Hamilton, T. F., Harms, I., Huynh-Ngoc, L., Liong Wee-Kwong, L., Osvath, I., Parsi, P., Petterssen, H., Povinec, P. P. & Sanchez, A. (1995). Marine radioactivity studies in the vicinity of sites with potential radionuclide releases. *International Symposium on Environmental Impact of Radioactive Releases, International Atomic Energy Agency* (IAEA). Vienna, Austria, 8–12 May, 1995.

Beck, H. L. & Krey, P. W. (1983). Radiation exposure in Utah from Nevada nuclear tests. *Science*, *220*, 18–24.

Bergman, R. & Baklanov, A. (1998). Radioactive sources of main radiological concern in the Kola–Barents region. Defense Research Establishment. FOA-B-98-00343-861-SE (78 pp.). Sweden.

Betis, J. (1993). Impact of the La Hague reprocessing plant on the surrounding environment. In H. Volke & S. Pretre (Eds), *Environmental Impact of Nuclear Installations* (pp. 33–36). Les Ulis Cedex A, France: Les Éditions de Physique.

Bol'shakov, V. N., Aleksakhin, R. M., Bol'shov, L. A., Chukanov, V. N., Kochetkov, L. A., Petukhov, V. I. & Tsyb, A. F. (1991). *Conclusion of the Commission for Evaluating the Ecological Situation in the Sphere of Influence of the Industrial Plant (IP) 'Mayak' of the USSR Ministry of Atomic Energy*. Organized by the Presidium of the USSR Academy of Sciences No. 1140-501, on June 12, 1990, Moscow, Russia.

Botov, N. G. (1992). ALWP-67: A Little-known big nuclear accident. *Proceedings of the Third International Conference on High Level Radioactive Waste Management*, April 12–16, 1992, Las Vegas, Nevada.

Bourlat, Y. & Martin, G. (1992). Precise determination of the concentration of radiocesium in the water of Mururoa Lagoon. *Journal of Environmental Radioactivity*, *17*, 13–29.

Bourlat, Y., Millies-LaCroix, J.-C. & Nazard, R. (1995). Determination of plutonium radioactivity in Mururoa Lagoon water. *Journal of Radioanalytical and Nuclear Chemistry*, *197*, 387–408.

Bourlat, Y., Millies-LaCroix, J.-C., Le Petit, G. & Bouriguignon, J. (1996). ^{90}Sr, ^{137}Cs and $^{239+240}$Pu in world ocean water samples collected from 1992 to 1994. In P. Guéguéniat, P. Gwermain & H. Métivier (Eds), *Radionuclides in the Oceans, Inputs and Inventories* (Ch. 2, pp. 75–93). Les Ulis Cedex A, France: Les Éditions de Physique.

Bowen, V. T., Noshkin, V. E., Livingston, H. D. & Volchok, H. L. (1980). Fallout radionuclides in the Pacific Ocean: Vertical and horizontal distributions, largely from GEOSECS stations. *Earth and Planetary Science Letters*, *49*, 411–434.

Bradley, D. J. (1997). *Behind the Nuclear Curtain: Radioactive Waste Management in the Former Soviet Union* (716 pp.). D. R. Payson (Ed.). Battelle Press.

Bradley, D. L. & Jenquin, U. P. (1995). Radioactive inventories and sources for contamination of the Kara Sea by rivers. PNWD-2316. Richland, Washington.

Bradley, D. J., Frank, C. W. & Mikerin, Y. (1996). Nuclear contamination from weapons complexes in the Former Soviet Union and the United States. *Physics Today*, *49*, 40–45.

Broecker, W. S., Peng, T. H., Ostland, G. & Stuiver, M. (1985). The distribution of bomb radiocarbon in the ocean. *Journal of Geophysical Research*, *90*, 6953–6970.

Broecker, W. S., Sutherland, S. & Smethie, W. (1995). Oceanic radiocarbon: Separation of the natural and bomb components. *Global Biogeochemical Cycles*, *9*, 263–288.

Buesseler, K. O. (1986). Plutonium isotopes in the North Atlantic (220 pp.). PhD Thesis. MIT/WHOI Joint Program in Oceanography.

Buesseler, K. O. (1997). The isotopic signature of fallout plutonium in the North Pacific. *Journal of Environmental Radioactivity*, *36*, 69–83.

Christensen, G. C., Malyshev, S. V., Strand, P., Salbu, B., Romanov, G. N., Selnaes, T. D., Oughton, D., Glagolenko, Y. V., Amundsen, I., Rudjord, A. L., Bjerk, T. O. & Lind, B. (1995). Radioactive contamination in the environment of the nuclear enterprise 'Mayak' PA. Results from the Joint Russian–Norwegian Field work in 1994. Presented at *The Second International Conference on Environmental Radioactivity in the Arctic*. August 21–25, 1995. Oslo, Norway.

Cochran, T. B., Norris, R. S. & Suokko, K. L. (1993). Radioactive contamination at Chelyabinsk-65 Russia. *Annual Review of Energy and the Environment*, *18*, 507–528.

Cochran, T. B., Norris, R. S. & Bukharin, A. O. (1995). *Making the Russian Bomb: From Stalin to Yeltsin*, Boulder: Westview Press.

Compton, A. A. (1956). *Atomic Quest, a Personal Narrative* (370 pp.). New York: Oxford University Press.

Cooper, L. W., Larsen, I. L., Beasley, T. M., Dolvin, S. S., Grebmeirer, J. M., Kelley, J. M., Scott, M. & Johnson-Pyrtle, A. (1998). The distribution of radiocesium and plutonium in sea ice-entrained Artic sediments in relation to potential sources and sinks. *Journal of Environmental Radioactivity*, *39*, 279–303.

Cooper, L. W., Kelley, J. M., Bond, L. A., Orlandini, K. A. & Grebmeirer, J. M. (2000). Sources of the transuranic elements plutonium and neptunium in arctic marine sediments. *Marine Chemistry*, *69*, 253–276.

Crocker, G. R., O'Connor, J. D. & Freiling, E. C. (1966). Physical and radiochemical properties of fallout particles. *Health Physics*, *12*, 1099–1104.

Dahlgaard, H. (1993). Where does all the Cs and Sr in Greenland come from? In P. Strand & E. Holm (Eds), *Environmental Radioactivity in the Arctic and Antarctic* (pp. 121–124). Østerås, Norway: Norwegian Radiation Protection Authority.

Dahlgaard, H., Aarkrog, A., Hallstadius, L., Holm, E. & Rioseco, J. (1986). Radiocesium transport from the Irish Sea via the North Sea and Norwegian Coastal Current to East Greenland. *Rapport et Proces-Verbaux Reunion Conseil Internationale Exploration de Mer*, *186*, 70–79.

Dahlgaard, H., Chen, Q., Nies, H., Ibbett, R. D. & Kershaw, P. J. (1988). On the background level of ^{99}Tc, ^{90}Sr, and ^{137}Cs in the North Atlantic. *Journal of Marine Systems*, *6*, 571–578.

Dahlgaard, H., Chen, Q. J. & Nielsen, S. P. (1991). Radioactive tracers in the Greenland Sea. In P. J. Kershaw & D. S. Woodhead (Eds), *Radionuclides in the Study of Marine Processes* (pp. 12–22). London: Elsevier.

Dahlgaard, H., Chen, Q., Nies, H., Ibbett, R. D. & Kershaw, P. J. (1995). On the background level of ^{99}Tc, ^{90}Sr, and ^{137}Cs in the North Atlantic. *Journal of Marine Systems*, *6*, 571–578.

Danilov-Danilyan, V. I. (1993). Statement (LC/16/J/4) made at the 16th Consultative Meeting of the Contracting Parties to the London Convention 1972. 10 November, 1993.

Degteva, M. O., Kozheurov, V. P. & Vorobiova, M. I. (1994). General Approach to dose reconstruction in the population exposed as a result of the release of radioactive waste into the Techa River. *Science of the Total Environment*. Special issue, *Radiation Exposure in the Southern Urals*.

Degteva, M. O., Vorobiova, M. I., Kozheurov, V. P., Tolstykh, E. I., Anspaugh, L. R. & Napier, B. A. (2000). Dose reconstruction system for the exposed population living along the Techa River. *Health Physics*, *78*, 542–554.

Dobry, T. J., Jr. (1980). Transuranic elements in space nuclear power systems. In W. C. Hanson (Ed.), *Transuranic Elements in the Environment* (pp. 83–85). DOE/TIC-22800.

DOE (1982). Enewetak Radiological Support Project, NVO-213 (p. 158). United States Department of Energy, Nevada: Nevada Operation Office.

DOE/EM (1997). Linking legacies: Connecting the Cold War nuclear weapons production processes to their environmental consequences (229 pp.). DOE/EM-0319. Washington, DC: US Department of Energy/Office of Environmental Management.

Drozhko, E. G., Nikipelov, B. V., Nikiforov, A. S., Suslov, A. P. & Tsarenko, A. F. (1989). Experience in radioactive waste management at the Soviet radiochemical plant and main approaches to waste reliable confinement development. In *The International Symposium on the Safety Assessment of Radioactive Waste Repositories*. Paris, France, October 9–12, 1989.

Eisenbud, M. & Gesell, T. (1997). *Environmental Radioactivity, from Natural, Industrial, and Military Sources* (4th edn) (656 pp.). San Diego, USA: Academic Press.

Fisher, N. S. & Fowler, S. W. (1983). The role of biogenic debris in the vertical transport of transuranic wastes in the sea. In T. P. O'Connor, W. V. Burt & I. W. Duedall (Eds), *Oceanic Processes in Marine Pollution*, Vol. 2 (Ch. 16, pp. 197–207). Florida: Robert E. Krieger Publishing Co.

Fowler, S. W., Ballestra, S., La Rosa, J. & Fukai, R. (1983). Vertical transport of particulate-associated plutonium and amercium in the Northwest Pacific. *Deep-Sea Research*, *7*, 35–47.

GESAMP (1990). The State of the UNEP Regional Seas Report and Studies No. 115. UNEP, Group of Experts on the Scientific Aspects of Marine Pollution (GESAMP).

Gosling, F. G. (1994). The Manhattan project: Making the atomic bomb. DOE/HR-0096. Washington, DC: US Government Printing Office (September, 1994).

Gray, J., Jones, S. R. & Smith, A. D. (1995). Discharges to the environment from Sellafield Site, 1951–1992. *Journal of Radiological Protion*, *15*, 99–131.

Guéguéniat, P., Kershaw, P., Hermann, J. & Bailly du Bois, P. (1997). New estimation of La Hague contribution to the artificial radioactivity of Norwegian water (1992–1995) and Barents Sea (1992–1997). *Science of Total Environment*, *202*, 249–266.

Hamilton, T. F., Milliès-LaCroix, J.-C. & Hong, G. H. (1996). ^{90}Sr (^{137}Cs) and Pu isotopes in the Pacific Ocean: Sources and Trends. In P. Guèguèniat, P. Germain & H. Mètivier (Eds), *Radionuclides in the Ocean: Inputs and Inventories* (Ch. 2, pp. 29–58). Les Ulis cedex A, France: Les Éditions de Physique.

Hardy, E. P., Krey, P. W. & Volchok, H. L. (1973). Global inventory and distribution of fallout plutonium. *Nature*, *241*, 444–445.

Hetherington, J. A. (1975). Some investigations into the behavior of plutonium in the marine environment. In *Impacts of Nuclear Releases into the Aquatic Environment*. IAEA symposium (SM-198/29) (pp. 193–212).

Hewlett, R. G. & Anderson Jr., O. (1962). *The New World, 1939–1946, Vol. I, History of the United States Atomic Energy Commission*. Pennsylvania State University Press, University Park.

Hicks, H. G. (1982). Calculation of the concentration of any radionuclide deposited on the ground by offset fallout from a nuclear detonation. *Health Physics*, *42*, 585–600.

Hicks, H. G. & Barr, D. W. (1984). Nevada test site fallout atom ratios: ^{240}Pu/^{239}Pu and ^{241}Pu/^{239}Pu. UCRL-53499/1 (4 pp.). Lawrence Livermore National Laboratory.

Hirose, K., Sugimura, Y. & Aoyama, M. (1992). Plutonium and ^{137}Cs in the western North Pacific: Estimation of residence time of plutonium in surface waters. *Applied Radiation and Isotopes*, *43*, 349–359.

Hirose, K., Amano, H., Baxter, M. S., Chaykovskaya, E., Chumichev, V. B., Hong, G. H., Isogai, K., Kim, C. K., Kim, S. H., Miyao, T., Torimoto, T., Nikitin, A., Oda, K., Pettersson, H. B. L., Povinec, P. P., Seta, Y., Tkjalin, A., Togawa, O. & Veletova, N. K. (1999). Anthropogenic radionuclides in seawater in the East Sea/Japan Sea: Results of the first-stage Japanese–Korean–Russian expedition. *Journal of Environmental Radioactivity*, *43*, 1–13.

Høibråten, S. & Thoresen, P. E. (1995). The sunken submarine Komsomolets and its effects on the environment. In P. Strand & A. Cooke (Eds), *Environmental Radioactivity in the Arctic* (pp. 2350–2355). Østerås, Norway: Norwegian Radiation Protection Authority.

Holm, E., Persson, B. R. R., Hallstadius, L., Aarkrog, A. & Dahlgaard, H. (1983). Radiocesium and transuranium elements in the Greenland and Barents Seas. *Oceanologica Acta*, *6*, 457–462.

Holm, E., Aarkrog, A., Ballestra, S. & Dahlgaard, H. (1986). Origin and isotopic ratios of plutonium in the Barents and Greenland Sea. *Earth and Planetary Science Letters*, *79*, 27–32.

Honjo, S. (1980). Material fluxes and modes of sedimentation in the mesopelagic and bathypelagic zones. *Journal of Marine Research*, *38*, 53–97.

IAEA (1991). Inventory of radioactive material entering the marine environment: Sea disposal of radioactive waste (83 pp.). IAEA-TECDOC-588. Vienna: International Atomic Energy Agency.

IAEA (1997). Predicted radionuclide release from marine reactors dumped in the Kara Sea (83 pp.). IAEA-TECDOC-938. Vienna: International Atomic Energy Agency.

IAEA (1998). The Radiological situation at the atolls of Mururoa and Fangataufa, main report (282 pp.). *Radiological Assessment Report Series*. Report by the International Advisory Committee. Vienna: International Atomic Energy Agency.

IAEA (1999). Inventory of radioactive waste disposal at sea. IAEA-TECDOC-1105 (121 pp.). Vienna: International Atomic Energy Agency.

IMO (1972). Final act of the conference with attachment including the convention on the prevention of marine pollution by dumping of wastes and other matter. *International Conference on the Convention on the Dumping of Wastes at Sea*. London 30 October–13 November 1972. London: International Maritime Organization.

IMO (1993). Resolution LC.51(16): Report of the 16th Consultative Meeting of the Contracting Parties to the Convention on the Prevention of Marine Pollution by Dumping of Wastes and Other Matter, LC 16/14. London: International Maritime Organization.

International Herald Tribune, October 12–13, 1996.

Iseki, K. (1981). Vertical transport of particulate organic matter in deep Bering Sea and Gulf of Alaska. *Journal of Oceanographical Society of Japan*, *37*, 101–110.

ITAR-TASS (1992). Nuclear reactors at Krasnoyarsk shut down after 30 years. June 30, 1992.

Jachmenev, V. (1995). *Problems of Radioactive Contamination of the Chelyabinsk Region and the Way to Solve Them. IIASA Meeting*. Luxemburg, Austria, March 6–10, 1995.

Jones, V. C. (1985). *The Making of the Atomic Bomb*. Washington, DC: US Government Printing Office.

Joseph, A. B., Gustafson, P. F., Russel, I. R., Schuert, E. A., Volchok, H. L. & Tamplin, A. (1971). Sources of radioactivity and their characteristics. In *Radioactivity in the Marine Environment* (Ch. 2, pp. 6–41). Washington: National Academy of Sciences.

Kathren, R. L. (1984). *Radioactivity in the Environment: Sources, Distribution and Surveillance* (397 pp.). Chur, Switzerland: Harwood Academic Publishers.

Kautsky, H. (1989). Content of radiocesium in the water of the northern North Atlantic during the year 1985. *Deutches Hydrographisches Zeitung*, *42*, 21–26.

Kennedy, J. W., Seaborg, G. T., Segre, E. & Wahl, A. C. (1946). Radioctive element 94 from deutrons on uranium. *Physical Reviews*, *70*, 555.

Kershaw, P. & Baxter, A. (1995). The transfer of reprocessing waste from north–west Europe to the Arctic. *Deep-sea Research II*, *42*, 1413–1448.

Kershaw, P., Woodhead, D. S., Lovett, M. B. & Leonard, K. S. (1995). Plutonium from European reprocessing operations-its behavior in the marine environment. *Applied Radiation and Isotopes*, *46*, 1121–1134.

Kershaw, K., McCubbin, D. & Leonard, K. S. (1999). Continuing contamination of North Atlantic and Arctic waters by Sellafield radionuclides. *Science of Total Environment*, *237/238*, 119–132.

Kershaw, P. J., Pentreath, R. J., Woodhead, D. S. & Hunt, G. J. (1992). A review of radioactivity in the Irish Sea (65 pp.). Aquatic Environment Monitoring Report, No. 32. MAFF Directorate of Fisheries Research, Lowestoft, UK.

Kiode, M., Bertine, K. K., Chow, T. J. & Goldberb, E. D. (1985). The $^{240}Pu/^{239}Pu$ ratio, a potential geochronometer. *Earth and Planetary Science Letters*, *72*, 1–8.

Kolstad, A. K. (1995). Expeditions to "Komsomolets" in 1993 and 1994. NPRA Stralevern Rapport 1995:7. Østerås, Norway: Norwegian Radiation Protection Authority.

Krey, P. W. (1967). Atmospheric burnup of a plutonium-238 generator. *Science*, *158*, 769–771.

Krey, P. W., Hardy, E. P., Pachucki, C., Rourke, F., Coluzza, J. & Benson, W. (1976). Mass isotopic composition of global fallout plutonium in soil. In *Transuranium Nuclides in the Environment*. IAEA-SM-199/39. Vienna: International Atomic Energy Agency.

Lassey, K. R., Mannning, M. R. & O'Brien, B. J. (1988). Assessment of the inventory of carbon-14 in the oceans: An overview. In *Inventories of Selected Radionuclides in the Oceans* (pp. 7–50). AEA-TECDOC-481. Vienna, Austria: International Atomic Energy Agency (IAEA).

Lee, Y. K. (1994). The study of the atmospheric deposition of anthropogenic radionuclides in the Korean Peninsula (43 pp.). MS Thesis. Hanyang University, Seoul (in Korean).

Livingston, H. D., Bowen, V. T. & Kupferman, S. L. (1982). Radionuclides from Windscale discharges II, their dispersion in Scottish and Norwegian coastal circulation. *Journal of Marine Research*, *40*, 1227–1258.

Livingston, H. D., Bowen, V. T., Casso, S. A., Volchok, H. L., Noshkin, V. E., Wong, K. M. & Beasley, T. M. (1985). Fallout nuclides in Atlantic and Pacific waters columns: GEOSECS Data, WHOI-85-19, Woods Hole, MA 02543 (73 pp.).

Martin, J. H. (1970). The possible transport of trace metals via moulted copopod exoskeletons. *Limnology and Oceanography*, *15*, 756–761.

Mitchell, P. I., Condren, O. M., Vintro, L. & McMahon, C. A. (1999). Trends in plutonium, americium and radiocesium accumulation and-long-term bioavailability in the western Irish Sea mud basin. *Journal of Environmental Radioactivity*, *44*, 223–251.

Monetti, M. A. (1996). Worldwide Deposition of Strontium-90 through 1990 (56 pp.). Environmental Measurements Laboratory, EML-579. New York: US DOE.

Moscow Interfax (1994), 21 January 1994.

Nagaya, Y. & Nakamura, K. (1976). ^{90}Sr and ^{137}Cs contents in the surface waters of adjacent seas of Japan and North Pacific during 1969 and 1973. *Journal of Oceanographycal Society of Japan*, *40*, 228–234.

Nagaya, Y. & Nakamura, K. (1981). Artificial radionuclides in the western Northwest Pacific (II): ^{137}Cs and ^{90}Sr in deep waters. *Journal of Oceanographycal Society of Japan*, *37*, 135–144.

Nagaya, Y. & Nakamura, K. (1984). ^{239,240}Pu, ^{137}Cs and ^{90}Sr in the central North Pacific. *Journal of Oceanographycal Society of Japan*, *40*, 416–424.

Nagaya, Y. & Nakamura, K. (1987). Artificial radionuclides in the western Northwest Pacific (II): ^{137}Cs and ^{239,240}Pu inventories in water and sediment columns observed from 1980 to 1986. *Journal of Oceanographycal Society of Japan*, *43*, 345–355.

Nakamura, K. & Nagaya, Y. (1975). Accumulation of radionuclides in coastal sediment of Japan (II): Contents of fission products in some coastal sediments collected in 1966–1972. *Journal of Radiation Research*, *16*, 184–192.

Nakamura, K. & Nagaya, Y. (1985). Accumulation of Cs-137 and Pu-239, 240 in sediments of the coastal sea and the North Pacific. In A. C. Sigleo & A. Hattori (Eds), *Marine and Estuarine Geochemistry* (Ch. 12, pp. 171–180). Chelsea: Lewis Pub.

Nakanishi, T., Tajima, M., Senaga, M., Takei, M., Ishikawa, A. & Sakamoto, K. (1984). Determination of ^{239,240}Pu in seawater. *Nuclear Instruments and Methods in Physics Research*, *223*, 239–242.

Nakanishi, T., Shiba, Y., Muramatsu, M. & Azizul Haque, M. (1995). Estimation of mineral aerosol fluxes to the Pacific by using environmental plutonium as a tracer. In H. Sakai & Y. Nozaki (Eds), *Biogeochemical Processes and Ocean Flux in the Western Pacific* (pp. 15–30). Tokyo: Terra Scientific (TERRAPUB).

NCRP (1936). X-ray Protection Report No. 3. Bethesda, Maryland, USA: National Council on Radiation Protection and Measurements.

NCRP (1938). Radium Protection Report No. 4. Bethesda, Maryland, USA: National Council on Radiation Protection and Measurements.

NEA/OECD (1996). Coordinated Research and Environmental Surveillance Programme Related to Sea Disposal of Radioactive Waste. CRESP Final Report 1991–1995. Nuclear Energy Agency, Organization for Economic Cooperation and Development, Paris, 1996.

Newell, R. W. (1971). The global circulation of atmospheric pollutants. *Scientific American*, *224*, 32–47.

Nikipelov, B. V. & Drozhko, E. G. (1990). Explosion in the south Ural Mountains, *PRIRODA*, *May 1990*, 48–49.

Nikipelov, B. V., Romanov, G. N., Buldakov, L. A., Babev, N. S., Kholina, Yu. B. & Mikerin, E. I. (1990). A radiation accident in the southern Urals in 1957. Original article submitted July 14, 1989, Plenum Publishing Co.

Nilsen, T., Kudrik, I. & Nitikin, A. (1996). The Russian northern fleet: Sources of radioactive contamination. Bellona Foundation, Oslo, Norway, 18 April 1996. Internet address: *http://www.grida.no./ngo/bellona/lhome.htm*.

Noshkin, V. E. & Bowen, V. T. (1973). Concentrations and distributions of long-lived fallout radionuclides in open ocean sediments. In *Radioactive Contamination of the Marine Environment* (pp. 671–686). Vienna, Austria: International Atomic Energy Agency (IAEA).

Noshkin, V. E. & Wong, K. M. (1979). Plutonium mobilization from sedimentary sources to solution in the marine environment. In *Proceedings of the Third NEA Seminar on Marine Radioecology* (pp. 165–178). Tokyo.

Noshkin, V. E. & Wong, K. M. (1980). Plutonium in the North Equatorial Pacific. In *Processes Determining the Input Behavior and Fate of Radionuclides and Trace Elements in Continental Shelf Environments*. Report 790382 (pp. 11–28). U.S. DOE Conference.

NRPA (1996). Dumping of radioactive waste and investigation of radioactive contamination in the Kara Sea. Results from 3 years of investigation (1992–1994) in the Kara Sea. Joint Norwegian–Russian Expert Group for Investigation of Radioactive Contamination in the Northern Areas. Østerås, Norway: Norwegian Radiation Protection Authority (NRPA).

Nuclear Fuel (1993). Tomsk-7 environmental Pu emission about 23 grams, Fact-Finder Says (p. 13).

Nuclear Waste News (1993). Tomsk-7 caused minimal radiation hazards (pp. 157–158). IAEA Team Reports.

Oi, N. (1998). Plutonium Challenges: Changing dimensions of global cooperation (pp. 12–16). IAEA Bulletin, 40/1/1998.

OTA (1995). Spent Fuel and Waste Management, Presented-by the Russian Delegation to the Congressional Office of Technology Assessment Meeting, January 17–18, 1995. Washington, DC: Office of Technology Assessment (OTA).

Pentreath, R. J. (1985). General review of literature relevant to water discharges (pp. 19–66). IAEA TECDOC-239. Vienna, Austria: International Atomic Energy Agency.

Petersen, K. R. (1970). An empirical model for estimating worldwide deposition from atmosphere nuclear detonations. *Health Physics*, *18*, 357–378.

Pettersson, H. B. L., Amano, K., Berezhnov, V. I., Chaykovskaya, E., Chumichev, V. B., Chung, C. K., Gastaud, J., Hirose, K., Hong, G. H., Kim, C. K., Kim, S. H., Lee, S. H., Morimoto, T., Nikitin, A., Oda, K., Povinec, P. P., Suzuki, E., Tkalin, A., Togawa, O., Veletova, N. K., Volkov, Y. & Yishida, K. (1999). Anthropogenic radionuclides in sediments in the NW Pacific Ocean and its marginal seas: Results of the 1994–1995 Japanese–Korean–Russian expeditions. *Science of the Total Environment*, *237/238*, 213–224.

Raisbeck, G. M., Yiou, F., Zhou, Z. Q., Kilius, L. R. & Dahlgaard, H. (1993). Antropogenic ^{129}I in the Kara Sea. In P. Strand & E. Holm (Eds), *Environmental Radioactivity in the Arctic and Antarctic* (pp. 125–128). Østerås, Norway: Norwegian Radiation Protection Authority.

Rhodes, R. (1986). *The Making of the Atomic Bomb*. New York: Simon and Schuster.

Robison, W. L., Noshkin, V. E., Hamilton, T. F., Conrado, C. L. & Bogen, K. (2001). *An Assessment of Current Day impact of Various Materials Associated with the U.S. Nuclear Test Program in the Marshall Islands* (21 pp.). UCRL-LR-143980. Lawrence Livermore National Laboratory.

Roether, W. (1974). The tritium and carbon-14 profiles at the GEOSECS I (1969) and GOGO I (1971) North Pacific stations. *Earth and Planetary Science Letters*, *23*, 108–115.

Sayles, F. L., Livingston, H. D. & Panteleyev, G. P. (1997). The history and source of particulate ^{137}Cs and $^{239+240}Pu$ deposition in sediments of the Ob River delta, Siberia. *Science of the Total Environment*, *202*, 25–71.

Schell, W. R., Lowman, F. G. & Marshall, R. P. (1980). Geochemistry of transuranic elements at bikini atoll. In W. C. Hanson (Ed.), *Transuranic Elements in the Environment* (pp. 541–577). U.S. Department of Energy.

Scott, M. R., Salter, P. F. & Halverson, J. E. (1983). Transport and deposition of plutonium in the ocean: Evidence from Gulf of Mexico sediments. *Earth and Planetary Science Letters*, *63*, 202–222.

SCRF (1995). Problems of arctic contamination. Materials prepared by a working group headed by Deputy Minister A. I. Volgin of the Ministry of Affairs for National and Regional Policy, Security Council of the Russian Federation, Moscow, Russia.

Seaborg, G. T. & Loveland, W. D. (1990). *The Elements Beyond Uranium* (359 pp.). New York: Wiley.

Sivintsev, Y. (1994a). Study of the nuclide composition and characteristics of the fuel in dumped submarine reactors and icebreaker "Lenin". Part 1 – Atomic icebreaker. Working materials of the International Arctic Seas Assessment Project. IAEA-IASAP-1. Vienna: International Atomic Energy Agency.

Sivintsev, Y. (1994b). Study of the nuclide composition and characteristics of the fuel in dumped submarine reactors and icebreaker "Lenin". Part 2 – Nuclear submarines. Working materials of the International Arctic Seas Assessment Project. IAEA-IASAP-5. Vienna: International Atomic Energy Agency.

Smith, J. N., Ellis, K. M. & Jones, E. P. (1990). Cesium-137 transport into the Arctic Ocean through Fran Strait. *Journal of Geophysical Research*, *95* (C2), 1693–1701.

Smith, J. N., Ellis, K. M., Aarkrog, A., Dahlgaard, H. & Holm, E. (1994). Sediment mixing and burial of the $^{239,240}Pu$ from the 1968 Thule, Greenland nuclear weapons accident. *Journal of Environmental Radioactivity*, *25*, 135–159.

Smith, J. N., Ellis, K. M., Naes, K., Dahle, S. & Matishov, D. (1995). Sedimentation and mixing rates of fallout radionuclides in Barents Sea sediments off Novaya Zemlya. *Deep-Sea Research II*, *42*, 1471–1493.

Smith, J. N., Ellis, K. M., Ployak, L., Ivanov, G., Forman, S. L. & Moran, S. B. (2000). $^{239,240}Pu$ transport into the Arctic Ocean from underwater nuclear tests in Chernaya Bay, Novaya Zemlya. *Continental Shelf Research*, *20*, 255–279.

Strand, P. & Cooke, A. (1995). *Environmental Radioactivity in the Arctic* (415 pp.). Scientific Committee of the Environmental Radioactivity in the Arctic, Østerås, Norway: Norwegian Radiation Protection Authority.

Strand, P. & Jolle, T. (1999). *Environmental Radioactivity in the Arctic* (353 pp). Scientific Committee of the Environmental Radioactivity in the Arctic, Østerås, Norway: Norwegian Radiation Protection Authority.

Tass World Service (1992). Two industrial nuclear reactors scheduled for closure. April 20, 1992.

UNSCEAR (1993). Sources and effects of ionizing radiation. United Nations Scientific Committee on the Effects of Atomic Radiation. 1993 Report to the General Assembly, with Scientific Annexes. United Nations Publication No. E.94.IX.82. New York: United Nations.

UNSCEAR (2000). Sources and effects of ionizing radiation. United Nations Scientific Committee on the Effects of Atomic Radiation. 2000 Report to the General Assembly, with Scientific Annexes, Vol. 1: Sources. United Nations Publication E.00.IX.3. New York: United Nations.

Vakulovsky, S. M. (1993). Transport of Artificial Radioactivity by the Ob to the Arctic seas. In P. Strand & E. Holm (Eds), *Environmental Radioactivity in the Arctic and Antarctic*. Østerås, Norway: Norwegian Radiation Protection Authority (August 1993).

Vakulovsky, S. M., Nikitin, S. & Chumichev, V. (1993). Radioactive contamination of the Barents and Kara Seas. *International Conference on Environmental Radioactivity in the Arctic*. June 7–9, 1993. Woods Hole Oceanographic Institution, Woods Hole, MA, USA.

Volchok, H. L., Bowen, V. T., Folson, T. R., Broecker, W. S., Schuert, E. A. & Bien, G. S. (1971). Oceanic distributions of radionuclides from nuclear explosions, sources of radioactivity and their characteristics. In *Radioactivity in the Marine Environment* (pp. 42–89). Washington: National Academy of Sciences.

Vorobiova, M. I., Degteva, M. O., Burmistrov, D. S., Safronova, N. G., Kozheurov, V. P., Anspaugh, L. R. & Napier, B. A. (1999). Review of historical monitoring data on Techa Review contamination. *Health Physics*, *76*, 605–618.

Weiss, W. & Roether, W. (1980). The rates of tritium input to the world oceans. *Earth and Planetary Science Letters*, *49*, 435–446.

White Book (1993). "*White Book*". *Facts and Problems Related to Radioactive Waste Disposal in Sea Adjacent to the Territory of the Russian Federation*. Materials for a report by the Government Commission on matters related to radioactive waste disposal at sea. Created by Decree No. 613 of the Russian Federation President of October 24, 1994. A. V. Yablokov, V. K. Karasev, V. M. Rumyantsev, M. E. Kokeev & O. J. Petrov (Eds). Moscow: Small World Publisher.

Yefimov, E. (1994). Radionuclide composition, characteristics of shielding barriers and analysis of weak points of the dumped reactors of submarine N. 601. Working materials of the International Arctic Seas Assessment Project. IAEA-IASAP-6. Vienna: International Atomic Energy Agency.

Yiou, F., Raisbeck, G. M., Zhou, Z. Q., Killius, L. R. & Kershaw, P. J. (1995). *Improved Estimates of Oceanic Discharges of* ^{129}I *from Sellafield and La Hague* (pp. 113–116). P. Strand & A. Cooke (Eds). Østerås, Norway: Norwegian Radiation Protection Authority.

Chapter 3

Transuranium nuclides in the world's oceans

Luis León Vintró [a], Peter I. Mitchell [a], Kilian J. Smith [a], Peter J. Kershaw [b], Hugh D. Livingston [c]

[a] *Department of Experimental Physics, University College Dublin, Belfield, Dublin 4, Ireland*
[b] *The Centre for Environment, Fisheries and Aquaculture Sciences (CEFAS), Lowestoft, Suffolk, NR33 0HT, UK*
[c] *International Atomic Energy Agency, Marine Environment Laboratory, MC 98012, Monaco*

For almost six decades, transuranium nuclides have been introduced to the world's oceans from a multiplicity of sources, both planned and accidental. During this time, much has been learned about the distribution and behaviour of these radionuclides in shelf, continental slope and deep ocean environments, as well as the oceanic physical and biochemical processes governing their dispersion, transfer and ultimate fate. In this chapter, the main sources of transuranium nuclides to the marine environment are reviewed, and an account given of the current state of knowledge regarding the main processes affecting their oceanic behaviour. Emphasis is given to the physico-chemical forms of transuranium nuclides upon entry to the ocean, the changes that take place upon dilution with seawater and the complex physical and biogeochemical processes that determine their long-term fate.

1. Introduction

Transuranium nuclides have been discharged to the environment from a multiplicity of sources, both planned and accidental. The first significant releases of transuranium nuclides to the marine environment began in the mid-1940s as a consequence of fallout from the first nuclear weapons tests and the discharge of effluents from weapons production facilities. Almost sixty years on since these first releases, the inventories of these elements in the world's oceans have been substantially increased by inputs from fallout from other nuclear detonations (mainly carried out in the 1950s and 1960s), authorised discharges from the nuclear industry, leakage from radioactive wastes of various kinds and a number of accidental releases.

During these six decades, much has been learned about the distribution and behaviour of these radionuclides in the oceans, as well as the oceanic physical and biochemical processes governing their dispersion, transfer and ultimate fate. This accumulated knowledge, gathered from a large number of studies, has allowed predictions to be made on the short- and long-term impact of transuranium nuclides on marine life and human health. It has also provided a conceptual framework with which to model and assess the possible impact of future releases.

In addition, the well-defined character of transuranium inputs (both in time and space), together with the relatively short history of these elements in the environment and the low solubility and high particle reactivity characterising their biogeochemical behaviour, have provided a unique opportunity to improve the general understanding of a wide range of marine processes, including particle dynamics in the water column, biological interactions, and mixing and burial within ocean sediments.

In this chapter, a brief overview is given of the main routes of production of transuranium nuclides. This is followed by an account of the main sources of transuranium nuclides to the marine environment, and a synthesis of the current state of knowledge regarding the key processes affecting their oceanic distribution and behaviour. Particular emphasis is given to the physical and chemical forms of transuranium nuclides before entering the oceans and following dilution with seawater, as well as to the physical and biogeochemical processes governing their fate.

2. Production of transuranium nuclides

Although traces of naturally-produced transuranium nuclides are to be found in nature, the overwhelming majority of present inventories of these elements derive from human activities, related mainly with power generation in nuclear fission reactors and with military applications. Whereas military stocks of transuranium nuclides have been maintained roughly constant in recent years, the quantities of transuranium nuclides generated from civil applications continue to increase steadily as a result of nuclear operations.

By 1995, a total of 437 nuclear reactors were operative worldwide with a total installed generating capacity of 344 gigawatts-electric (IAEA, 1996). Considering the net transuranium generation of a typical 1 GW_e light-water reactor (the dominant commercial reactor type), present annual world-wide production of transuranium nuclides can be estimated to be in the order of 4.6 tonnes of neptunium, 83 tonnes of plutonium, 1.3 tonnes of americium and 0.4 tonnes of curium. This to be added to the more than 1000 tonnes of plutonium and, to a lesser extent, other transuranium nuclides, which have been produced over the last few decades and are to be found in nuclear power reactors, waste disposal sites and nuclear weapons stockpiles around the world.

2.1. *Natural occurrence*

The occurrence of transuranium nuclides in nature was first established in 1942 by Seaborg & Perlman (1948), who were able to measure the small activities of plutonium in pitchblende and carnolite minerals. Further studies in a number of uranium ores established the amount of plutonium present in uranium minerals to be in the order of 1 atom of ^{239}Pu per 10^{11} atoms of uranium. Many of these early determinations gave only estimates of the plutonium concentrations, as no isotopic tracer was available for use as a chemical yield monitor of the plutonium sample purified from tonne quantities of mineral (Peppard et al., 1952). Precise determination of the concentrations of not only ^{239}Pu but also ^{237}Np was made possible with the development of high-sensitivity α spectrometry techniques and the production of suitable chemical yield tracers. Using mineral quantities under 10 g, Myers & Lindner (1971) found the $^{237}Np/U$ and $^{239}Pu/U$ atom ratios in samples of Belgian Congo pitchblende to be $(2.16 \pm 0.05) \times 10^{-12}$ and $(3.10 \pm 0.14) \times 10^{-12}$, respectively.

Assuming a concentration of uranium of 2.7×10^{-3} g kg^{-1} averaged over the world's crust, Taylor (2001) estimated an average ^{239}Pu concentration in the Earth's crust of approximately 2×10^{-14} g kg^{-1} ($\sim$50 μBq kg^{-1}). Plutonium concentrations in the oceans are probably some orders of magnitude lower, due to the lower concentrations of uranium and thorium in the ocean crust.

The presence of minute traces of another isotope of plutonium, ^{244}Pu ($T_{1/2} = 8.26 \times 10^7$ y), was confirmed by Hoffman et al. in 1971, who determined its abundance in Precambrian rare-earth minerals to be in the range 10^{-27}–10^{-25} g per gram. It has been estimated that, at present, $\sim$10 g of this primordial ^{244}Pu may still remain in the Earth's crust (Choppin et al., 1995; Taylor, 2001).

2.2. *Nuclear reactors*

The major routes for the production of transuranium nuclides in a nuclear reactor are depicted in Fig. 1. Excess neutrons produced by the fission of ^{235}U are captured in ^{238}U to yield, after decay of short-lived ^{239}U and ^{239}Np, the fissile ^{239}Pu. Non-fission capture of neutrons in ^{239}Pu results in the production of more massive plutonium isotopes such as ^{240}Pu, ^{241}Pu, ^{242}Pu and ^{243}Pu. Although this chain of reactions is responsible for most of the mass inventory of transuranium nuclides in the reactor, the highest α radioactivity in the generated plutonium results from ^{238}Pu, produced by neutron capture in ^{237}Np (Pigford & Ang, 1975). The latter in turn results from reaction chains initiated by non-fission neutron capture in ^{235}U and by fast neutron (n, 2n) reactions with ^{238}U. Some ^{238}Pu is also produced by fast neutron (n, 2n) reactions with ^{239}Pu.

Radioactive decay of ^{241}Pu and ^{243}Pu result in the formation of ^{241}Am and ^{243}Am, respectively. Although americium is initially the third most prevalent transuranium nuclide in a reactor after plutonium and neptunium, it becomes increasingly important some years after the end of irradiation due to the ingrown ^{241}Am following the decay of ^{241}Pu (Kim, 1991). Neutron capture processes in ^{241}Am and ^{243}Am lead to ^{242}Am, ^{242m}Am and ^{244}Am, which decay to produce ^{242}Cm and ^{244}Cm. These curium isotopes, together with small amounts of ^{243}Cm formed by neutron capture in ^{242}Cm, are major contributors to the total α activity of spent nuclear fuels (Schneider & Livingston, 1984).

The material inventories of plutonium, americium and curium generated annually by different types of reactors show considerable variations in the relative quantities of transuranium nuclides produced, which can be attributed to differences in the propensities of uranium and plutonium isotopes to undergo fission and to capture neutrons at different incident neutron energies. The relative abundances of transuranium nuclides in a reactor depend not only on the type of fuel and the energy spectrum of the neutrons, but also on the degree of 'burn-up'

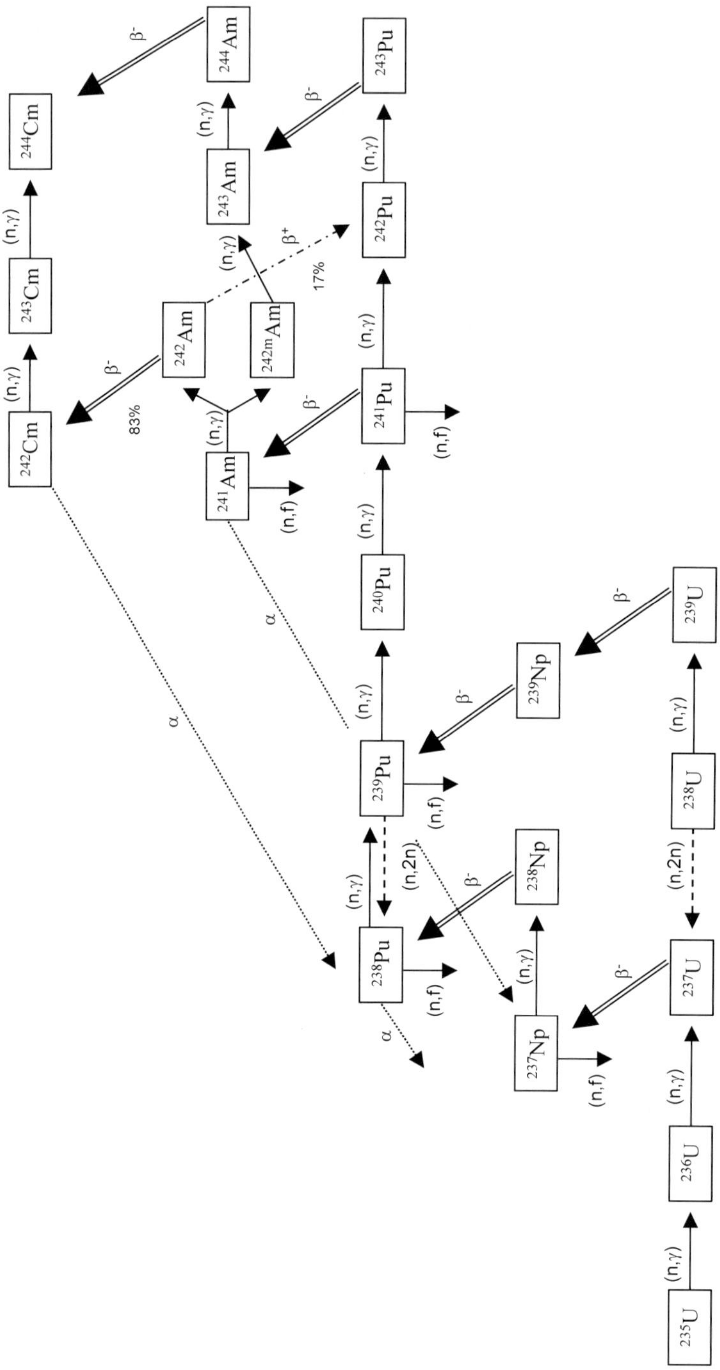

Fig. 1. Simplified reaction chain for the production of transuranium nuclides in a nuclear reactor (source: Schulz, 1976; NEA, 1989).

of the fuel, i.e. the neutron flux and the duration of the exposure to this flux. The process of neutron capture and the higher fissionability of transuranium nuclides with odd masses, lead to the accumulation of the less fissionable even mass isotopes if fuel is irradiated for longer periods (NEA, 1989).

In power generation, the burn-up of ^{238}U proceeds as far as economically feasible. Typical low-enriched uranium fuel discharged from a reliably operating light-water reactor contains approximately 58% ^{239}Pu and 11% ^{241}Pu (both fissile), plus 25% ^{240}Pu, 4% ^{242}Pu and 2% ^{238}Pu (Lovins, 1980). However, because burn-up and hence isotopic composition of discharged plutonium can vary enormously between and within reactors and with time, 'reactor-grade' plutonium is not a well-defined term, and is usually applied to fuel with a $^{240+242}$Pu content of $>19\%$ of the total plutonium (Dickson, 1981). In some circumstances, the $^{240+242}$Pu fraction can rise to as much as 49% (34% ^{240}Pu, 15% ^{242}Pu), although this is unusually high and approaches a practical limit.

By contrast, when plutonium is to be extracted for the production of nuclear weapons, the burn-up is kept low to reduce the presence of plutonium isotopes other than fissile ^{239}Pu (Burns et al., 1994). Typically, 'weapons-grade' plutonium contains no more than 8% ^{240}Pu (usually around 6%), perhaps 0.5% ^{241}Pu and negligible ^{242}Pu and ^{238}Pu (Lovins, 1980; Mitchell et al., 1997).

Information on the isotopic composition of the fuel and the relative proportions of the different transuranium nuclides is of importance, not only because it can be used to 'fingerprint' a particular source-term (as will be discussed later), but also because the radiotoxicity of the fuel (and, consequently, its radiological implications) may vary significantly depending on its composition. By way of example, it has been reported that the radiotoxicity of fuel from a pressured water reactor using enriched uranium oxide fuel can be up to ten times higher than that of weapons-grade plutonium or low burn-up fuel (Yamana et al., 2001).

2.3. *Nuclear explosions*

Transuranium nuclides are also produced in the course of nuclear explosions. In this case, very high neutron fluxes (10^{23}–10^{25} neutrons cm^{-2}) are produced for a very short period of time (10^{-8}–10^{-6} s), leading to the formation, by multiple neutron capture, of neutron-rich isotopes of the nuclear component (highly-enriched uranium or plutonium) which subsequently β-decay to more stable isobars (Fig. 2). The number of neutron captures and the proportions of transuranium nuclides formed depend on the neutron intensity and the duration of the flash (Keller, 1971).

Significant variations in the relative abundances of the transuranium elements take place depending on the yield and composition of the nuclear explosive. Prior to the first thermonuclear test in 1952, the fission weapons tested had a relatively low yield and the atomic masses of the transuranium nuclides produced did not exceed about 243. In subsequent thermonuclear devices, higher neutron yields resulted in the production of new transuranium nuclides with masses up to 255 (Diamond et al., 1960; Perkins & Thomas, 1980). For comparison, the relative mass abundance of the plutonium isotopes formed in the course of the first thermonuclear event and those measured in global fallout debris are presented in Table 1.

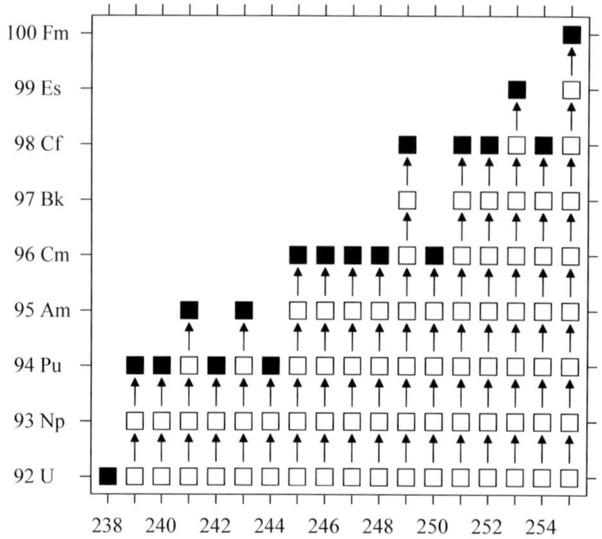

Fig. 2. Production of uranium isotopes in the 'Mike' thermonuclear event (1952), and their decay to β stability: (□) β-unstable nuclide, (■) β-stable nuclide (source: Diamond et al., 1960).

Table 1
Relative mass abundance of plutonium produced during the 'Mike' test compared with that measured in global fallout (the abundances are given relative to ^{239}Pu)

Mass number	Isobar	'Mike' event	Fallout
239	Pu	1.0	1.0
240	Pu	0.363	0.18
241	Pu	0.039	0.013
242	Pu	1.9×10^{-2}	3.4×10^{-3}

3. Transuranium nuclides in the oceans: sources and inventories

As a result of nuclear-related activities, small but significant amounts of transuranium nuclides have become widely dispersed over the world's oceans. Although the main source contributing to present-day concentrations remains global fallout from atmospheric nuclear weapons testing, authorised discharges from the nuclear industry, radioactive waste disposal and various accidental releases have resulted in a significant enhancement of transuranium nuclide concentrations at a local or regional scale.

3.1. *Atmospheric fallout*

Prior to the Limited Test Ban Treaty of 1963, agreed by the USA, the Former USSR and the UK, ~367 atmospheric explosions were conducted, resulting in a total explosive yield of 194 Mt from fission and 319 Mt from fusion (UNSCEAR, 1982). The most intense testing period took place between 1957 and 1963, with a total of 259 tests. From 1963, only the Peo-

Table 2
Estimated amounts of transuranium nuclides injected into the atmosphere. All activities are referred to 1961

Nuclide	Mass abundance (relative to ^{239}Pu)	Activity abundance (relative to ^{239}Pu)	Total injection (PBq)
^{238}Pu	0.00015	0.040	0.31
^{239}Pu	1	1	7.77
^{240}Pu	0.18	0.67	5.18
^{241}Pu	0.015	24.9	193
^{242}Pu	0.0044	2.8×10^{-4}	0.0022
^{242m}Am	2×10^{-7}	3×10^{-5}	0.00023
^{244}Cm	1.5×10^{-8}	2×10^{-5}	0.00016

ple's Republic of China and France conducted atmospheric tests, although on a much more limited scale. Since 1980, all testing has been confined to underground sites. Estimated activities of transuranium nuclides that have been injected into the atmosphere as a consequence of these tests are given in Table 2 (compiled from Hardy et al., 1973; Krey et al., 1976; Holm & Persson, 1977, 1978; Perkins & Thomas, 1980; UNSCEAR, 1982).

Most of the fallout produced in atmospheric tests has by now been transferred from the atmosphere to the Earth's surface. Considering that approximately 70% of the surface of the planet is ocean, it is no surprise that much of the fallout inventory has ended up there. The observed deposition pattern of transuranium and other radionuclides, with maxima at 40–50° N and S and minima at the Poles (Fig. 3), is a consequence of the location and yield of the nuclear detonations. Prior to 1952, devices with yields in the low-kiloton range, conducted at or near ground level, placed most of the debris in the troposphere. The low residence time for the debris in this layer confined the deposition to near the latitude of the test sites. The oceanic impact of these tests was most significant at the USA Pacific proving ground in the Marshall Islands, although the tests carried out by France in French Polynesia and by China at the Lop Nor test site also had a tropospheric component (Livingston & Povinec, 2002). After 1952, numerous multi-megatonne tests were carried out which injected most of the debris into the lower stratosphere, becoming globally dispersed (Perkins & Thomas, 1980).

This was the case at the Novaya Zemlya testing grounds, used by the Former USSR for large-scale atmospheric tests (mainly in 1961 and 1962). Although the contribution of these atmospheric tests to local contamination of the marine environment has been found to be negligible (Osvath et al., 1999), this is not so for a series of underwater nuclear tests conducted in the 1950s at Chernaya Bay, on the southwestern coast of the island. Plutonium-239, 240 levels exceeding 15,000 Bq kg^{-1} have been reported for sediments from the central region of the Bay, which are among the highest ever reported for the marine environment (Smith et al., 2000). The same authors estimate that approximately 11 TBq of ^{239,240}Pu are still retained in the sediments of Chernaya Bay, and that significant offshore transport of plutonium to the Barents Sea (2 TBq) occurred in the past, probably at the time of the tests.

Overall, it has been estimated that 12% of the atmospheric injection has been deposited close to the test sites as local fallout, 10% has been deposited in a latitude band around the test sites as tropospheric fallout and the remaining 78% has been deposited mainly on the same hemisphere as the test site as global fallout (UNSCEAR, 1982).

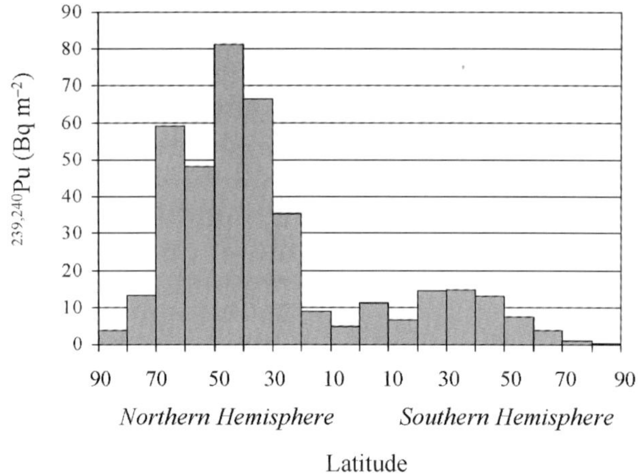

Fig. 3. Average latitudinal distribution of cumulative ^{239,240}Pu fallout to January 1971 (source: Hardy et al., 1973).

The isotopic composition of plutonium in global fallout derived from weapons testing has changed with time and has provided information on its origin. Koide et al. (1985) used the ^{240}Pu/^{239}Pu atom ratio in polar ice sheets to differentiate the particulate fallout from pre-moratorium atmospheric nuclear tests, dominated by the USA, and post-moratorium atmospheric tests, dominated by the Former USSR. Latitudinal variations of the plutonium isotopic composition have also been observed and have been attributed to the contribution of fallout debris from nuclear tests of different yields (Perkins & Thomas, 1980).

An additional source of atmospheric fallout was introduced in 1964 as a consequence of the burn-up in the Southern Hemisphere of the nuclear-powered SNAP-9A navigational satellite, containing 0.63 PBq ($\sim$1 kg) of ^{238}Pu, as it re-entered the atmosphere. The amount of ^{238}Pu introduced was twice the activity of this isotope injected in the atmosphere as a result of nuclear testing. About three-quarters of this activity was deposited in the Southern Hemisphere, and the remainder in the Northern Hemisphere. As a result, the mean ^{238}Pu/^{239,240}Pu activity ratio of $\sim$0.024 observed in both hemispheres prior to the event was altered to 0.034 ± 0.010 and 0.20 ± 0.05 in the Northern and Southern Hemispheres, respectively (Hardy et al., 1973).

Considerable amounts of ^{241}Am, mainly formed from the decay of ^{241}Pu, are also present in global fallout. Estimations of the deposited ^{241}Am, confirmed by measurements on soil samples, indicated that by 1974, ^{241}Am activity represented 22% of the ^{239,240}Pu activity (Krey et al., 1976). Further calculations, using parent–daughter nuclide relationships to account for ingrowth, indicated that the ^{241}Am content in soil will continue to increase until reaching a peak in the year 2037, when the ^{241}Am activity will represent 42% of that of ^{239,240}Pu (Perkins & Thomas, 1980).

3.2. *Authorised discharges from the nuclear industry*

In the course of normal operation of nuclear reactors some gaseous, liquid and solid wastes of low and intermediate level are generated, which can result in emissions to the environment. These emissions (which include transuranium radionuclides), however, are only small and their impact is confined to the immediate vicinity of the reactor. More significant are the discharges from coastal nuclear fuel reprocessing facilities, consisting of low-level liquid radioactive effluents arising from the treatment of spent nuclear fuel.

The most important radionuclide releases to the marine environment from this type of installation have taken place as a result of the operations at the nuclear fuel reprocessing facilities at Sellafield (UK) and La Hague (France), which have resulted in a significant increase in the inventories of anthropogenic radionuclides in the North Atlantic. Other installations, discharging radioactive waste into the ocean either directly or indirectly (through river discharges) have had a much smaller impact.

3.2.1. *Sellafield*

Low-level radioactive waste in liquid form has been discharged to the north-eastern Irish Sea from the Sellafield reprocessing plant since the early 1950s. Operations at Sellafield commenced in 1947 with the construction of two air-cooled reactors (Windscale Piles) for the production of military plutonium. The spent fuel from these reactors, until their shut-down in 1957, was reprocessed in a series of plutonium separation and plutonium purification plants which were in operation for the period 1952–1964. The reprocessing operations were transferred to a new integrated separation and purification plant in 1964, which has remained operative until the present (Gray, 1995). The main function of this plant has been the reprocessing of fuel from *Magnox* reactors. In 1985, the Site Ion Exchange Effluent Plant (SIXEP) and the Salt Evaporator waste treatment plant were commissioned, resulting in a reduction of radiocaesium discharges to relatively constant low levels. Since 1994, with the commissioning of the Enhanced Actinide Removal Plant (EARP), discharges of actinides were similarly reduced to relatively constant low levels. However, the processing of an accumulated backlog of waste by this plant resulted in increased discharges of certain radionuclides, most notably ^{99}Tc.

The liquid discharges from the Sellafield reprocessing plant, containing transuranium nuclides, arise mainly from water used as a coolant in spent fuel storage ponds and from wastes related to the reprocessing operations. The latter are routed through so-called 'sea-tanks' for treatment before discharge (Pentreath et al., 1985). Discharges of transuranium nuclides, (viz., Fig. 4) reached their peak in the early to mid-1970s due to increased throughputs and reprocessing of residues, and thereafter declined as new treatment facilities were introduced (Hunt, 1985). Overall, 0.12 PBq of ^{238}Pu, 0.61 PBq of ^{239,240}Pu, 22 PBq of ^{241}Pu, 0.54 PBq of ^{241}Am and smaller quantities of other transuranium nuclides have been discharged to the Irish Sea during the period 1952–2000 (BNFL, 1980–2001; Gray et al., 1995).

The radionuclide composition of the effluents has varied considerably over the years, depending on factors such as the degree of burn-up of the irradiated fuel, the cooling time between removal from the reactor and the reprocessing operation, the radionuclide content of the storage ponds due to corrosion of fuel, and the efficiency of the separation plant (McCarthy & Nicholls, 1990).

As an example, the isotopic composition of plutonium, obtained from reported discharges and environmental monitoring, indicate a steady increase of the ^{238}Pu/^{239,240}Pu activity ratio from a value of $\sim$0.03 in the 1960s to >0.3 in present discharges (Hetherington, 1976; Gray et al., 1995). Recent determinations of the ^{240}Pu/^{239}Pu atom ratio in dated sediment cores show a similar tendency, with values increasing from $\sim$0.05 in the 1960s up to $\sim$0.22 in present discharges (Kershaw et al., 1995a).

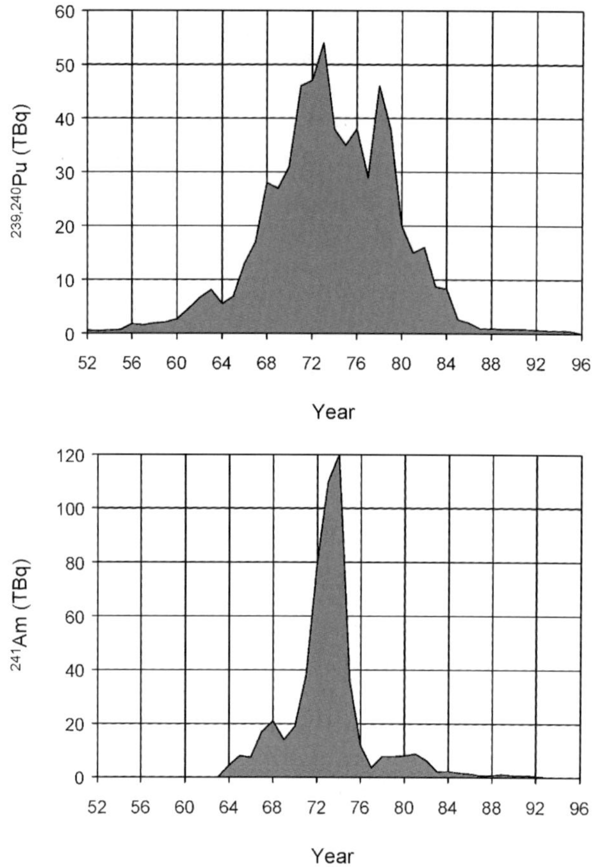

Fig. 4. Sellafield pipeline discharges of ^{239,240}Pu and ^{241}Am in the period 1952–1996 (source: BNFL, 1980–2001; Gray et al., 1995).

Changes in these ratios, with increasing proportions of ^{238}Pu and ^{240}Pu since the beginning of operations reveal, as discussed above, an increase in the degree of burn-up of the reprocessed fuel. This can be related to historical operations at the complex, initially dedicated to the production of low burn-up plutonium for the UK's weapons programme, as well as to improvements in fuel rod and reactor design, which have facilitated progressively longer burn-up in power-generating reactors (Kershaw et al., 1995b).

3.2.2. *Cap de La Hague*

Operations at La Hague commenced in 1966 with a plant for the reprocessing of spent fuel from commercial graphite–gas reactors. The complex was extended in 1976 to provide reprocessing facilities for spent fuel from light water reactors (SCOPE, 1993).

Although the activities discharged from La Hague represent a significant addition to existing inventories of certain radionuclides (e.g. ^{90}Sr, ^{125}Sb, ^{129}I), transuranium inputs from this source constitute only a small fraction of those released from Sellafield. Annual plutonium discharges have always been below 10^{11} Bq, a factor of 10^4 lower than those of Sellafield at their peak. As in the case of Sellafield, variations in the $^{238}Pu/^{239,240}Pu$ activity ratio in the discharges have occurred in the course of operation, with values of ~0.05 in the early 1970s increasing to ~0.65–1.2 in the period 1982–1988 and reaching values in the range 1.3–2.3 thereafter (Garcia, 1996).

3.2.3. *Other installations*

Over the years, small amounts of transuranium nuclides have been introduced to the eastern North Pacific Ocean from the Hanford plutonium production facility and to the western Atlantic from the Oak Ridge and Savannah River production plants in the USA. Beasley et al. (1981) estimated that 150–300 GBq of $^{239,240}Pu$ and 40–80 GBq of ^{241}Am had entered the eastern North Pacific Ocean between the 1950s and mid-1978 as a result of runoff from the Columbia River. Of this, less than 10% could be attributed to operations at Handford with the remainder originating from global fallout.

In the Former USSR, large amounts of radioactive waste have been discharged from the Krasnoyarsk, Tomsk and Mayak facilities into the Ob and Yenisey River systems, draining into the Arctic Ocean. Recent measurements have shown that the Techa River (belonging to the Ob River system) contains enhanced levels of transuranium nuclides (Trapeznikov et al., 1993). However, a significant fraction of the activity is deposited in riverine sediments, trapped in reservoirs and weak current areas, and is unlikely to reach the Arctic Ocean. Isotopic ratios ($^{137}Cs/^{239,240}Pu$, $^{238}Pu/^{239,240}Pu$) in sediments collected in the Ob River mouth indicate that the dominant source of plutonium in the delta is global fallout, with little (if any) input from the nuclear facilities located upstream (Panteleyev et al., 1995; Baskaran et al., 1995; Sayles et al., 1997).

Releases from other reprocessing plants, such as the Trombay and Tarapur plants in India and the Marcoule plant in southern France, have been relatively small compared with those from Sellafield and La Hague, and have only had a limited, local impact on transuranium concentrations near the points of discharge.

3.3. *Disposal of radioactive waste and nuclear waste repositories*

Most of the low- and intermediate-level wastes arising from reprocessing operations are treated, packaged and disposed of by shallow burial. Repositories for low-level radioactive waste exist, for example, in the USA (e.g. Hanford, Sheffield, West Valley, Morehead, Barnwell), UK (e.g. Drigg site in Cumbria) and the Former USSR (e.g. Kyshtym). However, until 1982, some of these packaged wastes were also dumped by western countries at deep sites in the ocean, mainly (98%) the North-east Atlantic. Although the majority of the activity disposed of consisted of β- and γ-emitters, small quantities of α-emitting nuclides were also included. It has been estimated that about 0.5 PBq of α activity was disposed by western countries at the disposal sites in the North-east Atlantic (Baxter et al., 1995). Follow-up measurements of radionuclide concentrations in sea water from the vicinity of these dump sites indicated that measurable leakage of transuranium elements (as evidenced by perturbations in fallout ratios) has occurred, although contamination is very localised and of no radiological significance (Table 3).

Radioactive waste was also dumped throughout the years by the Former Soviet Union in Arctic waters (Yablokov et al., 1993). The Kara Sea has been the major solid radioactive

Table 3
Concentrations of plutonium and isotopic ratios in seawater from the NE Atlantic dumping sites (source: Baxter et al., 1995)

Location	^{239,240}Pu (mBq m^{-3})	^{238}Pu/^{239,240}Pu
NE Atlantic dump sites ($n = 4$)	12 ± 5	0.10 ± 0.02
Control (fallout)	7.3 ± 0.3	0.029 ± 0.008

waste dumping site. Liquid and packaged solid wastes, together with a number of naval reactors from nuclear powered submarines, some with unloaded spent nuclear fuel, have added 0.1–0.5 PBq to the total actinide inventory of this Sea (Baxter et al., 1995; Baskaran et al., 1995), the highest contribution being due to ^{241}Pu (0.08–0.4 PBq). Enhanced levels of ^{239,240}Pu and other radionuclides have been reported in sediments collected close to dumped containers and reactor compartments in the Abrosimov and Stepovogo Ffjords (Salbu et al., 1997; Livingston & Povinec, 2000), demonstrating that leakage from the dumped object has already taken place. Nevertheless, the contamination appears to be very localised, with enhanced levels limited to a few meters from the objects, and no evidence (as yet) of any transfer of contamination from these sites to the overlaying water column or further afield (Mitchell et al., 1998; Livingston & Povinec, 2000; León Vintró et al., 2002).

Similar dumping practices also took place over the last three decades in the Sea of Japan, the Sea of Okhotsk and the western North Pacific Ocean, mainly by the Former Soviet Union and the Russian Federation, although the total activity reported to have been dumped is nearly an order of magnitude lower than that at the dumping sites in the Arctic Ocean (IAEA, 1999). To date, the distribution and inventories of transuranium nuclides at these sites are consistent with known weapons fallout sources (global and local) and with the oceanographic processes controlling the transfer of radionuclides in this part of the ocean (Livingston & Povinec, 2000).

3.4. *Accidental releases*

Besides planned discharges arising from operations associated with the nuclear fuel cycle, transuranium nuclides have entered the marine environment as a result of accidental releases from nuclear reactors and accidents involving military aircraft and submarines carrying nuclear weapons.

3.4.1. *Nuclear reactors*

The accident at Unit 4 of the Chernobyl nuclear power plant in April 1986 is the most serious accident to have occurred in the history of nuclear reactor operation. Severe violations of operating procedures in the course of an engineering test of the generator resulted in a prompt critical excursion of the 1 GW$_e$, graphite-moderated, light-water cooled reactor on April 26th, 1986. The energy generated led to a series of explosions that destroyed the outer containment, exposing the core to the environment and injecting highly radioactive debris into the atmosphere. The releases continued in the following days due to the subsequent fire in the graphite moderator. An attempt was made to cover the core with more than 5000 tons of boron, dolomite, sand, clay and lead. However, the blanketing of the core led to increased

Table 4
Estimated amounts of transuranium nuclides released to air from Chernobyl (source: IAEA, 1986; WHO, 1989)

Nuclide	Mass abundance (relative to ^{239}Pu)	Activity abundance (relative to ^{239}Pu)	Total injection (PBq)
^{239}Np	0.00045	1692	44
^{238}Pu	0.0042	1.15	0.03
^{239}Pu	1	1	0.026
^{240}Pu	0.42	1.54	0.04
^{241}Pu	0.12	200	5.2
^{242}Pu	0.042	0.0027	0.00007
^{241}Am	0.0041	0.23	0.006
^{242}Cm	0.00006	30	0.78
^{243,244}Cm	0.00017	0.23	0.006

temperatures and new releases from the 1st to the 5th of May. Emissions were terminated on May 6th upon cooling the reactor through tunnels constructed under the core (SCOPE, 1993).

Contamination spread throughout the Northern Hemisphere, and plumes of contaminated air described various trajectories following surface winds. Differences in the fallout deposition can be explained by the meteorological conditions prevailing during and after the accident. Although from a radiological point of view it was the 630 PBq of ^{131}I, 100 PBq of ^{137}Cs and 8 PBq of ^{90}Sr released which were of most concern, considerable quantities (3% of the reactor's content) of the transuranium nuclides were also released, most of which were deposited in the vicinity of the reactor. The estimated releases for these nuclides are given in Table 4.

Although the transuranium input to the oceans following the Chernobyl accident was comparatively small and localised, it represented a significant input to the Baltic Sea, as the first radioactive clouds from Chernobyl travelled to the north and caused high deposition over the Scandinavian region. Holm (1995) estimated that ~1.5 TBq of ^{239,240}Pu in the Baltic could be attributed to Chernobyl fallout, representing 9% of the total inventory (16.5 TBq). Given the isotopic composition of the plutonium in the reactor, the contributions of ^{238}Pu and ^{241}Pu were more significant, with ~52% of the ^{238}Pu and ~68% of the ^{241}Pu inventories (at 1.4 TBq and 189 TBq, respectively) being attributable to Chernobyl fallout.

The presence of short-lived ^{242}Cm (half-life = 163 days) in water samples collected in June and September of 1986 from the Black Sea (Livingston et al. 1988) suggested that Chernobyl-sourced transuranium nuclides were also deposited in this basin, a result confirmed by the measurement of elevated ^{238}Pu/^{239,240}Pu ratios in the same samples. Although this Chernobyl-sourced plutonium represented approximately half of the plutonium activity found in surface waters at the time of sampling, total Chernobyl inventories in the Black Sea are small when compared to global fallout plutonium, which has been preferentially removed to the deeper water column and sediments (Buesseler & Livingston, 1996).

3.4.2. *Aircraft accidents involving nuclear weapons*

In January 1966, four plutonium-bearing nuclear weapons were released at an altitude of 8500 m above the Mediterranean village of Palomares (Spain) following a mid-air collision between two U.S. aircraft, namely a B-52 bomber and a KC-135 refuelling tanker. Although two of the weapons were reported to have been recovered intact (one from the seabed in the nearby Gulf of Vera), the chemical explosive component of the other two detonated on impact on land, and plutonium was dispersed over an area of approximately 500 ha (NEA, 1981). To date, the quantity of plutonium released does not appear to have been published. However, allowing for clean-up, it has been estimated that the total residual ^{239,240}Pu inventory is in the order of 0.1 TBq (Aarkrog, 1995). A small fraction of this would appear to have found its way to the local marine environment as it has been reported that traces of plutonium of accident origin have been detected in sediments from the submarine canyon system situated south of the mouth of the (almost always) dry Almanzora River (Romero et al., 1991; Antón et al., 1994). Similar traces have been found in marine algae sampled in the vicinity of Palomares (Manjón et al., 1995).

Two years later, a second B-52 bomber, also carrying four plutonium-bearing weapons, crashed on Arctic ice in Bylot Sound (Greenland), 11 km west of the Thule Air Base. The plane and the chemical explosive component of all four weapons exploded on impact, causing the release of kilogramme quantities of insoluble plutonium oxide to the snow-pack in the locality (Risø, 1970; Aarkrog, 1977; Facer, 1980). Although it was initially estimated that the bulk of the plutonium was removed in the ensuing clean-up operation, leaving a residual contamination totalling 1 TBq (^{239,240}Pu) in a seabed area in the order of 1000 km^2 (Aarkrog et al., 1984; Smith et al., 1994), a recent re-assessment of this figure suggests 10 TBq as a more realistic estimate of the remaining inventory (Ericsson, 2002).

Measurement of ^{238}Pu/^{239,240}Pu ratios in filtered waters at Thule suggest that, in contrast with the seabed sediment, there is virtually no weapons-grade plutonium in the Thule water column at the present time (McMahon et al., 2000). A similar observation had been made in the early 1990s in coastal waters from the Palomares area (Mitchell et al., 1995), which suggests that plutonium dispersed from fractured nuclear weapons is present in a rather insoluble form (mainly as radioactive particles).

3.4.3. *Lost nuclear submarines*

Several nuclear-powered submarines (probably carrying nuclear weapons) have been lost by the US, the Former USSR and the Russian Federation, and now reside on the ocean floor at several locations. Overall, two US nuclear submarines and four Soviet/Russian submarines have been reported to have sunk after suffering some type of accident. The last two accidents have been of particular concern, as they took place in relatively shallow waters and in important fishing grounds.

In April 1989, a nuclear-powered submarine of the USSR Navy, the *Komsomolet*, sank to a depth of 1685 m in the Norwegian Sea, about 300 nautical miles from the Norwegian coast. The submarine contained an estimated 16 TBq of ^{239}Pu in the two nuclear warheads onboard and 5 TBq of actinides in the reactor core (Sivintsev, 1994; Høibråten & Thoresen, 1995). Studies of radionuclide concentrations in the vicinity of the site indicate that, to date, only minor contamination can be attributed to the submarine and its contents.

More recently, in August 2000, a Russian Oscar II class attack submarine, the *Kursk*, sank (with its 118 crew) to a depth of 116 m in the Barents Sea, 250 km from Norway and 80 km from the Kola Peninsula. According to Russian sources, the two reactors were shut down and the submarine was not carrying nuclear weapons. Results from two monitoring expeditions undertaken in August and October 2000, show that no radionuclide leakage from the submarine has taken place so far (Amundsen et al., 2001). Measurements carried out on debris from the submarine and on water and air collected by divers inside different compartments of the submarine show no increased levels of radioactivity, confirming that the reactors were shut down and that no flooding of the reactor had occurred at the time of sampling in October 2000. On October 8, 2001, the greater part of the *Kursk* was raised from the seabed and transported to a dry dock in Murmansk for examination.

4. Behaviour of transuranium nuclides in the marine environment

As highlighted above, a major part of the transuranium release to the environment has found its way into the world's oceans. Because of the variability of inputs both spatially and temporally, transuranium inventories in the oceans are not uniformly mixed, and activity concentrations vary widely between different marine zones. To illustrate, a summary of present $^{239,240}Pu$ concentrations in surface waters of the world's ocean is given in Fig. 5.

The state of knowledge regarding the behaviour of transuranium nuclides in the oceans has evolved progressively from a large number of studies. Various factors have been shown to be of importance in relation to transuranium behaviour. These include:

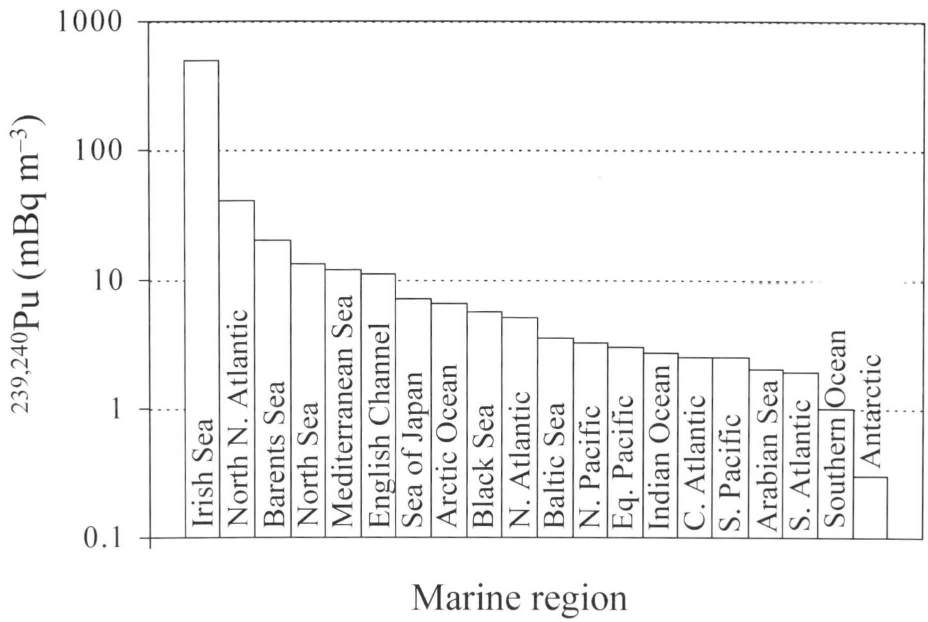

Fig. 5. Average $^{239,240}Pu$ concentrations in surface waters of the world's oceans in the year 2000 (source: Livingston & Povinec, 2002).

- the physical and chemical forms of these nuclides as they enter the ocean (dependent on the specific source);
- the physico-chemical properties of these nuclides upon contact with seawater (e.g. oxidation state distribution, complexation, sorption/desorption kinetics, etc.);
- the physical, chemical and biological processes that subsequently take place in the receiving medium (e.g. physical mixing of affected water masses, particle and biota removal, diagenetic processes in sediments, etc.).

4.1. *Physical and chemical forms of transuranium nuclides upon entry to the ocean*

The physico-chemical form of transuranium nuclides, which depends on the particular source term and release scenario, plays a significant role in their subsequent behaviour upon contact with seawater. Radionuclides released from a source may be present as fragments, particles, aerosols, pseudocolloids, colloids or low molecular mass forms (Salbu, 2001).

A significant fraction of transuranium nuclides originating in high temperature nuclear events (such as nuclear weapons tests and reactor fires) is released in the form of radioactive particles (Chamberlain & Dunster, 1958; Crocker et al., 1966; Cooper et al., 1994). Radioactive particles are also released under low temperature conditions (such as atmospheric emissions during normal operations in reactors and effluent discharges from nuclear reprocessing plants), and have often been observed in sediments contaminated by routine and accidental releases or leakage from dumped radioactive waste (Salbu, 2001).

4.1.1. *Weapons fallout*

In global fallout, plutonium was carried by relatively soluble sub-micron-sized particles (mainly composed of iron oxides) arising from the vaporisation and condensation of nuclear weapons material without any interaction with the Earth's surface (Adams et al., 1960; Joseph et al., 1971). In contrast, fallout from surface-based testing on coral atolls (e.g. Bikini, Enewetak) was in the form of calcium hydroxide-based particles resulting from the vaporisation of coral matrices during the blast (Adams et al., 1960; Buesseler, 1997). The partial solubilisation of these particles was accompanied by the release of hydroxyl ions and resulted in the formation of a shell of insoluble magnesium hydroxide on the fallout particle upon interaction with the magnesium ions in seawater (Buesseler, 1997). The difference in solubility between global stratospheric fallout particles and those from tropospheric fallout from the Pacific testing grounds is reflected in the enhanced $^{240}Pu/^{239}Pu$ ratios observed in the deep waters and sediments of the North-west Pacific (characteristic of the tropospheric fallout signal), which suggests that plutonium from the Pacific testing grounds was more rapidly removed from surface waters than was global fallout. A similar preferential removal of tropospheric fallout has been reported for silicate-based fallout debris from surface tests conducted at the Nevada Test Site (Buesseler & Sholkovitz, 1987).

4.1.2. *Effluents from reprocessing plants*

Despite complications associated with continuous changes in the chemical composition of effluent discharges and methodological difficulties in the separation techniques employed, a number of studies have been carried out in order to characterise the different chemical and physico-chemical forms present in effluents from reprocessing plants.

Analyses carried out in 1982 (and assumed to be typical of routine releases) showed that some 99% of the Pu (alpha), ^{241}Am and ^{243,244}Cm, and about 60% of the ^{237}Np present in sea tank releases from Sellafield were associated with particulate (>0.22 μm) material. In pond water effluents, the corresponding percentages were found to be somewhat lower (Pentreath et al., 1985). Differences in chemical speciation were also noted, with reduced plutonium (i.e. Pu(III, IV)) predominating in the filtrate of the effluent from the sea tanks and oxidised plutonium (i.e. Pu(V, VI)) predominating in the filtrate of the effluent from the cooling ponds. However, as the bulk of the effluent came from the sea tanks, only about 1% of the total would appear to have been in an oxidised form upon discharge. Nuclides of americium and curium were present only in the Am(III) and Cm(III) forms (i.e. chemically reduced), while for neptunium both reduced (Np(IV)) and oxidised (Np(V)) forms were present in approximately equal proportions in the total ^{237}Np discharge.

These observations were supported by a more recent study carried out in 1991 on sea tank and SIXEP effluent streams (Leonard et al., 1995). In the case of sea tank effluent, almost all of the ^{239,240}Pu and ^{241}Am activity present was found to be associated with the iron floc formed upon the neutralisation of acid liquors containing ferrous sulphamate – a chemical used to control the valency of plutonium during fuel reprocessing. What little plutonium was in the solution phase was determined to be in the reduced (tetravalent) form. Further, nuclides such as ^{137}Cs, ^{90}Sr and ^{99}Tc were found to be almost entirely in the solution phase. In SIXEP effluent, on the other hand, all of the radionuclides considered were almost entirely in a dissolved form, presumably as a result of the absence of particulate material in this waste stream.

Laboratory experiments to determine the colloidal size distribution of a suite of radionuclides in each of the effluent streams (SIXEP and seatank) were also carried out using ultrafiltration techniques (Leonard et al., 1995). Overall, the results suggest that colloidal forms of individual radionuclides, originating from the solution phase, are more likely to occur in the SIXEP rather than in the sea tank effluent, with significant fractions of the ^{239,240}Pu(V), ^{239,240}Pu(IV) and ^{241}Am in the former being in colloidally-bound form, as evidenced by the level of retention upon ultrafiltration (<3 kDa).

'Hot particles', likely to contain fuel fragments, have also been identified in the effluent (Pentreath et al., 1984) and have been shown to persist in the marine environment close to Sellafield for at least several months, with some being preserved in accreting estuarine sediments (Hamilton, 1981; Hamilton et al., 1991).

Recent speciation studies of radionuclides in effluents from the La Hague reprocessing plant have also shown that a major fraction of the transuranium nuclides are released in the form of particles and colloids (Salbu, 2001).

4.1.3. *Accidental releases and leakage from dumped radioactive waste*

The presence of radioactive particles containing anomalously high plutonium concentrations has been detected in seabed sediments from the Palomares and Thule accident sites (Risø, 1970; Mitchell et al., 1997). Studies carried out shortly after the accident at Thule showed the plutonium to be in the form of oxide particles with a very wide size distribution (mean size ~2 μm), associated with or adhered to particles of inert debris such as glass, metal and rubber (Holm, 1991). Localised heterogeneities have also been reported in the close vicinity of dumped objects in the Kara Sea (Salbu et al., 1997).

4.2. *Physical and chemical forms upon dilution with seawater*

Following entry to the ocean, the behaviour of transuranium nuclides is heavily influenced by the physical and chemical forms of each individual radionuclide. Of no element is this more true than plutonium, which can exist in solution in at least four distinct oxidation states simultaneously. These states include plutonium(III), plutonium(IV), plutonium(V) and plutonium(VI), of which plutonium(IV) and plutonium(V) predominate in seawater as the reduced (hydrolysed) $Pu(OH)_4$ and the oxidised PuO_2^+ forms, respectively (Morse & Choppin, 1986; Orlandini et al., 1986; Choppin & Kobashi, 1990). While plutonium(IV) is highly particle-reactive, being quickly adsorbed by suspended particles, sediments and natural colloids, plutonium(V) is more soluble and can be transported over greater distances within the dissolved phase before being scavenged.

Field studies have shown that, in the absence of complexing agents, oxidised forms of plutonium predominate in the dissolved (<0.45 μm) phase (Nelson & Lovett, 1978; Wahlgren & Orlandini, 1982; Mitchell et al., 1991, 1995). This observation applies to open oxygenated waters in oceans, semi-enclosed seas and shallow continental shelves, well removed from source-terms. The presence of significant levels of complexing agents (such as dissolved organic carbon) results in a redistribution in the partition of plutonium between the dissolved and particulate phases and in a change of the chemical species present in the dissolved phase. According to speciation calculations, Pu(III) should predominate in the dissolved phase under reducing conditions (Rai et al., 1980; Nash et al., 1988). Indeed, colloidal organic carbon at concentrations commonly encountered in natural waters (1–20 mg l^{-1}) has been shown to strongly interact with Pu(III, IV) species and inhibit their sorption onto suspended particles (Nelson et al., 1987; Orlandini et al., 1990).

Furthermore, speciation studies of plutonium adsorbed onto suspended particulate and sediment have shown plutonium to be almost entirely in a reduced chemical form (Nelson & Lovett, 1978; Mitchell et al., 1995).

In the case of americium and curium, it is well established that the aqueous solution chemistry of both elements is dominated by the trivalent Am(III) and Cm(III) forms. Indeed, the dominance of highly insoluble forms of americium and curium such as $Am(OH)_2^+$ and $Cm(OH)_2^+$ in alkaline solutions is predictable (Choppin, 1994). Measurements on the chemical speciation of americium in filtered water from the north-eastern Irish Sea appear to confirm that americium is present almost exclusively as Am(III) (Pentreath et al., 1986a). Accordingly, a much greater proportion of americium (and curium) than plutonium can be expected to be in a highly insoluble, hydrolysed form, leading to a stronger affinity for suspended particulate matter, colloids and sedimentary deposits. Neptunium, on the other hand, is normally present in its oxidised, relatively soluble, Np(V) form (Assinder, 1999).

4.2.1. *Weapons fallout*

Since global fallout was delivered in the form of sub-micron-sized particles that solubilised in a relatively short time upon contact with seawater, it can be assumed that, following deposition, transuranium elements originating from this source were present, at least initially, as dissolved species. Nevertheless, owing to the tendency to adsorb onto particles as outlined above, transuranium nuclides became preferentially removed from surface waters in comparison to other radionuclides (such as ^{3}H, ^{99}Tc, ^{137}Cs, ^{90}Sr) which, having a very small affinity for particle association, remained in solution.

Early ocean fallout studies confirmed the vertical separation of transuranium nuclides in the water column from the more soluble radionuclides that accompanied them as they entered the oceans. The removal of transuranium nuclides from surface water was controlled by the same physical processes active for the more soluble nuclides, but also by the association to suspended matter and subsequent sedimentation processes. Since the rate of removal is dependent on the particle load, this means that while in coastal regions transuranium nuclides were rapidly removed from the water column to the sediment compartment, in oligotrophic open oceans the rate of removal was much slower, with the major fraction of the inventory still residing in the water column at the present time (Livingston & Povinec, 2000).

By way of example, residence times for $^{239,240}Pu$ and ^{137}Cs in the surface mixed layer of the western North Pacific, estimated from long-term time-series data, were found to be 4 and 9 years, respectively (Hirose et al., 1992). The shorter residence time of plutonium with respect to radiocaesium is supported by the general finding that particulate $^{239,240}Pu$ concentrations range up to 21% of the total, compared with less than 1% for ^{137}Cs (Fowler et al., 1983; Hirose et al., 1992).

A longer residence time of about 15 years has been reported for plutonium in surface waters of the western Mediterranean (León Vintró et al., 1999). The longer residence time of plutonium in these waters reflects the lower productivity of this sea, and is supported by the smaller fraction of transuranium nuclides associated with particles, which is typically 5% for plutonium and 10% for americium. In continental shelf waters, the corresponding percentages in the particulate fraction increase to 10% and 45% for plutonium and americium, respectively (Papucci et al., 1996).

4.2.2. *Effluents from reprocessing plants*

Although, as discussed above, small amounts of transuranium elements in colloidal form were identified in effluents from the Sellafield reprocessing plant, further experiments by Leonard et al. (1995), following the dilution of effluent into seawater under laboratory conditions, indicated that these forms did not persist in seawater. The latter observation is supported by ultrafiltration analyses carried out by Mitchell et al. (2001) in the 1990s, which were unable to detect any colloidally-bound plutonium in seawater samples collected from the immediate vicinity of the discharge point.

Studies on the physico-chemical speciation of transuranium nuclides in open waters throughout the Irish Sea have shown that a significant proportion of the total plutonium and even more of the americium are associated with the particulate phase (Pentreath et al., 1986a; Kershaw et al., 1986; Mitchell et al., 1995). Measurements of the oxidation state distribution of plutonium in filtered water sampled throughout the Irish Sea show relatively little variation, spatially or temporally, with the bulk of the plutonium being in an oxidised, Pu(V), state (Nelson & Lovett, 1978; Lovett & Nelson, 1981; Pentreath et al., 1986a; Mitchell et al., 1991, 1995). A similar observation has been made in waters of the English Channel (Boust et al., 1996). Measurements on the oxidation state distribution of plutonium in these waters, carried out in 1994, showed no difference between the chemical speciation of plutonium in the near-field close to the La Hague reprocessing plant and that in mid-Channel waters, approximately

50 km away. This appears to confirm that an equilibrium in the partition of plutonium between the reduced and oxidised species is achieved relatively quickly following release to seawater.

In another study, Pentreath et al. (1986b) reported that only 1% of the neptunium in seawater from the vicinity of Sellafield is in its reduced, Np(IV), form, compared with approximately 50% in the original discharge. Further afield, the proportion of Np(IV) is even lower, with only 0.5% remaining in this form in the southern Scottish zone. It, thus, appears that neptunium becomes quickly oxidised upon contact with seawater, with only a very small proportion remaining un-oxidised.

4.3. *Transuranium nuclides and biogeochemical processes in the receiving medium*

4.3.1. *Water column*

One of the most important processes controlling the oceanic behaviour of transuranium nuclides in the water column is their association with solid phases. The extent of this association is very different when one compares shallow coastal areas, the margin areas of the major ocean basins and the deep ocean basins (Livingston & Povinec, 2002).

In shelf areas, high removal rates keep transuranium nuclide concentrations low in seawater and tend to accumulate them in areas of fine-grained sediment deposition. For example, most of the transuranium release from Sellafield has been reported to remain in a relatively narrow coastal zone close to the discharge point, incorporated in offshore and intertidal muddy sediments (Kershaw et al., 1995b). The high inventories of ^{239,240}Pu observed in continental shelf sediments from the western Mediterranean, often two to three times higher than the corresponding total cumulative fallout deposition ($\sim$ 82 Bq m^{-2}), have also been attributed to the preferential scavenging of this element in waters with a higher particle density (Papucci et al., 1996). This is in contrast to deep areas which do not receive significant particulate inputs from the shelf, and where sediment plutonium inventories as low as a few Bq m^{-2} have been reported. Intermediate inventory values are found along the continental slopes, which are not only themselves productive, but also represent particle sinks for material formed on adjacent shelves (Carpenter et al., 1987).

In open oceans, particle production depends primarily on productivity in surface waters, and particle removal of transuranium nuclides is mainly in association with biogenic particles (Fowler et al., 1983; Livingston & Andersen, 1983). For plutonium, an interesting feature characterises the vertical concentration profiles in most deep basins, namely the presence of a distinct sub-surface concentration maximum in the 200–1000 m depth range. Shallow, sub-surface maxima in plutonium concentrations have been observed in several ocean basins, including the Pacific Ocean (Bowen et al., 1980), the North Atlantic Ocean (Cochran et al., 1987; Nyffeler et al., 1996), the western Mediterranean Sea (Fukai et al., 1979), the Norwegian and Greenland Seas (Herrmann et al., 1998) and the central Arctic Ocean (Livingston et al., 1984; Smith & Ellis, 1995).

The presence of this maximum, first observed in vertical profiles throughout the North-west Pacific Ocean during the GEOSECS (Geochemical Sections) programme in 1973–1974, was initially attributed to the association of plutonium with particle populations sinking at different rates, coupled to horizontal and vertical water-mass movements acting on the slowest moving plutonium-bearing particles (Bowen et. al., 1980). Re-visit of some of these GEOSECS stations in expeditions undertaken in 1978, 1980, 1982 and 1997 (Nagaya & Nakamura, 1984;

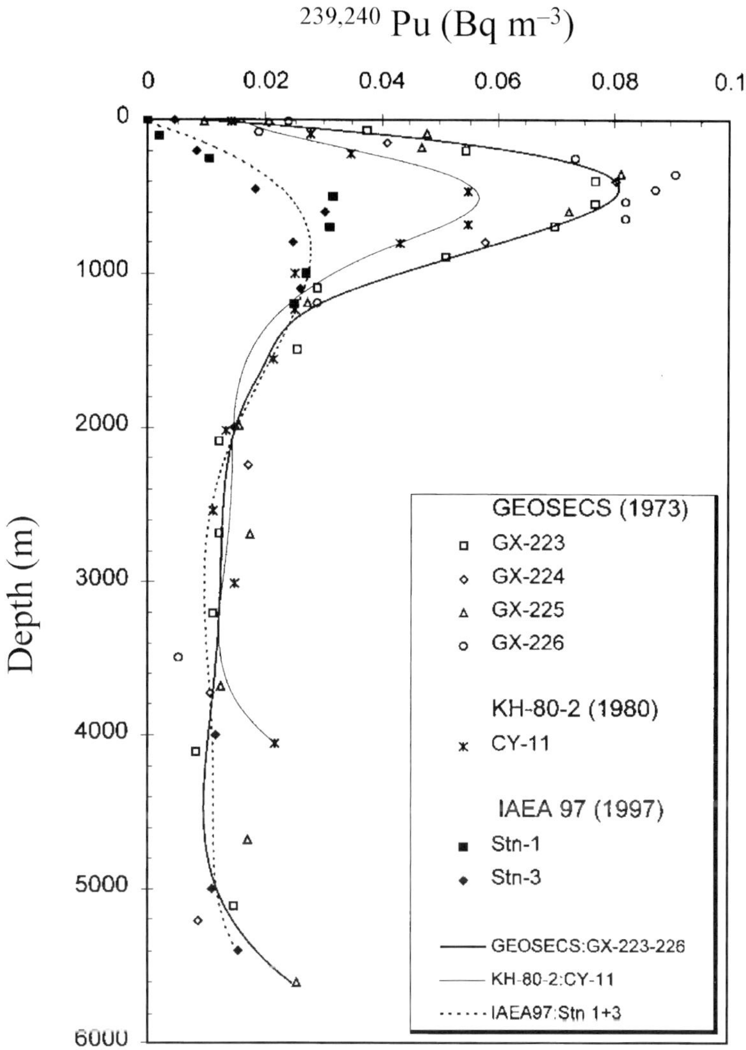

Fig. 6. ^{239,240}Pu profiles in water column of the GEOSECS (1973), Hakuho Maru (1980) and IAEA (1997) stations in the central NW Pacific Ocean (source: Livingston et al., 2001).

Livingston et al., 2001; Povinec et al., 2003), has shown that, after two decades, the maximum has become much smaller and less pronounced (a decrease by a factor of four), and that it has moved to deeper water layers, from 450 m to 850 m (Fig. 6). Comparison of the plutonium profile's evolution with those of the more soluble ^{237}Cs and ^{90}Sr nuclides over the same period, have confirmed that although vertical fluxes of plutonium in association with sinking particles clearly play a role in the formation of the sub-surface maximum, at least a major part of the observed plutonium changes over time has been caused by physical circulation processes in the upper water column (Livingston et al., 2001). The distribution of plutonium below depths ventilated physically from the surface results from the transport of plutonium in association with sinking particles which, on descent, may be subject to oxidation, bacterial degradation and demineralisation, with the consequent release of plutonium back into solution (Sholkovitz, 1983; Nyffeler et al., 1996; Livingston et al., 2001).

In contrast to the North-west Pacific stations, plutonium vertical profiles taken east and west of Bikini Atoll, and characterised by a similar sub-surface maximum, show no significant change over a period of 24 years. As these stations are in an area with very low biological productivity and oligotrophic conditions, the maintenance of the concentration profiles and inventories in the upper ocean has been related to the recirculation of these water-masses under near-permanent low productivity conditions (Livingston et al., 2001).

Another typical feature observed in ^{239,240}Pu water profiles in the NW Pacific Ocean is an enhanced ^{239,240}Pu concentration (by about a factor of two) near the bottom. This feature, which is not apparent in vertical profiles from the NE Pacific Ocean, has been associated with the presence of tropospheric fallout, as evidenced by the higher ^{240}Pu/^{239}Pu ratios measured in bottom water layers of the NW Pacific (Buesseler, 1987, 1997; Povinec et al., 2003). As indicated earlier, tropospheric fallout, associated to insoluble particles, is more rapidly removed from surface to deep waters than global fallout.

The presence and evolution of the plutonium sub-surface maximum has also been particularly well documented in waters of the western Mediterranean (Fukai et al., 1979, 1982; Ballestra, 1980; IAEA, 1991; Mitchell et al., 1995; Merino et al., 1997; León Vintró et al., 1999; Fowler et al., 2000; Lee et al., 2003). A comparison of the most recent studies, carried out in the mid-1990s, with those carried out in the late-1970s shows that while the depth of the sub-surface maximum has not changed significantly, the associated plutonium concentrations have been reduced by about half. In contrast, higher plutonium concentrations now characterise the deep Mediterranean waters, with values of about 25 mBq m^{-3} being reported in the early-1990s compared with values of 10 mBq m^{-3} in the late 1970s (Papucci et al., 1996). In a comprehensive study combining high-resolution water sampling with direct measurements of the vertical flux of transuranium nuclides using time-series sediment traps, Fowler et al. (2000) showed that the position of the sub-surface maximum in the north-western Mediterranean is closely controlled by the degree of primary production occurring in the upper water column. An assessment of ^{239,240}Pu inventories derived from the vertical profiles demonstrated that, while column inventories did not change substantially between the mid-1970s and the early 1990s, the fraction of the total inventory contained within the deeper layers increased significantly, confirming the slow, downward movement of plutonium in the western Mediterranean basin.

In contrast, a significant decrease (about 25%) in the water column inventory of ^{241}Am was observed over the same period. Enhanced scavenging of ^{241}Am, and a resultant more rapid removal from the water column relative to ^{239,240}Pu, has been supported by the measurement of enhanced ^{241}Am/^{239,240}Pu ratios in sinking particles and sediments throughout the western Mediterranean (Fukai et al., 1979; Merino et al., 1997; León Vintró et al., 1999; Fowler et al., 2000; Lee et al., 2003). Cochran et al. (1987) also reported water column data for ^{241}Am in the North-west Atlantic. The data show that this nuclide has a deeper water distribution than plutonium, and confirm that americium is being more rapidly transported towards ocean sediments on sinking particles.

Despite the clear role played by vertical fluxes of plutonium and americium in association with sinking particles in the evolution of vertical concentration profiles in the western Mediterranean, it should be emphasised that, as evidenced by the temporal evolution of vertical profiles for more soluble radionuclides, a major part of the observed transuranium changes over time can be attributed to physical circulation processes taking place in the water column. Indeed, studies on the vertical evolution of the soluble ^{137}Cs in this basin have shown that over the last twenty years, radiocaesium concentrations in surface waters have decreased by a factor of three, while those in deep waters have increased by about 30% (Papucci et al., 1996).

4.3.2. *The sediments*

As indicated above, the higher productivity and suspended particle loads found in shelf or continental slope environments result in a significant and rapid removal of transuranium nuclides from the water column to fine-grain sediment sinks. However, once delivered to the sediments, these radionuclides may be subject to post-depositional bioturbational downmixing and redistribution processes (Livingston & Bowen, 1979).

Much of the knowledge acquired about these processes comes from studies conducted in the north-eastern Irish Sea, where discharges from the Sellafield reprocessing plant have provided a unique opportunity to investigate the behaviour of transuranium nuclides in a particularly dynamic coastal environment. Plutonium concentrations in sediments from this region are dominated by the location of the point source and the nature of the seabed, in particular the zone of muddy sediments running up to the Scottish coast. Plutonium is transported northwards, predominantly attached to sediment particles (MacKenzie et al., 1987) and undergoes considerable mixing by physical and biological processes en route. The region is subject to frequent storms, and particle resuspension and transport is ubiquitous. The intertidal sediments along exposed beaches tend to be quite well mixed, to the depth of wave action, on a short time scale, while deposits in saltmarshes and floodplains such as those found along the Solway coast and the Esk Estuary are disturbed by channel migration on a scale of years or decades.

The offshore sediments are mixed by the macrobenthos to depths of over 1.5 m, in areas where the sedimentation rate is low ($\sim$1 mm y^{-1}; Kershaw et al., 1988). This results in mixing of plutonium to depth and a very inhomogeneous distribution over small space scales. The relatively high concentration of plutonium in the subtidal sediments has allowed the determination of ^{239,240}Pu concentrations in pore-waters. Lovett & Nelson (1981) demonstrated that plutonium in eastern Irish Sea sediments was primarily in a reduced, Pu(IV), form. In some areas, however, the presence of burrowing brittle stars in large numbers (up to 400 per square metre; Swift, 1993) results in the bio-irrigation of the upper few centimetres and allows the maintenance of high levels of oxidised Pu(V) (Kershaw et al., 1986).

The distribution and behaviour of plutonium in the estuaries bordering the eastern Irish Sea is complex. The migration of channels can lead to the re-exposure of relatively highly contaminated sediments, labelled with plutonium during the 1970s. The geochemical conditions in which the sediment finds itself may also change during the transfer of sediment to intertidal and saltmarsh zones, with the transition from a well-oxygenated, alkaline marine environment to a terrestrial location with slightly acidic conditions, variable redox potential and an increasing influence of fresh water (Pulford et al., 1998). In the Esk Estuary, for example, the tidal circulation results in the transport of particle-associated plutonium to the upper reaches, where the low salinity and pH changes cause the rapid desorption of a labile form of plutonium (Hamilton-Taylor et al., 1987).

In deep ocean basins, sediments accumulate very slowly and bioturbation processes are much less active than in shelf and continental slope sediments. Nevertheless, studies carried out using fallout isotopes have demonstrated that mixing to depths deeper than a few centimetres can still take place (Cochran, 1985; Livingston, 1986; Smith et al., 1986). The depth of the bioturbated layer is highly dependent on the characteristics of the benthic population, which in turn are strongly related to the variation of organic matter inputs and current velocities (Buffoni et al., 1992).

References

Aarkrog, A. (1977). Environmental behaviour of plutonium accidentally released at Thule, Greenland. *Health Physics*, *32*, 271–284.

Aarkrog, A. (1995). Inventory of nuclear releases in the World. In *Proc. NATO Advanced Study Institute Advanced Course on Radioecology* (12 pp.). Zarechny, Russia.

Aarkrog, A., Dahlgaard, H., Nilsson, K. & Holm, E. (1984). Further studies of plutonium and americium at Thule, Greenland. *Health Physics*, *46*, 29–44.

Adams, C. E., Farlow, N. H. & Schell, W. R. (1960). The composition, structures and origins of radioactive fall-out particles. *Geochimica et Cosmochimica Acta*, *18*, 42–56.

Amundsen, I., Lind, B., Reistad, O., Gussgaard, K., Iosjpe, M. & Sickel, M. (2001). The Kursk accident. Strålevern-Rapport 2001:5 (36 pp.). Østerås, Norway: Norwegian Radiation Protection Authority.

Antón, M. P., Gascó, C., Sánchez Cabeza, J. A. & Pujol, L. (1994). Geochemical association of plutonium in marine sediments from Palomares (Spain). *Radiochimica Acta*, *66/67*, 443–446.

Assinder, D. J. (1999). A review of the occurrence and behaviour of neptunium in the Irish Sea. *Journal of Environmental Radioactivity*, *44*, 335–347.

Ballestra, S. (1980). *Radioactivité Artificielle et Environment Marin. Etude Relative aux Transuraniens Pu-238, Pu-239+240, Pu-241 et Am-241 en Méditerranée* (130 pp.). PhD Thesis, University of Nice.

Baskaran, M., Asbill, S., Santschi, P., Davis, T., Brooks, J., Champ, M., Makeyev, V. & Khlebovich, V. (1995). Distribution of ^{239,240}Pu and ^{238}Pu concentrations in sediments from the Ob and Yenisey Rivers and the Kara Sea. *Applied Radiation and Isotopes*, *46* (11), 1109–1119.

Baxter, M. S., Fowler, S. W. & Povinec, P. P. (1995). Observations on plutonium in the oceans. *Applied Radiation and Isotopes*, *46* (11), 1213–1224.

Beasley, T. M., Ball, L. A. & Blakesley, B. A. (1981). Plutonium and americium export to the North-east Pacific Ocean by Columbia River runoff. *Estence Coastal, and Shelf Science*, *13*, 659–669.

BNFL (1980–2001). '*Annual Reports on Radioactive Discharges and Monitoring of the Environment, 1979; 1980; 1981; 1982; 1983; 1984; 1985; 1986; 1987; 1988; 1989; 1990; 1991; 1992; 1993; 1994; 1995; 1996; 1997; 1998, 1999, 2000*'. Risley: British Nuclear Fuels plc.

Boust, D., Mitchell, P. I., Garcia, K., Condren, O. M., León Vintró, L. & Leclerc, G. (1996). A comparative study of the speciation and behaviour of plutonium in the marine environment of two reprocessing plants. *Radiochimica Acta*, *74*, 203–210.

Bowen, V. T., Noshkin, V. E., Livingston, H. D. & Volchok, H. L. (1980). Fallout radionuclides in the Pacific Ocean: Vertical and horizontal distributions, largely from GEOSECS stations. *Earth and Planetary Science Letters*, *49*, 411–434.

Buesseler, K. O. (1987). The geochemistry of fallout plutonium in the North Atlantic: I. A pore water study in shelf, slope and deep-sea sediments. *Geochimica et Cosmochimica Acta*, *51*, 2623–2637.

Buesseler, K. O. (1997). The isotopic signature of fallout plutonium in the North Pacific. *Journal of Environmental Radioactivity*, *36* (1), 69–83.

Buesseler, K. O. & Livingston, H. D. (1996). Natural and man-made radionuclides in the Black Sea. In P. Guéguéniat, P. Germain & H. Métivier (Eds), *Radionuclides in the Oceans – Inputs and Inventories* (Ch. 1, pp. 199–217). Institut de Protection et de Surete Nucleaire. Les Ulis, France: Les Éditions de Physique.

Buesseler, K. O. & Sholkovitz, E. R. (1987). The geochemistry of fallout plutonium in the North Atlantic: II. ^{240}Pu/^{239}Pu ratios and their significance. *Geochimica et Cosmochimica Acta*, *51*, 2605–2623.

Buffoni, G., Delfanti, R. & Papucci, C. (1992). Accumulation rates and mixing processes in near-surface North Atlantic sediments: Evidence from C-14 and Pu-239, 240 downcore profiles. *Marine Geology*, *109*, 159–170.

Burns, P. A., Cooper, M. B., Johnston, P. N., Martin, L. J. & Williams, G. A. (1994). Determination of the ratios of ^{239}Pu and ^{240}Pu to ^{241}Am for nuclear weapons test sites in Australia. *Health Physics*, *67* (3), 226–232.

Carpenter, R., Beasley, T. M., Zahnle, D. & Somayajulu, B. L. K. (1987). Cycling of fallout (Pu, ^{241}Am, ^{137}Cs) and natural (U, Th, ^{210}Pb) radionuclides in Washington continental slope sediments. *Geochimica et Cosmochimica Acta*, *51*, 1897–1921.

Chamberlain, A. C. & Dunster, H. J. (1958). Environmental impact of particles emitted from Windscale Piles, 1954–57. *Science of the Total Environment*, *63*, 139–160.

Choppin, G. R. (1994). Aspects of nuclear waste disposal of use in teaching basic chemistry. *Journal of Chemical Education*, *71* (10), 826–829.

Choppin, G. R. & Kobashi, A. (1990). Distribution of Pu(V) and Pu(VI) in seawater. *Marine Chemistry*, *30*, 241–247.

Choppin, G. R., Liljenzin, J. O. & Rydberg, J. (1995). *Radiochemistry and Nuclear Chemistry* (2nd edn). Oxford: Butterworth-Heinemann.

Cochran, J. K. (1985). Particle mixing rates in sediments of the eastern equatorial Pacific: Evidence from ^{210}Pb, ^{239,240}Pu and ^{137}Cs distributions at MANOP sites. *Geochimica et Cosmochimica Acta*, *49*, 1195–1210.

Cochran, J. K., Livingston, H. D., Hirschberg, D. J. & Surprenant, L. D. (1987). Natural and anthropogenic radionuclide distributions in the Northwest Atlantic Ocean. *Earth and Planetary Science Letters*, *84*, 135–152.

Cooper, M., Burns, P., Tracy, B., Wilks, M. & Williams, G. (1994). Characterisation of plutonium contamination at the former nuclear weapons testing range at Maralinga in South Australia. *Journal of Radioanalytical and Nuclear Chemistry*, *177*, 161–184.

Crocker, G. R., O'Connor, J. D. & Freiling, E. C. (1966). Physical and radiochemical properties of fallout particles. *Health Physics*, *12*, 1099–1104.

Diamond, H., Fields, P. R., Stevens, C. S., Studier, M. H., Fried, S. M., Inghram, M. G., Hess, D. C., Pyle, G. L., Mech, J. F., Manning, W. M., Ghiorso, A., Thompson, S. G., Higgins, G. H., Seaborg, G. T., Browne, C. I., Smith, H. L. & Spencer, R. W. (1960). Heavy isotope abundances in Mike thermonuclear device. *Physical Review*, *119* (6), 2000–2004.

Dickson, D. (1981). US nuclear weapons: Lasers purify. *Nature*, *292*, 401.

Ericsson, M. (2002). *On Weapons Plutonium in the Arctic Environment* (150 pp.). PhD Thesis, Risø–R–1321(EUN). Risø National Laboratory, Roskilde.

Facer, G. (1980). Quantities of transuranic elements in the environment from operations relating to nuclear weapons. In Wayne C. Hanson (Ed.), *Transuranic Elements in the Environment* (p. 86). US DOE/TIC-22800. Office of Health and Environmental Research.

Fowler, S. W., Ballestra, S., La Rosa, J. & Fukai, R. (1983). Vertical transport of particulate-associated plutonium and americium in the Northwest Pacific. *Deep-Sea Research*, *7*, 35–47.

Fowler, S. W., Noshkin, V. E., La Rosa, J. & Gastaud, J. (2000). Temporal variations in plutonium and americium inventories and their relation to vertical transport in the Northwestern Mediterranean Sea. *Limnology and Oceanography*, *45* (2), 446–458.

Fukai, R., Holm, E. & Ballestra, S. (1979). A note on vertical distribution of plutonium and americium in the Mediterranean Sea. *Oceanologica Acta*, *2*, 129–132.

Fukai, R., Ballestra, S. & Vas, D. (1982). Characteristics of the vertical transport of transuranic elements through the Mediterranean water column. In *VI Workshop on Marine Pollution of the Mediterranean Sea* (pp. 95–101). Monaco: CIESM.

Garcia, K. (1996). *Distribution et comportament du plutonium dans les eaux de la Manche et de l'estuarie de la Seine* (323 pp.). PhD Thesis, Universite de Paris Sud U.F.R. Scientifique d'Orsay, Paris.

Gray, J., Jones, S. R. & Smith, A. D. (1995). Discharges to the environment from the Sellafield site, 1951–1992. *Journal of Environmental Protection*, *15* (2), 99–131.

Hamilton, E. I. (1981). Alpha-particle radioactivity of hot particles from the Esk Estuary. *Nature*, *290*, 690–693.

Hamilton, E. I., Williams, R. & Kershaw, P. J. (1991). The total alpha particle for some components of marine ecosystems. In P. J. Kershaw & D. S. Woodhead (Eds), *Radionuclides in the Study of Marine Processes*. London: Elsevier.

Hamilton-Taylor, J., Kelly, M., Mudge, S. & Bradshaw, K. (1987). Rapid remobilization of plutonium from estuarine sediments. *Journal of Environmental Radioactivity*, *5*, 409–423.

Hardy, E. P., Krey, P. W. & Volchok, H. L. (1973). Global inventory and distribution of fallout plutonium. *Nature*, *241*, 444–445.

Herrmann, J., Nies, H. & Goroncy, I. (1998). Plutonium in the deep layers of the Norwegian and Greenland Seas. *Radiation Protection Dosimetry*, *75* (1–4), 237–245.

Hetherington, J. A. (1976). The behaviour of plutonium nuclides in the Irish Sea. In W. Miller & J. N. Stannard (Eds), *Environmental Toxicity of Aquatic Radionuclides* (pp. 81–106). Michigan: Ann Arbor Science.

Hirose, K., Sugimura, Y. & Aoyama, M. (1992). Plutonium and ^{137}Cs in the western North Pacific: Estimation of residence time of plutonium in surface waters. *Applied Radiation and Isotopes*, *43*, 349–359.

Hoffman, D. C., Lawrence, F. O., Mewherter, J. L. & Rourke, F. M. (1971). Detection of plutonium-244 in nature. *Nature*, *234*, 132–134.

Høibråten, S. & Thoresen, P. E. (1995). The sunken submarine Komsomolets and its effects on the environment. In P. Strand & A. Cooke (Eds), *Proc. Second Int. Conf. on Environmental Radioactivity in the Arctic* (pp. 350–355). Østerås, Norway: Norwegian Radiation Protection Authority.

Holm, E. (1991). Anthropogenic radionuclides in marine polar regions. In M. García León & G. Madurga (Eds), *Low-Level Measurements of Man Made Radionuclides in the Environment* (pp. 425–445). Huelva: World Scientific.

Holm, E. (1995). Plutonium in the Baltic Sea. *Applied Radiation and Isotopes*, *46* (11), 1225–1229.

Holm, E. & Persson, R. B. R. (1977). Radiochemical studies of ^{241}Pu in Swedish Reindeer lichens. *Health Physics*, *29*, 43–51.

Holm, E. & Persson, R. B. R. (1978). Global fallout of curium. *Nature*, *273*, 289–290.

Hunt, G. J. (1985). Timescales for dilution and dispersion of transuranics in the Irish Sea near Sellafield. *Science of the Total Environment*, *46*, 261–278.

IAEA (1986). *USSR State Committee in the Utilisation of Atomic Energy: The Accident at the Chernobyl Nuclear Power Plant and Its Consequences*. Vienna: International Atomic Energy Agency.

IAEA (1991). *Biennial Report 1989–91* (79 pp.). Monaco: International Atomic Energy Agency.

IAEA (1996). Nuclear Power Plant Status. In *Reports on Global Activities of the International Atomic Energy Agency*, *IAEA-Newsbriefs* 11(2:71) (p. 10). April–May 1996, Vienna.

IAEA (1999). Inventory of radioactive waste disposals at sea. IAEA-TECDOC-1005. Vienna: International Atomic Energy Agency.

Joseph, A. B., Gustafson, P. F., Russell, I. R., Schuert, E. A., Volchok, H. L. & Templin, A. (1971). Sources of radioactivity and their characteristics. In *Radioactivity in the Marine Environment* (pp. 6–41). National Academy of Sciences.

Keller, C. (1971). *The Chemistry of the Transuranium Elements*, Vol. 3 (655 pp.). K. H. Lieser (Ed.). Germany: Verlag Chemie GmbH.

Kershaw, P. J., Pentreath, R. J., Lovett, M. B. & Boggis, S. J. (1986). Apparent distribution coefficients of transuranium elements in U.K. coastal waters. In T. H. Sibley & C. Myttenaere (Eds), *Application of Distribution Coefficients to Radiological Assessment Models* (pp. 277–287). London: Elsevier.

Kershaw, P. J., Swift, D. J. & Denoon, D. C. (1988). Evidence of recent sedimentation in the eastern Irish Sea. *Marine Geology*, *85*, 1–14.

Kershaw, P. J., Sampson, K. E., McCarthy, W. & Scott, R. D. (1995a). The measurement of the isotopic composition of plutonium in an Irish Sea sediment by mass spectrometry. *Journal of Radioanalytical and Nuclear Chemistry (Articles)*, *198*, 113–124.

Kershaw, P. J., Woodhead, D. S., Lovett, M. B. & Leonard, K. S. (1995b). Plutonium from European reprocessing operations – its behaviour in the marine environment. *Applied Radiation and Isotopes*, *46* (11), 1121–1134.

Kim, J. I. (1991). Chemical behaviour of transuranic elements in the natural environment. In *Proc. Third International Symposium on Advanced Nuclear Energy Research – Global Environment and Nuclear Energy* (14 pp.). March 1991. Mito, Japan.

Koide, M., Bertine, K. K., Chow, T. J. & Goldberg, E. D. (1985). The ^{240}Pu/^{239}Pu ratio, a potential geochronometer. *Earth and Planetary Science*, *72*, 1–8.

Krey, P. W., Hardy, E. P., Pachucky, C., Rourke, F., Coluzza, J. & Benson, W. K. (1976). Mass isotopic composition of global fallout plutonium in soil. In *Transuranium Nuclides in the Environment* (pp. 671–678). IAEA-SM-199/39. Vienna: International Atomic Energy Agency.

Lee, S., La Rosa, J. J., Levy-Palomo, I., Oregioni, B., Khanh Pham, M., Povinec, P. & Wyse, E. (2003). Recent inputs and budgets of ^{90}Sr, ^{137}Cs, ^{239,240}Pu and ^{241}Am in the Northwest Mediterranean Sea. *Deep Sea Research II, 50*, 2817–1834.

Leonard, K. S., McCubbin, D. & Lovett, M. B. (1995). Physico-chemical characterisation of radionuclides discharged from a nuclear establishment. *Science of the Total Environment, 175*, 9–24.

León Vintró, L., Mitchell, P. I., Condren, O. M., Downes, A. B., Papucci, C. & Delfanti, R. (1999). Vertical and horizontal fluxes of plutonium and americium in the western Mediterranean and the Strait of Gibraltar. *Science of the Total Environment, 237/238*, 77–91.

León Vintró, L., McMahon, C. A., Mitchell, P. I., Josefsson, D., Holm, E. & Roos, P. (2002). Transport of plutonium in surface and sub-surface waters from the Arctic shelf to the North Pole via the Lomonosov Ridge. *Journal of Environmental Radioactivity, 60*, 73–89.

Livingston, H. D. (1986). Fallout plutonium in western North Pacific Sediment. IAEA-TECDOC-368. Vienna: International Atomic Energy Agency.

Livingston, H. D. & Andersen, R. F. (1983). Large particle transport of plutonium and other fallout radionuclides to the deep ocean. *Nature, 303*, 228–231.

Livingston, H. D. & Bowen, V. T. (1979). Pu and ^{137}Cs in coastal environments. *Earth and Planetary Science Letters, 43*, 29–45.

Livingston, H. D. & Povinec, P. P. (2000). Anthropogenic marine radioactivity. *Ocean & Coastal Management, 43*, 689–712.

Livingston, H. D. & Povinec, P. P. (2002). A millennium perspective on the contribution of bomb fallout radionuclides to ocean science. *Health Physics, 82* (5), 656–668.

Livingston, H. D., Kupferman, S. L., Bowen, V. T. & Moore, R. M. (1984). Vertical profiles of artificial radionuclide concentrations in the Central Arctic Ocean. *Geochimica et Cosmochimica Acta, 48*, 2195–2203.

Livingston, H. D., Buesseler, K. O., Izdar, E. & Konuk, T. (1988). Characteristics of Chernobyl fallout in the Southern Black Sea. In J. C. Guary, P. Guéguéniat & R. J. Pentreath (Eds), *Radionuclides: A Tool for Oceanography* (pp. 204–216). Essex: Elsevier.

Livingston, H. D., Povinec, P. P., Ito, T. & Togawa, O. (2001). The behaviour of plutonium in the Pacific Ocean. In A. Kudo (Ed.), *Plutonium in the Environment. Radioactivity in the Environment*, Vol. 1 (pp. 267–292). Amsterdam: Elsevier.

Lovett, M. B. & Nelson, D. M. (1981). Determination of some oxidation states of plutonium in sea water and associated particulate matter. In *Techniques for Identifying Transuranic Speciation in Aquatic Environments* (pp. 27–35). STI/PUB/613. Vienna: International Atomic Energy Agency.

Lovins, A. B. (1980). Nuclear weapons and power-reactor plutonium. *Nature, 283*, 817–823.

MacKenzie, A. B., Scott, R. D. & Williams, T. M. (1987). Mechanisms for northwards dispersal of Sellafield waste. *Nature, 299*, 613–616.

Manjón, G., García León, M., Ballestra, S. & López, J. J. (1995). The presence of man-made radionuclides in the marine environment in the south of Spain. *Journal of Environmental Radioactivity, 28* (2), 171–189.

McCarthy, W. & Nicholls, T. M. (1990). Mass-spectrometric analysis of plutonium in soils near Sellafield. *Journal of Environmental Radioactivity, 46*, 261–278.

McMahon, C. A., León Vintró, L., Mitchell, P. I. & Dahlgaard, H. (2000). Oxidation state distribution of plutonium in surface and subsurface waters at Thule, NW Greenland. *Applied Radiation and Isotopes, 52*, 697–703.

Merino, J., Sánchez-Cabeza, J. A., Bruach, J. M., Masqué, P. & Pujol, L. (1997). Artificial radionuclides in a high resolution water column profile from the Catalan Sea (the North Western Mediterranean). *Radioprotection–Colloques, 32* C2, 85–90.

Mitchell, P. I., Vives Batlle, J., Ryan, T. P., Schell, W. R., Sánchez-Cabeza, J. A. & Vidal-Quadras, A. (1991). Studies on the speciation of plutonium and americium in the western Irish Sea. In P. J. Kershaw & D. S. Woodhead (Eds), *Radionuclides in the Study of Marine Processes* (pp. 37–51). London: Elsevier.

Mitchell, P. I., Vives i Batlle, J., Downes, A. B., Condren, O. M., León Vintró, L. & Sánchez-Cabeza, J. A. (1995). Recent observations on the physico-chemical speciation of plutonium in the Irish Sea and the western Mediterranean. *Applied Radiation and Isotopes, 46* (11), 1175–1190.

Mitchell, P. I., León Vintró, L., Dahlgaard, H., Gascó, C. & Sánchez-Cabeza, J. A. (1997). Perturbation in the ^{240}Pu/^{239}Pu global fallout ratio in local sediments following the nuclear accidents at Thule (Greenland) and Palomares (Spain). *Science of the Total Environment, 202*, 147–153.

Mitchell, P. I., McMahon, C. A., León Vintró, L. & Ryan, R. W. (1998). Plutonium in Arctic surface and sub-surface waters at the St Anna and Voronin troughs. *Radiation Protection Dosimetry*, *75* (1–4), 247–252.

Mitchell, P. I., Downes, A. B., León Vintró, L. & McMahon, C. A. (2001). Studies of the speciation, colloidal association and remobilisation of plutonium in the marine environment. In A. Kudo (Ed.), *Plutonium in the Environment. Radioactivity in the Environment*, Vol. 1 (pp. 175–200). Amsterdam: Elsevier.

Morse, J. W. & Choppin, G. R. (1986). Laboratory studies of plutonium in marine systems. *Marine Chemistry*, *20*, 73–89.

Myers, W. A. & Lindner, M. (1971). Precise determination of the natural abundance of ^{237}Np and ^{239}Pu in Katanga pitchblende. *Journal of Inorganic and Nuclear Chemistry*, *33*, 3233–3238.

Nagaya, Y. & Nakamura, K. (1984). ^{239,240}Pu, ^{137}Cs and ^{90}Sr in the Central North Pacific. *Journal of the Oceanographical Society of Japan*, *40*, 416–424.

Nash, K. L., Cleveland, J. M. & Rees, T. F. (1988). Speciation patterns of actinides in natural waters: A laboratory investigation. *Journal of Environmental Radioactivity*, *7*, 131–157.

NEA (1981). The environmental and biological behaviour of plutonium and some other transuranium elements (116 pp.). Report by an Expert Group. Nuclear Energy Agency.

NEA (1989). Plutonium fuel: An assessment (160 pp.). Report by an Expert Group, Nuclear Energy Agency.

Nelson, D. M. & Lovett, M. B. (1978). Oxidation states of plutonium in the Irish Sea. *Nature*, *276*, 599–601.

Nelson, D. M., Larsen, R. P. & Penrose, W. R. (1987). Chemical speciation of plutonium in natural waters. In J. E. Pinder III, J. J. Alberts, K. W. McLeod & R. G. Scheckhise (Eds), *Environmental Research on Actinide Elements* (pp. 27–48). U.S. Department of Energy.

Nyffeler, F., Cigna, A. A., Dahlgaard, H. & Livingston, H. D. (1996). Radionuclides in the Atlantic Ocean: A survey. In P. Guéguéniat, P. Germain & H. Métivier (Eds), *Radionuclides in the Oceans – Inputs and Inventories* (Ch. 1, pp. 1–28). Institut de Protection et de Surete Nucleaire. Les Ulis, France: Les Éditions de Physique.

Orlandini, K. A., Penrose, W. R. & Nelson, D. M. (1986). Pu(V) as the stable form of oxidised plutonium in natural waters. *Marine Chemistry*, *18*, 49–57.

Orlandini, K. A., Penrose, W. R., Harvey, B. R., Lovett, M. B. & Findlay, M. W. (1990). Colloidal behaviour of actinides in an oligotrophic lake. *Environmental Science and Technology*, *24* (5), 706–712.

Osvath, I., Povinec, P. P. & Baxter, M. S. (1999). Kara Sea radioactivity assessment. *Science of the Total Environment*, *237/238*, 167–169.

Panteleyev, G. P., Livingston, H. D., Sayles, F. L. & Medkova, O. N. (1995). Deposition of plutonium isotopes and Cs-137 in sediments of the Ob delta from the beginning of the nuclear age. In P. Strand & A. Cooke (Eds), *Proc. Second International Conference on Environmental Radioactivity in the Arctic* (pp. 57–64). Oslo & Østerås, Norway: Norwegian Radiation Protection Authority.

Papucci, C., Charmasson, S., Delfanti, R., Gascó, C., Mitchell, P. I. & Sánchez-Cabeza, J. A. (1996). Time evolution and levels of man-made radioactivity in the Mediterranean Sea. In P. Guéguéniat, P. Germain & H. Métivier (Eds), *Radionuclides in the Oceans – Inputs and Inventories*. (Ch. 8, pp. 177–197). Institut de Protection et de Surete Nucleaire. Les Ulis, France: Les Éditions de Physique.

Pentreath, R. J., Lovett, M. B., Jefferies, D. F., Woodhead, D. S., Talbot, J. W. & Mitchell, N. (1984). The impact on public radiation exposure of transuranium nuclides discharged in liquid wastes from fuel reprocessing at Sellafield, UK. In *Symp. Radioactive Waste Management* (pp. 315–329). Vienna: International Atomic Energy Agency.

Pentreath, R. J., Harvey, B. R. & Lovett, M. B. (1985). Chemical speciation of transuranium nuclides discharged into the marine environment. In R. A. Bullman & J. R. Cooper (Eds), *Seminar on Speciation of Fission and Activation Products in the Environment* (pp. 312–325). Oxford: Elsevier Applied Science.

Pentreath, R. J., Woodhead, D. S., Kershaw, P. J., Jeffries, D. F. & Lovett, M. B. (1986a). The behaviour of plutonium and americium in the Irish Sea. *Rapport et Proces-Verbaux Reunion Conseil Internationale Exploration de Mer*, *186*, 60–69.

Pentreath R. J., Kershaw, P. J., Harvey, B. R. & Lovett, M. B. (1986b). The behaviour of certain long-lived radionuclides in the marine environment. In *Behaviour of Long-Lived Radionuclides Associated with the Deep-Sea Disposal of Radioactive Wastes* (pp. 101–114). IAEA-TECDOC-265. Vienna: International Atomic Energy Agency.

Peppard, D. F., Mason, G. W., Gray, P. R. & Mech, J. F. (1952). Occurrence of the $(4n + 1)$ series in nature. *Journal of the American Chemical Society*, *74*, 6081–6084.

Perkins, R. W. & Thomas, C. W. (1980). Worldwide Fallout. In Wayne C. Hanson (Ed.), *Transuranic Elements in the Environment* (pp. 53–82). US DOE/TIC-22800. Office of Health and Environmental Research.

Pigford, T. H. & Ang, K. P. (1975). The plutonium fuel cycles. *Health Physics*, *29*, 451–468.

Povinec, P., Livingston, H. D., Shima, S., Aoyama, M., Gastaud, J., Goroncy, I., Hirose, K., Huynh-Ngoc, L., Ikeuchi, Y., Ito, T., La Rosa, J., Wee Kwong, L. L., Lee, S., Moriya, H., Mulsow, S., Oregioni, B., Pettersson, H. & Togawa, O. (2003). IAEA'97 expedition to the NW Pacific Ocean – results of oceanographic and radionuclide investigations of the water column. *Deep Sea Research II*, *50*, 2607–2637.

Pulford, I. D., Allan, R. L., Cook, G. T. & MacKenzie, A. B. (1998). Geochemical associations of Sellafield-derived radionuclides in saltmarsh deposits of the Solway Firth. *Environmental Geochemistry and Health*, *20*, 95–101.

Rai, D., Serne, R. J. & Swanson, J. L. (1980). Solution species of plutonium in the environment. *Journal of Environmental Quality*, *9*, 417–420.

Risø (1970). Project Crested Ice: A Joint Danish–American Report on the Crash Near Thule Air Base on 21 January 1968 of a B-52 Bomber Carrying Nuclear Weapons (97 pp.). Risø Report No. 213.

Romero, L., Lobo, A. M., Holm, E. & Sánchez, J. A. (1991). Transuranics contribution off Palomares coast: Tracing history and routes to the marine environment. In P. J. Kershaw & D. S. Woodhead (Eds), *Radionuclides in the Study of Marine Processes*. London: Elsevier.

Salbu, B. (2001). Speciation of radionuclides in the environment. In R. A. Meyers (Ed.), *Encyclopedia of Analytical Chemistry: Instrumentation and Applications* (pp. 12993–13016). Chichester: Wiley.

Salbu, B., Nikitin, A., Strand, P., Christensen, G. C., Chumichev, V. B., Lind, B., Fjelldal, H., Bergan, T. D., Rudjord, A. L., Sickel, M., Valetova, N. K. & Foyn, L. (1997). Radioactive contamination from dumped nuclear waste in the Kara Sea – results from the joint Russian–Norwegian expeditions in 1992–1994. *Science of the Total Environment*, *202*, 185–198.

Sayles, F. L., Livingston, H. D. & Panteleyev, G. P. (1997). The history and source of particulate ^{137}Cs and ^{239,240}Pu deposition in sediments of the Ob delta, Siberia. *Science of the Total Environment*, *202*, 25–41.

Schneider, D. L. & Livingston, H. D. (1984). Measurement of curium in marine samples. *Nuclear Instruments and Methods in Physics Research*, *223*, 510–516.

Schulz, W. (1976). *The Chemistry of Americium* (290 pp.). Technical Information Center, Energy Research and Development Administration. Springfield: U.S. Department of Commerce.

SCOPE (1993). *Radioecology after Chernobyl: Biogeochemical Pathways of Artificial Radionuclides* (367 pp.). Sir F. Warner & R. M. Harrison (Eds). Scientific Committee on Problem on the Environment. Chichester: Wiley.

Seaborg, G. T. & Perlman, M. L. (1948). Search for elements 93 and 94 in nature. Presence of 94^{239} in pitchblende. *Journal of American Chemical Society*, *70*, 1571–1573.

Sholkovitz, E. R. (1983). The geochemistry of plutonium in fresh and marine water environments. *Earth Science Reviews*, *19*, 95–161.

Sivintsev, Y. (1994). Study of nuclide composition and characteristics of fuel in dumped submarine reactors and atomic icebreaker 'Lenin', Part I – Atomic Icebreaker. IAEA-IASAP-1. Vienna: International Atomic Energy Agency.

Smith, J. N. & Ellis, K. M. (1995). Radionuclide tracer profiles at the CESAR Ice Station and Canadian Ice Island in the western Arctic Ocean. *Deep Sea Research II*, *42* (6), 1149–1470.

Smith, J. N., Boudreau, B. P. & Noshkin, V. E. (1986). Plutonium and ^{210}Pb distributions in Northeast Atlantic sediments: Subsurface anomalies caused by non-local mixing. *Earth and Planetary Science Letters*, *81*, 15–28.

Smith, J. N., Ellis, K. M., Aarkrog, A., Dahlgaard, H. & Holm, E. (1994). Sediment mixing and burial of the ^{239,240}Pu pulse from the 1968 Thule, Greenland nuclear weapons accident. *Journal of Environmental Radioactivity*, *25*, 135–159.

Smith, J. N., Ellis, K. M., Polyak, L., Ivanov, G., Forman, S. L. & Moran, S. B. (2000). ^{239,240}Pu transport into the Arctic Ocean from underwater nuclear tests in Chernaya Bay, Novaya Zemlya. *Continental Shelf Research*, *20*, 255–279.

Swift, D. J. (1993). The macrobenthic infauna off Sellafield, Cumbria (north-east Irish Sea) with special reference to bioturbation. *Journal of the Marine Biological Association of the United Kingdom*, *73*, 143–162.

Taylor, D. M. (2001). Environmental plutonium – creation of the Universe to twenty-first century mankind. In A. Kudo (Ed.), *Plutonium in the Environment. Radioactivity in the Environment*, Vol. 1 (pp. 1–14). Amsterdam: Elsevier.

Trapeznikov, A. V., Pozolotina, V. N., Chebotina, M. Y., Chukanov, V. N., Trapeznikova, V. N., Kulikov, N. V., Nielson, S. P. & Aarkrog, A. (1993). Radioactive contamination of the Techa River, the Urals. *Health Physics*, *65*, 481–488.

UNSCEAR (1982). *Ionizing Radiation: Sources and Biological Effects* (773 pp.). New York: United Nations.

Wahlgren, M. A. & Orlandini (1982). Comparison of the geochemical behaviour of plutonium, thorium and uranium in selected North American lakes. In *Environmental Migration of Long-Lived Radionuclides* (pp. 757–774). IAEA-SM-257/89. Vienna: International Atomic Energy Agency.

WHO (1989). Health hazards from radiocaesium following the Chernobyl nuclear accident. *Journal of Environmental Radioactivity*, *10*, 107–121.

Yablokov, A. V., Kasarev, V. K., Rumyantsev, V. M., Kokeyev, M. Ye., Petrov, O. I., Lystsov, V. N., Yemelyanenkov, A. F. & Rubtsov, P. M. (1993). "*White Book*". *Facts and Problems Related to Radioactive Waste Disposal in Seas Adjacent to the Territory of the Russian Federation*. Moscow: Office of the President of the Russian Federation.

Yamana, H., Yamamoto, T. & Moriyama, H. (2001). Isotopic ratio of Pu released from fuel cycle facilities – importance of radiochemically pure ^{236}Pu as a tracer. In A. Kudo (Ed.), *Plutonium in the Environment. Radioactivity in the Environment*, Vol. 1 (pp. 31–46). Amsterdam: Elsevier.

MARINE RADIOACTIVITY
Hugh D. Livingston (Editor)

Chapter 4

Overview of point sources of anthropogenic radionuclides in the oceans

G. Linsley, K.-L. Sjöblom, T. Cabianca

International Atomic Energy Agency, Marine Environment Laboratory, MC 98012, Monaco

1. Introduction

This chapter reports and discusses the sources and amounts of radioactive material introduced into the marine environment by human activities such as radioactive waste disposal at sea, accidents at sea involving radioactive materials and discharges of radioactive effluents from coastal sites and rivers. It does not include radioactive materials that have entered the marine environment due to nuclear weapons testing or due to nuclear accidents such as the Chernobyl accident.

Although much of the information has been provided or approved by national governments it cannot be regarded as being complete. In particular, the information provided on discharges to the marine environment is only indicative. The information on radioactive waste disposal events at sea is probably the most reliable although the data on radioactive content of the individual disposals is less so. In relation to the reported accidents at sea, it is likely that the reported information does not represent a complete picture of the accidents that have actually occurred.

2. Radioactive waste disposal at sea – a history

For many years the oceans were used for the disposal of industrial waste, including waste from the nuclear industry. In the 1970s, the practice of sea dumping became subject to an international convention (IMO, 1972), which had the aim of regularising procedures and preventing activities that could lead to marine pollution. As time went on pressure mounted especially from smaller countries not engaged in ocean disposal, for waste disposal activities to be further restricted. In November 1993, it was finally decided that the disposal of industrial and radioactive wastes at sea should be prohibited. In the following paragraphs the main events leading to the prohibition are described.

2.1. *Law of the sea*

In 1958, the United Nations Conference on the Law of the Sea concluded, inter alia, that "every State shall take measures to prevent pollution of the sea from dumping of radioactive wastes, taking into account standards and regulations which may be formulated by competent international organisations".

Pursuant to its responsibilities as one of the identified "competent international organisations", the International Atomic Energy Agency (IAEA) set up successive scientific panels to provide guidance and recommendations to ensure that the disposal of radioactive waste at sea would not result in unacceptable hazards to man. The first of these meetings was held in 1957 and it subsequently led to the publication of IAEA Safety Series No. 5, *Radioactive Waste Disposal into the Sea* (IAEA, 1961). The purpose of the report was to offer recommendations which could be used as a basis of international agreement to ensure that "any disposal of radioactive waste into sea involves no unacceptable degree of hazard to man". Already this report regarded high level radioactive waste unsuitable for disposal at sea. The justification for disposal of low and intermediate level disposal was based on the large dilution capacity of the sea and the concept of maximum permissible concentrations.

2.2. *London Convention 1972*

The Convention on the Prevention of Marine Pollution by Dumping of Wastes and Other Matter (London Convention, 1972, formerly referred to as the London Dumping Convention) entered into force in 1975. For the regulation of materials to be disposed of in the marine environment, 'black' and 'grey' lists were established. The disposal of substances on the 'black list' (Annex I to the Convention) was prohibited except in trace quantities. Substances on the 'grey' list (Annex II to the Convention) were subject to 'special care' measures to ensure that their disposal, which had to be carried out under the provision of a 'special permit', would not have adverse effects on the marine environment.

High-level radioactive waste (HLW) was included in the 'black' list. The IAEA, which was recognised by the Contracting Parties to the London Convention as the competent international body in matters relating to radioactive waste disposal and radiation protection, was entrusted with the responsibility for defining HLW unsuitable for dumping at sea.

Other types of radioactive waste and hazardous material, not on the 'black' list – low and intermediate level wastes – were included in the 'grey' list. In issuing the special permits for the dumping of these types of radioactive waste, countries were advised to take the recommendations of the IAEA fully into account.

2.3. *Role of the international organisations*

In fulfilment of its obligations to the London Convention, the IAEA formulated and periodically reviewed its definition of HLW and recommendations for the use of national authorities on the issuance of 'special permits' for the dumping of radioactive waste other than HLW. In 1974, the IAEA presented the first provisional definition and recommendations to the London Convention. The last revision, published as IAEA Safety Series No. 78, was issued in 1986 (IAEA, 1986).

IAEA recommendations include, among other things, the requirement that the secretariat of the London Convention – the International Maritime Organization (IMO) headquartered in London – be notified prior to dumping operations. Selection criteria for dump sites and guidance for performing the required environmental assessment are also included. Revisions of the definition and recommendations between 1974 and 1986 took into account improvements in the understanding of the dispersion and behaviour of radionuclides in the marine environment and of changes in international radiation protection criteria.

The dumping of radioactive waste at sea took place solely under national authority until 1977. At that time, the Organization for Economic Co-operation and Development (OECD) established a 'Multilateral Consultation and Surveillance Mechanism' to co-ordinate the ocean disposal of its member states. Later, the OECD also established a Co-ordinated Research and Environmental Surveillance Programme (CRESP) to provide additional information for assessing the suitability of the Northeast Atlantic dumpsite, which was used by the OECD member states.

2.4. *Regional conventions*

After the institution of the London Convention, which is a global convention, several regional conventions for the protection of the sea were established, either under the umbrella of the United Nations Environment Programme (UNEP) or independently.

Many of these, while promoting the objectives of the London Convention, adopted more restrictive approaches to the regulation of dumping. Thus, the sea disposal of radioactive waste was totally prohibited in the Baltic Sea (1974), Mediterranean Sea (1976), Black Sea (1992), and in certain areas of the South Pacific (1985) and Southeast Pacific (1989).

2.5. *Temporary moratorium and inter-governmental review*

By the early 1980s, there was increasing disquiet among many of the Contracting Parties to the London Convention over the continuing practice of sea dumping of low-level radioactive waste. This led to a proposal being made at the Convention's 1983 Consultative Meeting for Contracting Parties to voluntarily suspend all sea dumping of radioactive waste pending a review of the safety of the practice, which was to be carried out by an independent panel of scientific experts.

In the following decade, expert groups reviewed the practice of sea dumping of radioactive waste from scientific and technical, as well as political, legal, economic and social perspectives. Their reports were presented to and discussed at annual meetings of the Contracting Parties. No conclusive decision could be reached by the Contracting Parties during this time and the voluntary moratorium was successively extended until 1993.

2.6. *Prohibition of sea dumping of radioactive wastes*

The discussions at the Consultative Meeting of Contracting Parties in November 1993 were inflamed by reports of the illicit dumping of liquid radioactive waste by the Russian Federation in the Sea of Japan in October 1993. The meeting adopted, by a majority vote, the prohibition of dumping of all types of radioactive waste to come into effect on 20 February 1994. The meeting also adopted the prohibition of dumping of industrial waste to come into effect by 1 January 1996.

The prohibitions were brought about by amending the Annexes to the Convention. As a result of the amendments, all types of radioactive wastes and radioactive matter are at the present included in the 'black' list.

The Russian Federation made a declaration not accepting the amendments associated with radioactive waste dumping, though stating that it will continue its endeavours to ensure that the sea is not polluted by the dumping of wastes and other matter. For it, the old Annexes of the Convention concerning this specific issue are still, in 2002, in force, and so too are the IAEA's definition and recommendations.

3. Coastal discharges – international control guidelines

After the termination of solid industrial and radioactive waste disposal into the oceans, the only remaining route by which waste can legally enter the marine environment is by effluent discharges to rivers and from coastal locations.

A non-binding instrument, the Montreal Guidelines for the Protection of the Marine Environment Against Pollution from Land-based Sources, was agreed in 1985. It was replaced in 1995 by the Global Programme of Action for Protection of the Marine Environment from Land-based Sources (GPA), an agreement under the auspices of UNEP. Like the Montreal Guidelines the GPA does not have the status of an international convention, it is rather a recommendation to governments. The GPA aims at preventing the degradation of the marine environment from land-based activities by assisting Governments to identify and assess the problems.

The IAEA has provided a series of guidance documents to national authorities on the limitation and regulation of discharges to the environment, in harmony with developments in the area of radiation protection as recommended by the International Commission on Radiological Protection (ICRP). The first guidance was issued as Safety Series No.45, *Principles for Establishing Limits for the Release of Radioactive Material into the Environment* in 1978 (IAEA, 1978). The most recent guidance was issued in 2000 (IAEA, 2000).

At the regional level an important legal instrument is the Convention for the Protection of the Marine Environment of the Northeast Atlantic (OSPAR Convention). The OSPAR Convention commits the Contracting Parties to take all possible steps to prevent and eliminate pollution of the marine environment of the Northeast Atlantic by applying the precautionary approach and best environmental technologies and environmental practices.

The 1998 Ministerial Meeting of the OSPAR Commission held in Sintra adopted strategies to direct its future work in the following four main areas:

(a) protection and conservation of ecosystems and biological diversity,
(b) hazardous substances,
(c) radioactive substances and
(d) eutrophication.

The Sintra Ministerial Statement represents a commitment to progressive and substantial reductions of discharges, emissions and losses of radioactive substances, with the ultimate aim of achieving concentrations in the environment near background values for naturally occurring radioactive substances and close to zero for artificial radioactive substances. Similar objectives are established for synthetic chemical substances of a hazardous character.

4. Inventory of radioactive waste disposals at sea

4.1. *Background*

In 1946, the first sea disposal operation took place at a site in the Northeast Pacific Ocean, some 80 km off the coast of California. Over the next thirty-five years, most sea disposal operations were performed under national authority approval and, in many cases, under an international consultative mechanism: Organization for Economic Co-operation and Development/Nuclear Energy Agency (OECD/NEA) Consultation Mechanism for Northeast Atlantic Dump Sites.

Some disposal operations continued until 1993 and included disposal of liquid and solid waste and nuclear reactor vessels, with and without fuel, into the oceans. Liquid waste was mainly released into the Arctic and Pacific Oceans. Solid waste, mostly packaged, was dumped into the Atlantic, Arctic and Pacific Oceans. Nuclear reactor vessels without fuel were dumped in the Atlantic, Arctic and Pacific Oceans, while nuclear reactor vessels with fuel were dumped only in the Arctic Ocean (Kara Sea).

The OECD/NEA has kept records of the disposal operations of packaged low level radioactive waste carried out by its Member States (NEA/OECD, 1983). In addition, the OECD/NEA has developed specific guidelines for waste package design (NEA/OECD, 1974) and site operational procedure (NEA/OECD, 1979). With respect to waste package performance requirements, the NEA/OECD has specifically stated that: "... the packages should be designed to ensure that their content is retained within them during descent to the sea-bed. This should normally ensure that the packages will remain intact for a period of time after they have reached the sea bed" (NEA/OECD, 1974).

4.2. *Data base on disposal at sea*

In response to requests from the London Convention, 1972, the IAEA has developed a global inventory of radioactive materials entering the marine environment due to the disposal at sea of radioactive waste and due to accidents and losses involving radioactive materials. The inventory was first established on the basis of the results of a questionnaire sent by the IAEA to Contracting Parties to the London Convention in 1986. Confirmation that the data were accurate was obtained separately from official sources in each of the countries responsible for the disposals. The first report on the global inventory of radioactive waste disposal at sea was issued in 1991 (IAEA, 1991). Subsequently the report was revised mainly to take account of the information received on sea disposal operations carried out by the Former Soviet Union and the Russian Federation and reissued in 1999 (IAEA, 1999).

Radioactive waste disposed of at sea comprises liquid and solid, low- and high-level waste. Low-level solid waste such as paper and textiles from decontamination processes, resins and filters etc. was usually solidified with cement or bitumen and packaged in metal containers. Large parts of nuclear installations such as steam generators, main circuit pumps, lids of reactor pressure vessels were disposed of unpackaged. In addition, nuclear reactor pressure vessels, without fuel or containing damaged spent nuclear fuel were dumped mainly into the Arctic seas, and without fuel also to the Pacific and Northwest Atlantic. The reactor pressure vessels containing spent fuel were usually filled with a polymer-based solidification agent (furfural) to provide an additional protective barrier.

The information in the IAEA inventory database is heterogeneous due to the various ways in which records on disposal operations have been kept in different countries. Usually an indication of the date of the disposal operation as well as of the location of the disposal site, in geographical co-ordinates, is given. The type, number and weight or volume of the disposed containers is reported. The weight or volume is representative of the disposed containers but not of the radioactive waste itself. Total alpha and total beta–gamma activities of disposed wastes are reported. In addition, some countries have provided more detailed information on radionuclide composition of the waste.

4.3. *Summary of global information on radioactive waste disposals at sea*

4.3.1. *Distribution of disposal operations – geographical and temporal*

The first reported sea disposal operation of radioactive waste took place in 1946 and the latest in 1993. During the 48 years history of sea disposal, 14 countries have used more than 80 sites to dispose of approximately 85 PBq[1] (2.3 MCi) of radioactive waste (Fig. 1, Table 1).

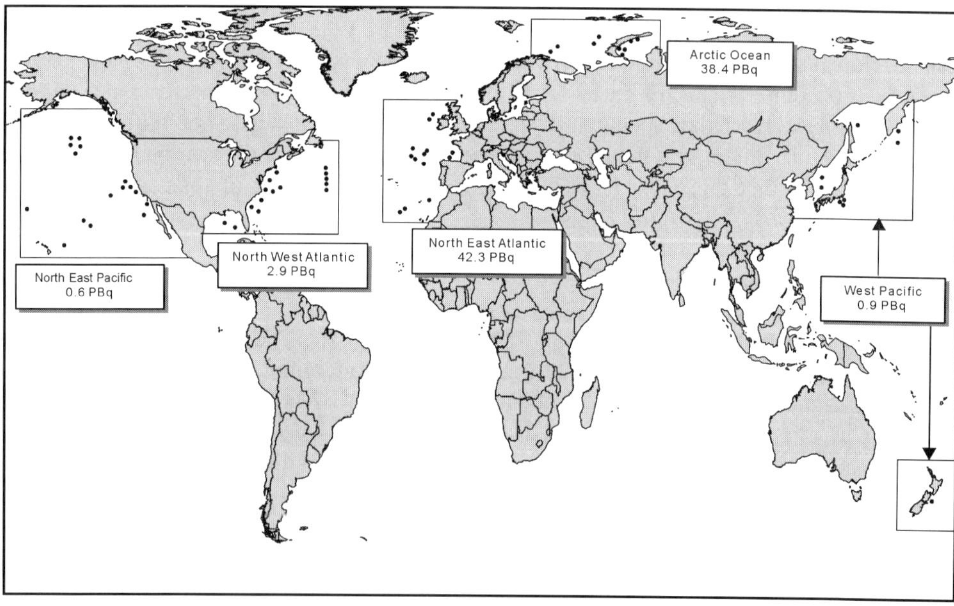

Fig. 1. Worldwide disposal at sea of radioactive waste.

[1] 1 PBq = 10^{15} Bq.

Table 1
Activity of alpha and beta–gamma emitters and tritium disposed of in the Atlantic, Pacific and Arctic Oceans between 1946 and 1993

Country	Activity (TBq[2])				Percentage of total activity (%)
	Alpha	Beta–gamma[a]	Tritium	Total	
Atlantic sites:					
Belgium	29	2091	787	2120	2.5
France	8.5	345		353	0.4
Germany	0.02	0.2		0.2	–
Italy	0.07	0.1		0.2	–
Netherlands	1.1	335	99	336	0.4
Sweden	0.94	2.3		3.2	–
Switzerland	4.3	4415	3902	4419	5.2
United Kingdom	631	34,456	10,781	35,088	41.2
United States		2942		2942	3.5
Total (Atlantic sites)	675	44,587	15,569	45,262	53.2
Arctic sites:					
Former Soviet Union		38,369		38,369[b]	45.1
Russian Federation		0.7		0.7	
Total (Arctic sites)		38,370		38,370	45.1
Pacific sites:					
Japan	0.01	15		15	0.02
Republic of Korea				NI[c]	
New Zealand	0.01	1.0		1.0	–
Russian Federation		2.1		2.0	–
Former Soviet Union		874		874[b]	1.0
United States		554		554	0.7
Total (Pacific sites)	0.02	1446		1446	1.7
Total (all sites)				85,078	100

[a]Tritium activities are included in the beta–gamma values.
[b]For solid packaged low-level waste, activity is expressed as ^{90}Sr equivalents.
[c]No information (NI) available in terms of activity disposed of by the Republic of Korea.

An examination of the quantities of waste disposed of by each country involved shows that some countries used this waste management option only for small amounts of waste. Two countries conducted only one disposal operation each and one country conducted only two disposal operations. On the other hand, five countries used the sea disposal option regularly for the disposal of large quantities of waste. Figure 2 summarises, as a percentage, the total activity of radioactive waste disposed of by involved countries in the Atlantic, Pacific and Arctic Oceans.

The past dumping operations of radioactive waste can be summarised broadly as follows (see Table 2): around 53% of the activity in the disposed radioactive waste is associated with the disposal of low level packaged solid waste, of which more than 93% was disposed of at the Northeast Atlantic dumping sites by eight countries, predominantly by the United Kingdom.

[2] 1 TBq = 10^{12} Bq.

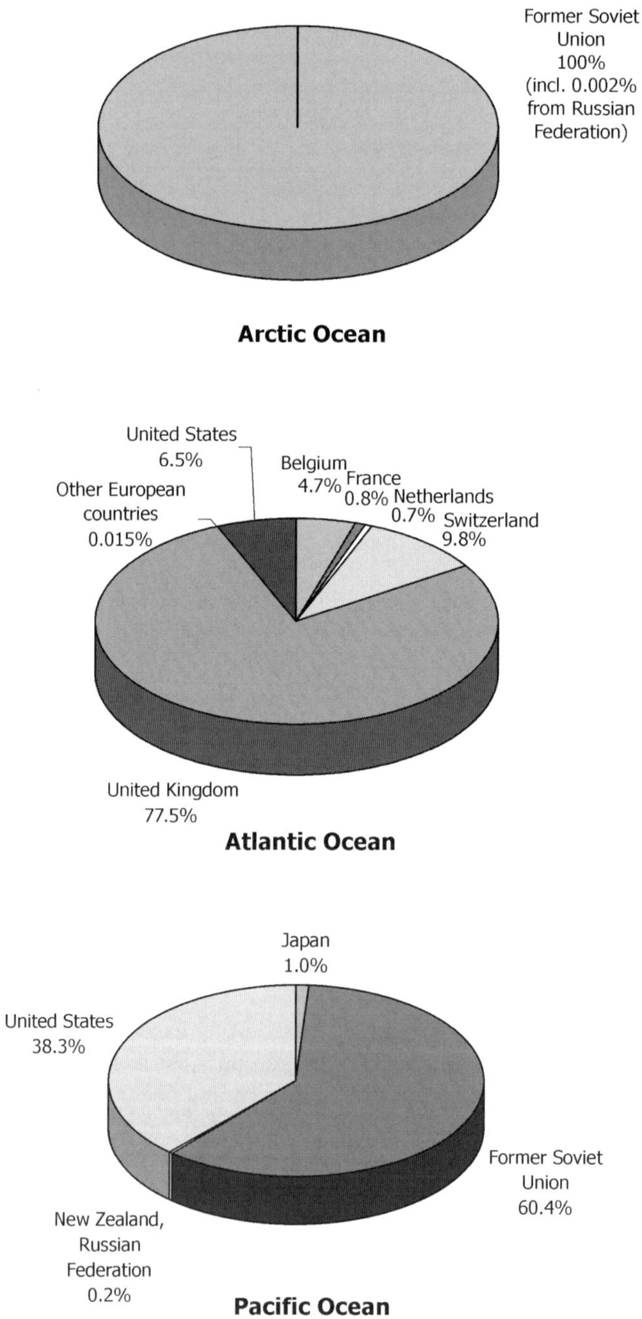

Fig. 2. Percentage by country of total activity of radioactive material disposed of in the Arctic, Atlantic and Pacific Oceans.

Table 2
Distribution of activity (TBq) for different types of waste dumped in the world oceans

Waste type	Activity (TBq)				Percentage (%)
	Atlantic	Pacific	Arctic	Total	
Reactors with spent nuclear fuel	0	0	36,876	36,876	43.3
Reactors without spent nuclear fuel	1221	166	143	1530	1.8
Low level solid waste	44,043	821	585	45,449	53.4
Low level liquid waste	<0.001	459	765	1223	1.5
Total	45,265	1445	38,369	85,078	
Percentage (%)	53.2	1.7	45.1		100

Some 43% of the activity in the disposed radioactive waste is associated with the disposal of reactors with spent nuclear fuel by the Former Soviet Union in the Kara Sea (Arctic Ocean). The disposal of low-level liquid and solid waste in the Arctic Ocean makes up less than 1.6% of the total global disposed activity. The inventory of all wastes disposed of into the Pacific Ocean amounts to about 1.7% of the total global disposed activity.

Initial information on the dumping of naval reactors with and without spent fuel to the Kara Sea by the Former Soviet Union was given in 1993 in the '*White Book*' (Office of the President of the Russian Federation, 1993). In all six reactors with spent fuel, a special container with fuel and ten reactors from which the fuel was removed were dumped between 1965 and 1988 in the shallow fjords of Novaya Zemlya and in the Novaya Zemlya Depression. The total inventory at the time of disposal was estimated to be 89 PBq. A further study indicated that the total activity at the time of disposal was lower, 37 PBq, and that in 1994, due to the radioactive decay, this was reduced to 4.7 PBq (IAEA, 1997). Current activity in the reactors is estimated to be about 3.7 PBq. The Former Soviet Union dumped four reactors altogether without spent nuclear fuel in the Sea of Japan in the years 1971 and 1979. A separate study related to the dumping operations and the inventory suggests that the total activity of those reactors, mainly activation products, was, at the time of dumping, 166 TBq, and in 2002 about 22 TBq (Sivintsev & Kiknadze, 1998).

The United States dumped the reactor shell of submarine Seawolf to the Northwest Atlantic in 1959. The radioactivity involved was estimated to be 1.2 PBq at the time of dumping (IAEA, 1999).

The chronological distribution of the dumping operations of low and intermediate level waste is shown in Figs 3–5. The dumping at the Northeast Atlantic site (Fig. 3) started in 1954, when 20 TBq were disposed of; annual inventories increased gradually and reached their highest level of almost 7000 TBq in 1980, shortly before the moratorium on low level radioactive waste disposal was introduced.

The temporal distribution of disposals of solid packaged and unpackaged low and intermediate level waste (excluding nuclear reactors and spent fuel) into the Arctic Seas (Fig. 4) they reached in 1964 and the annual inventories remained at less than 40 TBq until 1982 when a peak of more than 75 TBq. They peaked again in 1988, when 70 TBq were dumped in the same area.

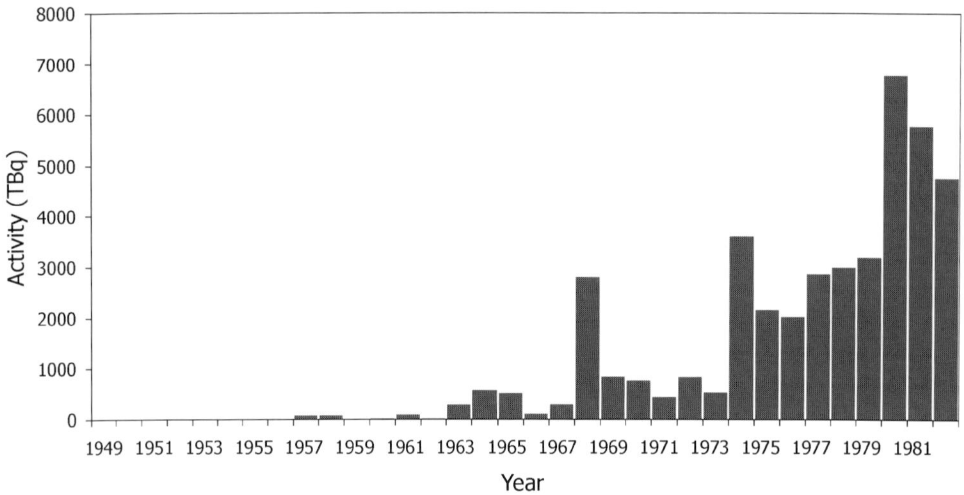

Fig. 3. Temporal distribution of radioactive waste disposals in the Northeast Atlantic Ocean.

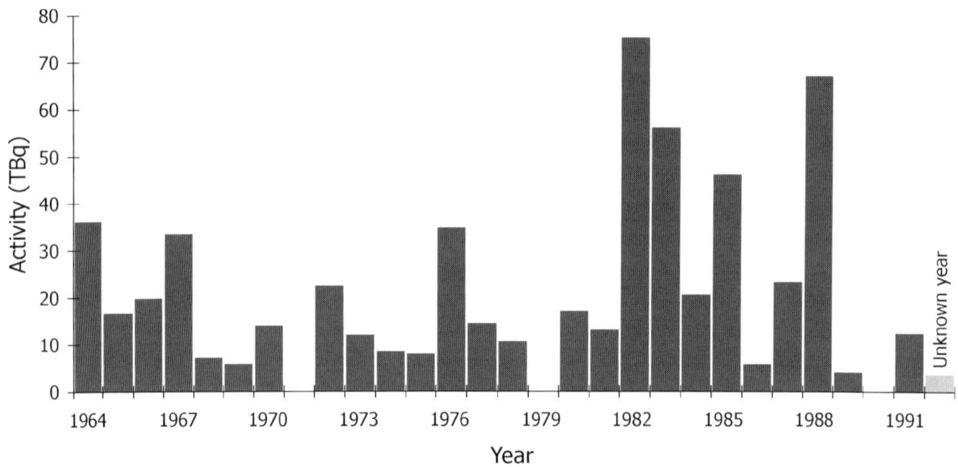

Fig. 4. Temporal distribution of low and intermediate level solid waste disposals in the Arctic Ocean (spent nuclear fuel not included).

The temporal distribution of disposals of liquid low-level waste into the Arctic Ocean is given in Fig. 5. The operations started in 1959 and continued until 1992. Two major peaks, of 350 TBq in 1976 and 195 TBq in 1988, dominate the picture. The last disposal operation in the Arctic Ocean was carried out in 1992 when low-level liquid radioactive waste was released by the Russian Federation into the Barents Sea (Office of the President of the Russian Federation, 1993).

The last recorded case of ocean disposal occurred in 1993 when the Russian Federation released low-level liquid radioactive waste into the Sea of Japan (Danilov-Danilyan, 1993).

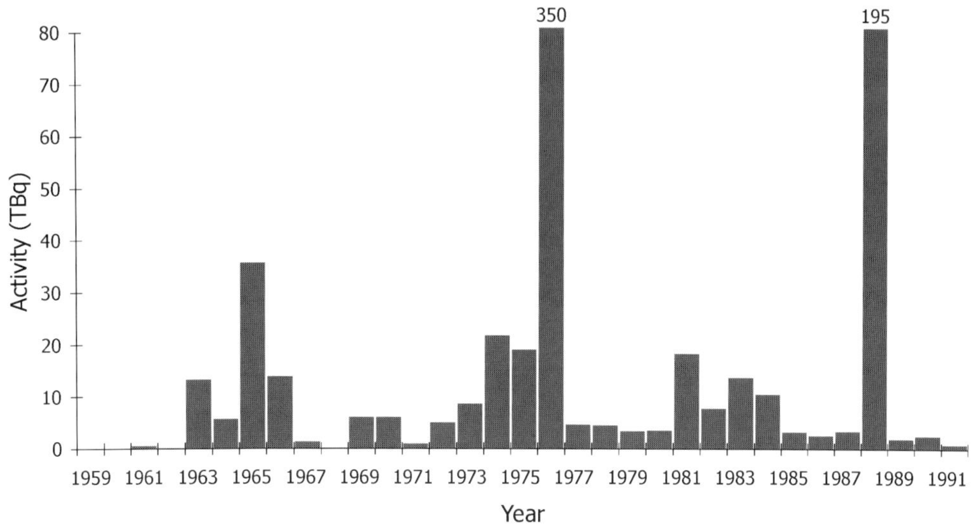

Fig. 5. Temporal distribution of low-level liquid waste disposals in the Arctic Ocean.

4.3.2. *Radionuclide composition of wastes*

At the North Atlantic dumping sites, tritium alone represents one third of the total activities (see Table 1). Tritium, together with other beta–gamma emitters such as ^{90}Sr, ^{134}Cs, ^{137}Cs, ^{55}Fe, ^{58}Co, ^{60}Co, ^{125}I and ^{14}C, contributes more than 98% to the total activity of the waste. The waste also contains low quantities (less than 2%) of alpha-emitting radionuclides, with plutonium and americium isotopes representing 96% of the alpha-emitters present (NEA/OECD, 1985).

The initial information on the radionuclide composition of waste disposed of by the Former Soviet Union presented in the 1993 '*White Book*' concerns only the reactor compartment of the nuclear icebreaker *Lenin* at the time of disposal. Radionuclides include ^{60}Co, ^{137}Cs, ^{90}Sr, ^{238}Pu, ^{241}Am and ^{244}Cm. The results of the analysis of radionuclide composition in all reactors dumped in the Kara Sea performed in the later study is summarised in Table 3 (IAEA, 1998).

4.3.3. *Environmental impact of disposed waste*

The dumpsites in Northeast Atlantic have been periodically surveyed by the countries involved in the disposal operations since 1977. Generally these surveys did not detect any radioactivity associated with the dumping operations in the water samples, but in the survey carried out in 1992 elevated concentrations of ^{238}Pu were detected in water samples collected at the dumpsites indicating leakages from the packages. At some locations also the concentrations of $^{293+240}$Pu, ^{241}Am and ^{14}C in the water were enhanced. It should be noted that the design of packages for the dumped waste was not intended to confine the radionuclides for tens of years but rather to secure that the wastes are transported intact to the sea bottom (Baxter et al., 1995).

The dumpsites used by the United States in Northeast Pacific and Northwest Atlantic have been surveyed form time to time. Some samples taken in the immediate vicinity of waste

Table 3
Estimated inventory of individual isotopes in the marine reactors dumped in the Kara Sea in 1994 (IAEA, 1998)

Isotope	Activity in 1994 (Bq)	Percentage of group (%)	Percentage of total (%)
Fission products:			86
^{3}H	5.2×10^{13}	1.3	
^{85}Kr	3.9×10^{13}	0.97	
^{90}Sr	9.5×10^{14}	24	
^{90}Y	9.5×10^{14}	24	
^{99}Tc	2.8×10^{11}	0.01	
^{125}Sb	1.7×10^{11}	<0.01	
^{129}I	3.9×10^{8}	<0.01	
^{137}Cs	1.0×10^{15}	25	
^{137m}Ba	9.9×10^{14}	25	
^{147}Pm	2.5×10^{12}	0.06	
^{151}Sm	2.4×10^{13}	0.60	
^{155}Eu	9.6×10^{10}	<0.001	
Activation products:			12
^{14}C	4.9×10^{11}	0.09	
^{60}Co	1.4×10^{14}	25	
^{59}Ni	6.3×10^{12}	1.1	
^{63}Ni	3.4×10^{14}	61	
^{152}Eu	6.0×10^{13}	11	
^{154}Eu	1.1×10^{13}	2.0	
^{205}Pb	1.9×10^{8}	<0.001	
^{207}Bi	1.7×10^{10}	<0.001	
^{208}Bi	6.2×10^{9}	<0.001	
^{210}Bi	3.4×10^{9}	<0.001	
Actinides:			2
^{238}Pu	1.6×10^{12}	1.7	
^{239}Pu	6.2×10^{12}	6.4	
^{240}Pu	2.7×10^{12}	2.8	
^{241}Pu	7.8×10^{13}	81	
^{241}Am	8.3×10^{12}	8.6	
Total	4.7×10^{15}		100

containers have shown slightly elevated levels of caesium and plutonium isotopes (IAEA, 1975; USNOOA, 1995).

In the studies carried out in 1992–1994 at the Arctic dump sites, in the fjords of Novaya Zemlya, ^{60}Co, ^{137}Cs and $^{239+240}Pu$ were detected in sediment samples indicating slight, non-radiologically significant leakages from the waste disposed in metal containers. No clear indication of leakages from the major objects containing nuclear fuel was found (Joint Norwegian–Russian Expert Group for Investigation of Radioactive Contamination in the Northern Areas, 1996).

Studies carried out at the dumpsites of the Sea of Japan have so far not revealed any indication of leakages from the dumped waste (Baxter et al., 1995; Petterson et al., 1995, 1996).

It should be noted that the leakages from the dumped waste packages are insignificant, although in some cases measurable. The localised increase in radionuclide concentrations in seawater and sediment represent less than 0.1% of the natural radioactivity in those media.

4.3.4. *Estimation of the current inventory of disposed waste*

The cumulative inventory of radioactive wastes in the world's oceans is shown in Fig. 6; the dark line in the graph shows the estimated actual cumulative inventory, which takes account of the radioactive decay of the waste. The decay corrected cumulative inventory of radioactive waste in the world oceans was estimated using information on the radionuclide composition of the low-level radioactive waste disposed of in the Northeast Atlantic sites, the high level radioactive waste disposed by the Former Soviet Union in the Arctic Ocean and in the Pacific Ocean. These wastes represent more than 93% of the radioactive material dumped at sea. No information on the isotopic composition of the remaining radioactive material disposed of in the world's oceans is available and therefore no correction for radioactive decay was included for this waste.

The maximum inventory (45 PBq) was reached in 1982, the last year when dumping in the Northeast Atlantic sites took place. The graph in Fig. 6 also shows two other peaks, in 1965 (20 PBq) and in 1967 (31 PBq) corresponding to the dumping of reactors with nuclear fuel in the Arctic Ocean. Currently the total inventory of radioactive material disposed of at sea can be estimated to be 21 PBq. By 2050 it will be further reduced to less than 10 PBq or around 11% of the total radioactive disposed of at sea. This has to be considered a maximum value as it includes around 6 PBq of waste for which no isotopic composition is known and no correction for radioactive decay was made.

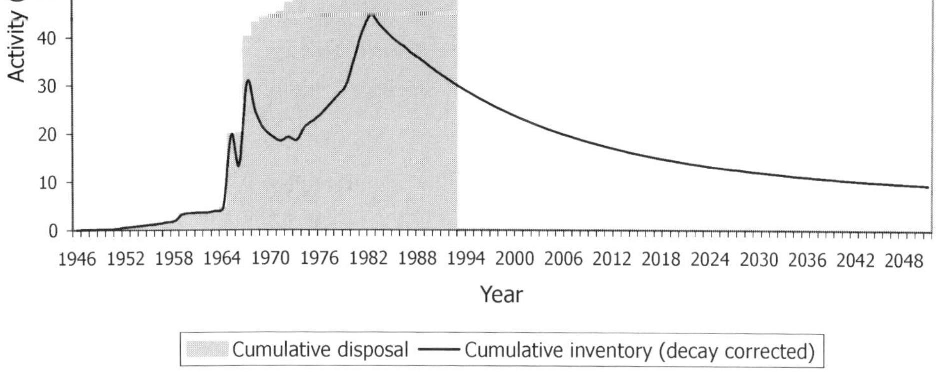

Fig. 6. Cumulative inventory of radioactive waste disposed in the world oceans.

5. Inventory of accidents and losses at sea

5.1. *Collection of information*

In 1989, pursuant to the recommendations of the Contracting Parties of the London Convention 1972 (IMO, 1989), the IAEA started to gather information on accidents and losses at sea, initially using data available from the extensive open literature existing on this subject (Ringot et al., 1987; Eriksen, 1990; NEA/OECD, 1990; Gladkov & Sivintsev, 1994; Sharpe, 1994; Kasatonov, 1996). Subsequently, IAEA's Member states reviewed and, as appropriate, officially confirmed the information and provided details on accidents to be included in the inventory database. A report on the information was issued in 2001 (IAEA, 2001).

5.2. *Sources of radioactivity in the marine environment resulting from accidents and losses at sea*

Seven possible sources of radioactive material entering the marine environment as a result of accidents and losses were identified for inclusion in the inventory:

1. Nuclear powered military surface or underwater vessels.
2. Nuclear weapons and military vessels capable of carrying such weapons.
3. Nuclear powered civilian ships.
4. Nuclear energy sources used in spacecraft, satellites and in the deep sea as acoustic signal transmitters.
5. Radioisotope thermoelectric generators (RTG) used, for instance, to supply power to lighthouses.
6. Cargos of nuclear material in transit.
7. Sealed radiation sources.

Information on losses of minor sources, such as depleted uranium used as ballast, or alloys with very low radioactive content, is not included in the IAEA inventory. Radionuclide sources associated with nuclear weapon losses are included for completeness. However, no detailed information on inventories of radioactive material associated with these sources is available.

5.3. *Summary of global information*

The global information is summarised in Table 4 and Fig. 7 and discussed below.

5.3.1. *Nuclear powered military vessels*

In general, one or two nuclear reactors are used to propel nuclear vessels and are usually of the small pressurised-water design not dissimilar to the larger version used in nuclear power plants. Information available up to October 2000 indicates that there were about 400 reactors in nuclear powered vessels around the world (Eriksen, 1990; Sharpe, 1994; OTA, 1996). The number of nuclear powered vessels is smaller than the number of reported reactors because most of the Russian submarines had two reactors while submarines of other countries typically have one reactor on board. In 2000, the Russian Federation had around 75 active military vessels including 3 surface ships and 72 submarines representing about 150 reactors. As of

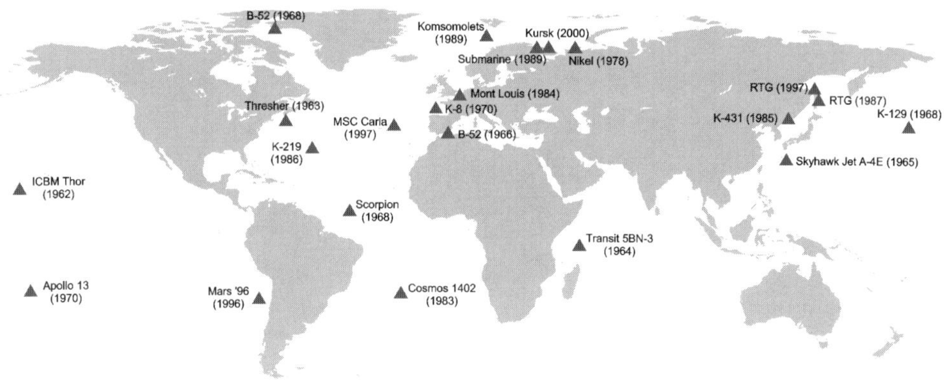

Fig. 7. Location of confirmed accidents at sea resulting in actual or potential release to the marine environment.

1994 it was estimated that the USA had about 117 nuclear vessels, the UK had 16, France had 11 and China had one nuclear military vessel (Sharpe, 1994; OTA, 1996). It has been confirmed that six nuclear submarines have been lost due to accidents since 1963 (Fig. 7, Table 4) at various sites in the Atlantic Ocean: two from the USA Navy, *Thresher* in 1963 and *Scorpion* in 1968; three others from the Soviet Union Navy, K-8 in 1970, K-219 in 1986 and K-278 *Komsomolets* in 1989, and one from the Russian Federation, K-141 *Kursk* in 2000 (Kasatonov, 1996; Office of the President of the Russian Federation, 1993; U.S. Department of the Navy, 1984; Amundsen et al., 2001). With the exception of the Russian submarine *Kursk*, the depth at the sites of the accidents, below 1500 meters, has not permitted the recovery of any of the submarines or their nuclear reactors. The *Kursk*, which sunk at a depth of 108 m in the Barents Sea was recovered in October 2001 with its reactors intact.

The primary barrier to prevent radionuclide release in case of an accident involving a nuclear submarine is the reactor pressure vessel which is designed to contain radioactive substances during either normal or accidental conditions and is expected to limit or delay radionuclide release into the marine environment. Radiological surveys on samples of seawater, sediments and deep sea organisms collected near the various sites of past accidents have been carried out. So far, monitoring has not shown any elevated levels of radionuclides above those due to nuclear weapons fallout except for some ^{60}Co detected in sediment samples collected close to the submarines *Scorpion* and *Thresher* (U.S. Department of the Navy, 1984; Knolls Atomic Power Laboratory, 1993a, b, 2000); and ^{137}Cs in water and sediment samples near the wreck of the *Komsomolets* (Gladkov & Sivintsev, 1994).

In a few cases accidents have occurred at military bases during the maintenance or defuelling of nuclear submarines resulting to releases to the environment. The largest reported accidental release of liquid radioactive waste (74 TBq) occurred in 1989 during the anchorage of a submarine of the Soviet Northern Fleet in Ara Bay. The accident led to the radioactive contamination of a sea area of about 1 km^2 (Office of the President of the Russian Federation, 1993; Petrov, 1995). In 1985, during refuelling work on the nuclear submarine K-431 at a pier in the Navy Shipyard of Chazhma Bay (Russian Far East) a prompt explosive criticality accident occurred in the reactor compartment. As a result, radionuclides with an activity of about 185 TBq

Table 4
Summary of confirmed accidents at sea resulting in actual or potential release to the marine environment

Date	Vessel involved	Country	Geographical area	Coordinates		Depth (m)	Radioactive material involved	Recovered	Total activity	Estimated activity released
				Latitude	Longitude					
2/6/1962	ICBM Thor Rocket	USA	Pacific Ocean, Johnston Island	16°45′N	169°32′W	–	Nuclear test device	No	–	–
19/6/1962	ICBM Thor Rocket	USA	Pacific Ocean, Johnston Island	16°45′N	169°32′W	–	Nuclear test device	No	–	Release occurred
10/4/1963	Nuclear submarine SSN-593 '*Thresher*'	USA	Atlantic Ocean, 100 miles east of Cape Cod	41°46′N	65°3′W	2590	Nuclear reactor	No	1.15 PBq[a]	0.04 GBq
21/4/1964	Satellite 'Transit 5BN-3'	USA	West Indian Ocean, North of Madagascar	–	–	–	SNAP-9A generator	No	630 TBq	630 TBq
5/12/1965	Skyhawk Jet A-4E	USA	Pacific Ocean, 250 miles South of Kyushu, 70 miles east of Okinawa	27°35′N	131°19′E	4800	1 nuclear weapon	No	–	–
17/1/1966	B-52 Bomber	USA	Mediterranean Sea, 5 miles off Palomares Spain	37°12′N	1°41′W	914	4 nuclear weapons[b]	Yes	–	1.37 TBq
21/1/1968	B-52 Bomber	USA	Arctic Ocean, Thule, Greenland	76°32′N	69°17′W	247	4 nuclear weapons	Partial	–	3.12 TBq
11/4/1968	Diesel submarine K-129	Former Soviet Union	Pacific Ocean, 1230 miles from Kamchatka	40°6′N	179°57′E	6000	2 Nuclear warhead(s)	Yes	37 GBq	–
22/5/1968	Nuclear submarine SNN-583 '*Scorpion*'	USA	Atlantic Ocean, 400 miles South West Azores	8°N[c]	40°6′W[c]	>3000	Nuclear reactor 2 nuclear warheads	No	1.3 PBq[a]	0.04 GBq

Table 4
(Continued.)

Date	Vessel involved	Country	Geographical area	Co-ordinates		Depth (m)	Radioactive material involved	Recovered	Total activity	Estimated activity released
				Latitude	Longitude					
8/4/1970	Nuclear submarine K-8	Former Soviet Union	Bay of Biscay	–	–	4000	2 reactors	No	9.25 PBq	–
							Nuclear warhead(s)		30 GBq	
11/4/1970	Spacecraft 'Apollo 13'	USA	South Pacific Ocean, Tonga Trench south of Fiji	21°38′S	165°22′W	6000	SNAP-27 generator	No	1.63 TBq	–
1978	Lighter '*Nikel*'	Former Soviet Union	South-eastern Barents Sea, off Kolguyev Island	69°31′N	47°56.03′E		Unenclosed solid radioactive LLW and ILW	No	1.5 TBq	–
8/2/1983	Satellite 'Cosmos 1402'	Former Soviet Union	South Atlantic Ocean, 1600 km East of Brazil	–	–	–	Reactor core U-235, Sr-90, Cs-137	No	1 PBq	–
25/8/1984	Surface vessel '*Mont Louis*'	France	North Sea, 20 km off Zeebrugge	51°24.2′N	2°50′E	25	Containers of uranium hexafluoride	Yes	6 TBq	–
10/8/1985	Nuclear submarine K-431	Former Soviet Union	Soviet Pacific Coast, Chazhma Bay Shkotovo-22	43°N	132°E	–	Reactor core	Yes	185 TBq	Release occurred
6/10/1986	Nuclear submarine K-219	Former Soviet Union	Atlantic Bermudas	31°29′N	54°42′W	5500	2 reactors	No	9.25 PBq	–
20/8/1987	RTG power supply	Former Soviet Union	Sea of Okhotsk, off Sakhalin Island	50°2′N[d]	144°E	~30	Sr-90 sealed source	No	25.3 PBq	–

Table 4
(Continued.)

Date	Vessel involved	Country	Geographical area	Co-ordinates		Depth (m)	Radioactive material involved	Recovered	Total activity	Estimated activity released
				Latitude	Longitude					
7/4/1989	Nuclear submarine K-278 '*Komsomolets*'	Former Soviet Union	Norwegian Sea, 180 km SW Bear Island	73°46.3′N	13°15.9′E	1680	Reactor core	No	3.59 PBq	<370 GBq
1989	Nuclear submarine	Former Soviet Union	Ara Bay, Kola Peninsula	69°30′N	33°E	–	Liquid radioactive waste	No	74 TBq	Release occurred
16/11/1996	Interplanetary station 'Mars' 96'	Russian Federation	Near west coast of Chile	25°6′ S[e]	75°24′W	–	Pu-238	No	174 TBq	–
8/8/1997	RTG power supply	Russian Federation	Sea of Okhotsk, off Sakhalin Island	54°19′N[f]	142°15′E	–	Sr-90 sealed source	No	1.3 PBq	–
24/11/1997	Surface vessel 'MSC Carla'	France	Atlantic Ocean, 70 nautical miles off the Azores	40°3′N	22°50′W		Three packages containing ^{137}Cs sealed sources	No	326 TBq	No release occurred
12/8/2000	Nuclear submarine K-141 '*Kursk*'	Russian Federation	Barents Sea, off Rybatschi Peninsula	69°37′N	37°35′E	116	Two nuclear reactors	No	1–2 EBq	No release occurred

[a] Estimates as of 1984.
[b] Of the 4 nuclear weapons, one fell into Mediterranean Sea and was recovered intact, one was recovered intact from fields near village of Palomares and the other two were destroyed on impact. Release of radioactivity into the Mediterranean Sea was from these two nuclear weapons destroyed on land.
[c] Co-ordinates are estimates of the location where debris were found.
[d] Estimated. The area where the RTG sank is between latitude 49°59′N and 50°5′5″N, and between longitude 144°3′6″E and the coastline.
[e] Estimated. The location where the Mars' 96 Interplanetary Station fell is a 800 × 200 km area with the centre at 25°6′S, 75°24′W.
[f] Estimated. The area where the RTG sank is within the following coordinates: 54°18′54.18″N, 142°14′59.17″E; 54°19′05.18″N, 142°15′32.37″E; 54°19′19.53″N, 142°14′09.06″E; 54°19′43.51″N, 142°15′21.91″E.

(mainly short-lived radionuclides) were released into the atmosphere. A fraction of these radionuclides was deposited in the waters of the Bay in an area of approximately 0.1 km^2. The total activity of ^{60}Co in bottom sediments of the radioactive contaminated part of the bay was estimated to be 185 GBq (Sivintsev & Vysotskii, 1994; Petrov, 1995).

5.3.2. *Nuclear weapons and military vessels capable of carrying such weapons*

Nuclear weapons have been designed to be carried by submarines, surface ships, aircraft, and rockets. In January 2000 the world stock of nuclear weapons was estimated to be about 20000 (SIPRI, 2001). There are seven recorded accidents listed in the IAEA inventory that have resulted in the confirmed loss of one or more nuclear weapons. In two cases the nuclear weapons were destroyed, spreading radioactive contamination in the environment.

In 1966, a U.S. B-52 bomber and a refuelling tanker collided in mid-air near the village of Palomares in Spain: the B-52 crashed. Of the four thermonuclear weapons carried by the aircraft, one fell into the ocean and was retrieved some months later another was recovered intact from the fields where it had landed, while the remaining two were destroyed on impact. The pyrophoric plutonium metal was ignited, creating a cloud of oxide fume that contaminated soil over an area of 2.26 km^2. The top 10 cm of soil was removed immediately after the accident but some areas remained partially contaminated by plutonium and americium. Heavy rains in the region followed by floods washed a fraction of the contaminated soil into the Mediterranean continental shelf. Analyses of sediment cores collected in the period 1985–1991 in the marine area near Palomares show enhanced levels of transuranics derived from the accident. The input of the Palomares accident into the Mediterranean has been estimated to be a maximum of 1.37 TBq of plutonium (USEPA, 1992; Papucci et al., 1996; Eisenbud & Gesell, 1997).

In 1968, a B-52 bomber crashed on the ice near the Thule airbase in northern Greenland. The bomber carried four nuclear weapons, which were destroyed, spreading plutonium over a large area of the ice. The contamination was mainly confined to the marine environment. The monitoring surveys carried out between 1968 and 1979 showed that most of the activity released is confined to the sediments and the benthic environment within a distance of 50 km of the crash site. The total inventory of plutonium estimated by these surveys was 3.1 TBq (Aarkrog et al., 1987; Smith et al., 1994).

In March 1968, the diesel submarine K-129, which carried two torpedoes with nuclear warheads and three IBM rockets, was lost in Pacific Ocean near the Hawaiian Islands. In August 1974, the bow part of the submarine with nuclear warheads was raised (Office of the President of the Russian Federation, 1993; Shirokorad, 1997).

In cases where the nuclear weapons have sunk intact with the submarine no releases to the environment have been detected in the environment of the submarine. The NS *Scorpion* carried two nuclear weapons. Analysis of sediment, water and marine life samples at the *Scorpion* site using sensitive methods has revealed no evidence of leakage of plutonium from the nuclear weapons (Knolls Atomic Power Laboratory, 1993b, 2000). In the case of the NS *Komsomolets* (Kasatonov, 1996), a detailed survey was carried out around the two nuclear warheads aboard the submarine. To impede washout of plutonium as the warhead corrodes, large holes in the hull of the sunken submarine were covered with special plates to reduce the flow of water through the torpedo compartment.

5.3.3. *Civilian nuclear powered vessels*
Since the launch of the first nuclear powered civilian ship in 1959, eleven additional civilian nuclear powered vessels have been commissioned (Juurmaa & Laukia, 1989; Nielsen & Bøhmer, 1994; Khlopkin & Zotov, 1997). The first nuclear powered civilian ship, the USSR's icebreaker *Lenin* was launched in 1959, and decommissioned in 1990. The first full-scale prototype cargo vessel was the NS *Savannah* commissioned by the USA in 1962 and decommissioned in 1970. In 2001 seven civilian nuclear powered vessels were operational, six of them being icebreakers and one a lighter-carrier.

In the first years of operation, the power on the icebreaker *Lenin* was generated by three nuclear reactors. In 1965, during routine repair, an operational error allowed the core of one of the reactors to be left without water for some period of time. As a consequence a part of the reactor core was damaged. The damaged fuel and all three reactor vessels from which fuel had been removed were subsequently dumped into the Kara Sea (see Section 4). Later two more powerful reactors were installed and the icebreaker operated until decommissioning in 1990 (IAEA, 1997).

5.3.4. *Nuclear energy sources used in spacecraft, satellites and in the deep sea*
Nuclear energy sources are used in some types of spacecraft, satellites and deep sea acoustic signal transmitters for generation of heat or electricity. Two types of nuclear energy sources are available: radioisotope thermoelectric generators (RTGs) and nuclear reactors. In RTGs the most commonly used radionuclide is ^{238}Pu, which has a half-life of 87.7 years. RTGs containing ^{90}Sr (half-life of 28.3 years) have also been used. A typical RTG contains approximately 1 PBq of ^{238}Pu or about 10 PBq of ^{90}Sr. RTGs containing plutonium have been used in deep sea acoustic beacon signal transmissions. Today they are mainly used for outer space missions (NEA/OECD, 1990).

For higher energy demands, nuclear reactors containing up to 90% enriched ^{235}U are used. For example, small nuclear reactors have been extensively used by the Soviet Union in some of the Cosmos series of satellites. More than thirty nuclear powered satellites in the Cosmos series have been launched. At the end of the operational time, the normal procedure is to boost the satellite to a higher orbit, with a lifetime of at least 500 years, to allow for the decay of the fission products before the satellite with its nuclear reactor re-enters the Earth's atmosphere and burns up.

There have been four recorded accidental re-entries of nuclear powered satellites, and three recorded accidental re-entries of a spacecraft. Five of these accidents resulted in the actual or potential release of radionuclides into the environment.

Two United States ICBM Thor Rockets were unsuccessfully launched from Johnston Island in June 1962. The first missile with a nuclear test device was lost from the tracking system and destroyed for security reasons. No nuclear detonation occurred. It is believed that special nuclear materials were vaporised or finely fragmented as a result of the detonation of the high explosive material. No information is available concerning the extent to which, if any, radioactive material entered the marine environment (USEPA, 1992). The second one with a nuclear test device and two experimental re-entry vehicles flew initially on a normal course but after 10,000 m the rocket motor stopped and the missile and warhead were deliberately destroyed. One of the re-entry vehicles, the instrument pod and the missile fell on Johnston Island. A substantial amount of debris fell on to the water around Johnston Island. Approximately 250 pieces of the system, some of which were contaminated with plutonium, were recovered from the lagoon waters around the island (USEPA, 1992).

In 1964 the United States navigational satellite 'Transit 5BN-3' with a SNAP-9A radioisotope generator[3] containing 630 TBq of ^{238}Pu in metallic form failed to achieve orbit and re-entered the atmosphere at 120 km altitude and burned up over the West Indian Ocean north of Madagascar. The nuclear fuel was vaporised during re-entry and was dispersed worldwide (Sandia National Laboratories, 1964; United Nations, 1980; NEA/OECD, 1990).

After completion of its mission in late December 1982, the USSR radar imaging satellite 'Cosmos 1402' failed to boost its nuclear reactor into a higher orbit. The spacecraft was split into three parts. The part containing the reactor core, containing an estimated 1 PBq of fission products, re-entered and broke up over the South Atlantic at about 1600 km east of Brazil in February 1983. It is not known to what extent the reactor core was vaporised during re-entry (NEA/OECD, 1990; Petrov & Schlayhter, 1992).

In 1968 the United States spacecraft Nimbus B-1 containing two SNAP-19 radioisotope generators did not reach orbit due to a booster failure at launch. The booster was destroyed and the spacecraft fell with the generator into the Santa Barbara Channel near to the coast of California. The two capsules containing a total of 1265 TBq of ^{293}Pu were recovered intact from a depth of 100 m (NEA/OECD, 1990; United Nations, 1980).

After a successful launch in April 1970 of the manned spacecraft 'Apollo 13' a malfunction of the oxygen support forced the crew to use the lunar landing module as a survival facility during the flight around the moon and to return to earth with the lunar landing module attached. The landing module, with a SNAP-27 radioisotope generator containing 1.63 Bq of ^{239}Pu, re-entered the atmosphere intact and landed in the deep ocean south of Fiji Islands. No attempt was made to recover the generator from a depth of 6000 m because the exact location was unknown (NEA/OECD, 1990; Sandia National Laboratories, 1964).

The Russian automatic Interplanetary Station 'Mars' 96' was launched in 1996 but as a result of an unsuccessful burn, the booster block entered the Earth's atmosphere and fell into the Pacific Ocean to the West of Chile. The 'Mars' 96' probe contained 18 ^{238}Pu RTGs, with a total activity 174 TBq. Potential local radioactive contamination of the marine environment cannot be excluded (United Nations, 1997a, b).

5.3.5. *Nuclear powered lighthouses*

Lighthouses in Former Soviet Union waters were often powered by radionuclide thermoelectric generators (RTGs), which may contain up to several petabequerels of ^{90}Sr. The Russian Federation is currently the only user of this type power source in lighthouses. It was been estimated that 500 RTGs are in use (NATO, 1995), with a very large total ^{90}Sr activity, perhaps of the order of 5000 PBq. There have been two recorded incidents where RTGs have been lost at sea, both occurring near the eastern coast of Sakhalin Island in the Sea of Okhotsk and involving emergency disposals of the RTGs during transportation by helicopter (Ivanzhin, 1999, 2000; Sivintsev & Kiknadze, 1999). In the first incident, which occurred in 1987, the RTG contained about 25 PBq of ^{90}Sr. The second RTG accident occurred in 1997 and it contained 1.3 PBq of ^{90}Sr.

[3]SNAP = Systems for Nuclear Auxiliary Power.

5.3.6. *Cargos of nuclear material in transit*
Transport by sea is a common practice for radioactive material within the nuclear fuel cycle; these materials include uranium hexafluoride, enriched uranium, spent nuclear fuel and solidified high level waste. The IAEA has defined criteria for packaging design and performance for the various classes of radioactive material and has developed regulations for safe transport of radioactive materials (IAEA, 1996b). There are three records of cargo ship accidents involving radioactive material, but none of them have resulted in recorded release of radionuclides into the environment.

During transportation of encapsulated solid low and medium level radioactive waste with a total activity of 40 Ci ^{90}Sr equivalent,[4] the lighter 'Nikel' was lost in 1978 in a storm, 20 miles north–west of Kolguyev Island in the south-eastern Barents Sea (Office of the President of the Russian Federation, 1993). No attempt has been made to recover the lighter.

In 1984, the cargo carrier 'Mont Louis' collided with a car ferry 20 km off Zeebrugge and sank in shallow waters. Among the cargo, 30 containers of uranium hexafluoride enriched to less than 1% were present. The type 48Y containers were cylindrical and weighed 15 tons each and were all recovered. One container showed signs of being breached. Intensive environmental monitoring was carried out but did not reveal any release of radionuclides to have occurred (Ringot et al., 1987).

In the accident on 24 November 1997 involving the vessel 'MSC Carla', 70 nautical miles off the Azores, north San Miguel Island, Atlantic Ocean, 3 type B packages containing caesium chloride ^{137}Cs sealed sources (total activity: 326 TBq) were lost, no release of activity was reported (IPSN, 1998).

5.3.7. *Sealed radiation sources*[5]
Sealed radiation sources are used widely in the marine environment in association with oil and gas prospecting and extraction. In some instances the logging tool and drill string containing the sealed source becomes stuck in the drill hole and recovery is not feasible. The equipment is generally left in place and the hole is cemented. This results in situations where radioactive material could potentially enter the marine environment.

In general, these losses have occurred deep in the sediment. Radionuclides involved in these losses of sealed sources have included tritium, ^{55}Fe, ^{60}Co, ^{109}Cd, ^{137}Cs, ^{192}Ir, ^{226}Ra, ^{232}Th, ^{241}Am–Be and ^{252}Cf. The nature of the containment as well as the location of the loss is such that, in general, radionuclide release could occur only after a long period of time (NATO, 1995).

Worldwide, more than half a million sealed radiation sources are estimated to be in commercial use (IAEA, 1996b), only a small fraction of them is being used in activities related to marine applications. The current IAEA inventory includes information of more than one

[4]Sr-90 equivalent activities are calculated by converting the gamma radiation dose rate of each waste package using an empirical relationship which is based on the radionuclide content of a standard package and the ratios of the maximum permissible concentration of these radionuclides in drinking water to the maximum permissible concentration of Sr-90.

[5]The term 'sealed radiation source' indicates radioactive material that is either permanently sealed in a capsule or closely bounded and in a solid form. The capsule bonded and in a solid form. The capsule or material of a sealed source shall be strong enough to maintain leak tightness under the conditions of use and wear for which the source was designed, also under foreseeable mishaps (IAEA, 1996a).

hundred losses of sealed sources. However, it is clear that the inventory is far from complete and it is not presented here.

6. Radioactive discharges to the marine environment

6.1. *An historic perspective*

From the mid-1940s, anthropogenic radionuclides were introduced into the marine environment as a result of the nuclear weapons production programmes of countries such as the USA, Former Soviet Union and the United Kingdom.

In the USA, the first two nuclear reactors became operational at Hanford in 1944. Water from the Columbia River was used to cool the reactors and activation products and some fission products were released to the river and from there to the Pacific Ocean at a distance of about 600 km from the plant. Precise information on the early releases is not available; annual releases to the Pacific Ocean in the mid-1960s have been estimated at about 10 PBq (National Academy of Sciences, 1971). In the 1940s and early 1950s the release rates were estimated to be about 2 orders of magnitude lower (Robertson et al., 1973).

At the end of 1948, the Mayak plant in the Russian region of Chelyabinsk started to produce plutonium as part of the Former Soviet Union's nuclear programme. Over the next 2 years about 108 PBq of intermediate and low-level liquid radioactive waste, consisting mainly of medium- and long-lived beta emitting radionuclides, such as ^{90}Sr (12 PBq), ^{95}Zr/Nb (14 PBq), ^{103,106}Ru (28 PBq) and ^{137}Cs (13 PBq) were released into the Techa River (NRPA, 1997), a tributary of the Ob River, which flows into the Kara Sea. From 1951, releases were significantly reduced as a consequence of discharges of intermediate-level liquid radioactive waste being diverted to Lake Karachai and the construction of a series of dams and reservoirs, between 1956 and 1963. In the period between 1962 and 1992 a total of 108 TBq were released into the Techa River (NRPA, 1997). Owing to the distance between the Mayak plant and the Arctic Sea only a small fraction of the more mobile radionuclides discharged from Mayak, such as ^{90}Sr reached the Kara Sea (Christensen et al., 1995). The other two major facilities for the production of nuclear weapons in the Former Soviet Union, Tomsk-7 and Krasnoyarsk-26 started to operate in the 1950s. These two facilities are located on the Tom River, a tributary of the Ob River, and on the Yenisey River. Although reservoirs were built at both sites to collect large volumes of liquid waste, and although low-level liquid wastes were released into the Tom River, deep underground injection has been the main method to store liquid radioactive wastes at these two facilities, resulting in little radioactivity reaching the Arctic Oceans. Tsaturov & Polikarpov (1993) have estimated that the total input of ^{90}Sr to the Arctic Seas by rivers in the period 1961–1994 was 2.4 PBq (1.4 PBq to the Kara Sea). Recent studies on sediment profiles from the Ob Estuary have indicated that the contribution of land-based nuclear installations from the Russian Federation and the Former Soviet Union to the radioactivity in the Arctic Seas is small compared to fallout (Panteleyev et al., 1995).

In the United Kingdom, the nuclear weapons production programme started in the late 1940s with the establishment of the Atomic Energy Research Establishment at Harwell and the construction of the nuclear installations at Windscale (later known as Sellafield) and Springfields. While liquid radioactive effluents from Harwell and Springfields are released into riverine systems (the Thames River and the Ribble Estuary), Windscale was the first major nuclear installation to release liquid radioactive effluents directly into the sea. Discharges of radioactive material into the Irish Sea started in 1952, when the nuclear installation became operational. In that year about 700 TBq of liquid radioactive effluents were discharged to the north-eastern Irish Sea (Gray et al., 1995). Throughout the 1950s, 1960s and 1970s discharges from Sellafield steadily increased to reach a peak in 1975 when 11 PBq were discharged (Gray et al., 1995). The isotopic composition of the effluents released have changed markedly in time; with the quantities of shorter-lived fission product nuclides such as ^{95}Zr + Nb, ^{106}Ru and ^{144}Ce declining steadily since the early 1970s, longer-lived nuclides such as ^{137}Cs peaking in the mid to late 1970s and declining thereafter, and the major transuranics, ^{241}Am and $^{239+240}$Pu, peaking in the early to mid 1970s and also declining thereafter. By the early 1990s, discharges were roughly two orders of magnitude or more lower than they were at their peak (Gray et al., 1995).

Since 1992, with the commissioning of the Enhanced Actinide Removal Plant (EARP), discharges of Pu and Am have been further reduced. However, as a result of the processing of medium-level radioactive liquors previously stored on site and the commissioning of the new Thermal Oxide Reprocessing Plant (THORP) in 1994, there have been increases in the quantities of ^{99}Tc, ^{129}I, ^{60}Co and ^{14}C released. Current annual discharges of low-level radioactive liquid are of the order of 100 TBq. The radionuclides contributing most to the overall releases from Sellafield, excluding ^{3}H, are ^{137}Cs (41 PBq), ^{106}Ru (28 PBq), ^{95}Zr + Nb (24 PBq), ^{241}Pu (22 PBq), ^{90}Sr (6.4 PBq), ^{144}Ce (6.2 PBq) and ^{134}Cs (5.8 PBq); amongst alpha-emitters the main contributors are ^{239}Pu (0.61 PBq), ^{241}Am (0.54 PBq) and ^{238}Pu (0.13 PBq). Figure 8 shows the estimate of the releases of tritium, alpha and beta–gamma emitters from Sellafield for the period 1952–1999.

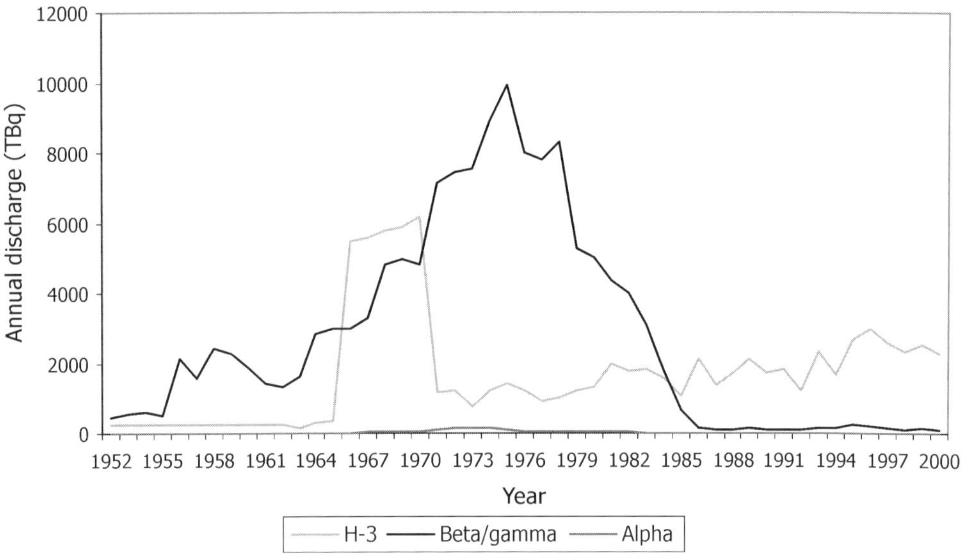

Fig. 8. Annual liquid discharges from Sellafield from 1952 to 2000.

In addition to releases from Sellafield, the most significant contributions to the input of radioactivity into the North European waters are the radioactive discharges from the nuclear fuel reprocessing plants at Cap de la Hague, in France, and to a lesser extent, Dounreay, in United Kingdom. The nuclear fuel reprocessing plant at Cap de la Hague started to operate in 1965. A total of around 19 PBq of low-level liquid radioactive waste, excluding tritium, have been released from Cap de la Hague into the English Channel. The most important contributing radionuclides are ^{106}Ru + Rh (13 PBq), ^{90}Sr (2.2 PBq), ^{125}Sb (1.7 PBq) and ^{137}Cs (1 PBq); current annual releases, excluding ^{3}H, are of the order of 30 TBq. Discharges of ^{3}H have risen steadily since the start of operations at the site; current annual releases are of the order of 10–15 PBq, 3–4 times higher than discharges of the same radionuclides from Sellafield. The first recorded discharges from Dounreay into the North Sea are for 1958. Historical releases of low-level liquid radioactive waste, excluding ^{3}H, total around 11 PBq. The main contributing radionuclides have been ^{95}Nb (3.5 PBq), ^{95}Zr (2.4 PBq), ^{144}Ce (1.8 PBq) and ^{106}Ru (1 PBq); releases of ^{3}H have been low (0.2 PBq in total). Discharges peaked in 1967 when 1.2 PBq of alpha and beta–gamma emitters were released; current annual discharges are less than 1 TBq.

6.2. *Available information on discharges*

Published information on worldwide discharges of radioactive materials to the marine environment is not comprehensive. The United Nations Scientific Committee on the Effects of Atomic Radiation (UNSCEAR) publishes information on releases of radioactive material to the environment from nuclear power plants and nuclear fuel reprocessing plants (United Nations, 1977, 1982, 1988, 1993, 2000), but this information is incomplete. For nuclear power plant only liquid discharges of ^{3}H are reported separately, while discharges of all other radionuclides are generally reported collectively as total alpha or total beta. For reprocessing plants, the information provided includes a partial isotopic composition of the material released. Figures 9a and b show liquid discharges of ^{3}H and other radionuclides from all nuclear power plants from 1975 as reported in UNSCEAR reports. The two graphs show that while releases of radionuclides other than ^{3}H show a downward trend with discharges in the 1990s being typically 3 times lower than those in the 1970s, discharges of ^{3}H have increased over the last quarter of a century and are now a factor of 4–5 higher.

Because of the incompleteness of the information it not possible to determine the radioactive discharges from land based nuclear installations into the marine environment for some regions of the world, such as Central America, the Indian sub-continent, West Asia and the Russian Federation.

More detailed information is available for Western European countries. In the second half of the 1980s the Commission of the European Communities established the MARINA project (CEC, 1990), to estimate the radiological exposure of the population of the European Community from radioactivity in North European waters. As part of this project a comprehensive survey of the discharges of radionuclides into the North Atlantic Ocean was carried out. Data from a total of 72 sites releasing directly or indirectly into the Northern European waters were collected. Information is provided on discharges from the start of the practices until the end of 1984 for all sites.

The MARINA project was followed by two similar projects, MARINA-MED (CEC, 1994) and MARINA-BALT (Nielsen, 2000), during which liquid discharges into the Mediterranean

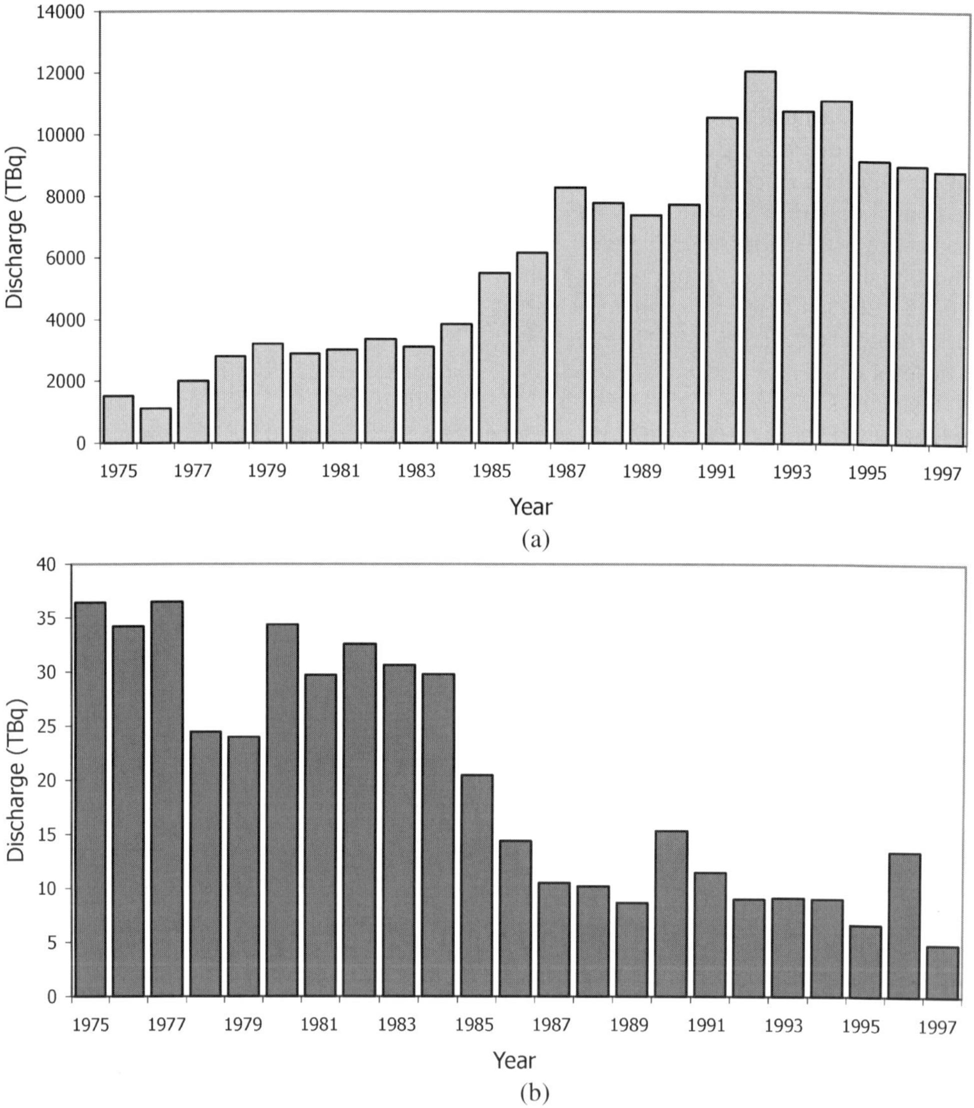

Fig. 9. World-wide annual liquid discharges from nuclear power plants.

Sea and the Baltic Sea from 24 nuclear sites (including some installations already included in the MARINA project) were collected. More recently a follow-up to the MARINA project, called MARINA II (CEC, 2002) has been carried out, which has updated discharges from European nuclear facilities into the North European waters. Table 5 provides a summary of the total liquid discharges by type of facility into the North European Waters up to 2000 as given in the MARINA II report.

Table 5
Percentage contribution of alpha emitters, beta emitters and ^3H to the total liquid discharges from nuclear sites releasing into the Northern European waters

Type of facility	Alpha emitters	Beta emitters	^3H	Total
Reprocessing plants	95.0%	91.4%	78.0%	83.2%
Power stations	0.2%	0.6%	20.4%	12.5%
Research reactors	1.9%	6.5%	1.3%	3.4%
Other (Fuel fabrication)	2.9%	1.4%	0	0.6%
Total (PBq, to 2000)	1.4	150	230	380

References

Aarkrog, A. et al. (1987). Studies of transuranics in an Arctic marine environment. *Journal of Radioanalytical and Nuclear Chemistry, 115* (1), 39–50.

Amundsen, I. et al. (2001). The Kursk accident. Strålevern Rapport 2001:5. Østerås, Norway: Norwegian Radiation Protection Authority.

Baxter, M. S. et al. (1995). Marine radioactivity studies in the vicinity of sites with potential radionuclide releases. *Proc. Int. Symp. on Environmental Impact of Radioactive Releases, 1995.* Vienna: IAEA.

Christensen, G. C., Malyshev, S. V., Strand, P., Salbu, B., Romanov, G. N., Selnæa, T. D., Oughton, D., Glakolenko, Y. U., Amundsen, I., Rudjord, A. L., Bjerk, T. O., Lind, B. (1995). Radioactive contamination in the environment of the nuclear enterprise Mayak PA. Results from the joint Russian–Norwegian field work in 1994. In P. Strand & A. Cooke (Eds), *Environmental Radioactivity in the Arctic.* Østerås, Norway: NRPA.

Commission of the European Communities (1990). The radiological exposure of the population of the European Community from radioactivity in North European Marine waters. Project "Marina". Report EUR-12483-EN. Luxembourg: CEC.

Commission of the European Communities (1994). The radiological exposure of the population of the European Community from radioactivity in the Mediterranean Sea. Project "Marina-Med". Report EUR-15564-EN. Luxembourg: CEC.

Commission of the European Communities, Marina II (2002). Update of the MARINA project on the radiological exposure of the population of the European Community from radioactivity in North European Marine waters. Radiation Protection 132. Luxembourg: CEC.

Danilov Danilyan, V. I. (1993). Statement (LC/16/I/4) made on 10 November 1993 by Mr. V. I. Danilov-Danilyan, Minister of Environmental Protection and Natural Resources of the Russian Federation, at the 16th Consultative Meeting of the Contracting Parties to the London Convention 1972.

Eisenbud, M. & Gesell, T. (1997). *Accidents Involving Military Aircraft Carrying Nuclear Weapons, in Environmental Radioactivity from Natural, Industrial and Military Sources* (4th edn) (pp. 429–432). San Diego, CA: Academic Press.

Eriksen, V. O. (1990). *Sunken Nuclear Submarines*, Oslo, Norway: Norwegian University Press.

Gladkov, G. A. & Sivintsev, Y. (1994). Radiation conditions in the region of the sunken nuclear submarine "Komsomolets", *Atomnaya Energiya, 77* (5), 379–386 (in Russian).

Gray, J., Jones, S. R. & Smith, A. D. (1995). Discharges to the environment from the Sellafield Site. *Journal of Radiological Protection, 1* (2), 99–131.

Institut de Protection de Surete Nucleaire (1998). Communication dated 26 May 1998 from Institut de Protection de Surete Nucleaire, France.

International Atomic Energy Agency (1961). *Radioactive Waste Disposal into the Sea.* Safety Series No. 5. Vienna: IAEA.

International Atomic Energy Agency (1975). Impacts of nuclear releases into the aquatic environment. *Proc. Symp.*, Otaniemi, 1975. IAEA.

International Atomic Energy Agency (1978). *Principles for Establishing Limits for the Release of Radioactive Material into the Environment.* Safety Series No. 45. Vienna: IAEA.

International Atomic Energy Agency (1986). *Definition and Recommendations for the Convention for the Prevention of Marine Pollution by Dumping of Wastes and Other Matter*. Safety Series No. 78. Vienna: IAEA.

International Atomic Energy Agency (1991). Inventory of radioactive material entering the marine environment: Sea disposal of radioactive waste. IAEA-TECDOC-588. Vienna: IAEA.

International Atomic Energy Agency (1996a). *International Basic Safety Standards for Protection Against Ionizing Radiation and for the Safety of Radiation Sources*. Safety Series No. 115. Vienna: IAEA.

International Atomic Energy Agency (1996b). Regulations for the safe transport of radioactive materials (1996 edn). IAEA Safety Standards Series No. ST-1. Vienna: IAEA.

International Atomic Energy Agency (1997). Predicted radionuclide release from marine reactors dumped in the Western Kara Sea. Report of the Source Term Working Group of the International Arctic Seas Assessment Project (IASAP). IAEA-TECDOC-938. Vienna: IAEA.

International Atomic Energy Agency (1998). Radiological conditions of the Western Kara Sea, assessment of the radiological impact of the dumping of radioactive waste in the Arctic Seas. Report on the International Arctic Seas Assessment Project (IASAP). Radiological Assessment Reports Series. Vienna: IAEA.

International Atomic Energy Agency (1999). Inventory of radioactive waste disposal at Sea. IAEA-TECDOC-1105. Vienna: IAEA.

International Atomic Energy Agency (2000). *Regulatory Control of Radioactive Discharges to the Environment*. Safety Standards Series WS-G-2.3. Vienna: IAEA.

International Atomic Energy Agency (2001). Inventory of accidents and losses at sea involving radioactive material. IAEA-TECDOC-1242. Vienna: IAEA.

International Maritime Organization (1972). Final act of the conference with attachment including the convention for the prevention of marine pollution by dumping of wastes and other matter. *International Conference on the Convention on the Dumping of Wastes at Sea*. London 30 October–13 November 1972. London: IMO.

International Maritime Organization (1989). Report of the Twelfth Consultative Meeting of the Contracting Parties to the Convention on the Prevention of Marine Pollution by Dumping of Wastes and Other Matter. LDC 12/16. London: IMO.

Ivanzhin, A. P. (1999). Inventory of radioactive wastes and objects in the pacific resulting from nuclear fleet activities. In *Proc. Int. Sem. on ISTC Projects, Radiation Legacy in the CIS*. Brussels 8–9 March 1999. Luxembourg: CEC.

Ivanzhin, A. P. (2000). Personal communication.

Joint Norwegian–Russian Expert Group for Investigation of Radioactive Contamination in the Northern Areas (1996). Dumping of radioactive waste and investigation of radioactive contamination in the Kara Sea. Results from 3 years of investigations (1992–1994) in the Kara Sea. Østerås, Norway: Norwegian Radiation Protection Authority.

Juurmaa, K. & Laukia, A. (1989). Nuclear-powered merchant ships today and tomorrow. *Nuclear Europe*, *78*, 28–29.

Kasatonov, I. V. (Ed.) (1996). *Russian Navy in Post War Period, in Three Centuries of the Russian Fleet*, Vol. 3 (pp. 208–431) (in Russian).

Khlopkin, N. S. & Zotov, A. P. (1997). Merchant marine nuclear-powered vessels. *Nuclear Engineering and Design*, *173*, 201–205.

Knolls Atomic Power Laboratory (1993a). Deep sea radiological environmental monitoring conducted at the site of the nuclear-powered submarine Thresher sinking. Report KAPL-4748.

Knolls Atomic Power Laboratory (1993b). Deep sea radiological environmental monitoring conducted at the site of the nuclear-powered submarine Scorpion sinking. Report KAPL-4749.

Knolls Atomic Power Laboratory (2000). Deep sea radiological environmental monitoring performed during September 1998 at the sites of the sunken submarines USS Thresher and USS Scorpion. Report KAPL-4842.

National Academy of Sciences (1971). *Radioactivity in the Marine Environment*. Washington, DC: National Academy of Sciences.

Nielsen, S. P. (2000). The Radiological Exposure of the Population of the European Community to Radioactivity in the Baltic Sea. Project "Marina-Balt". Report EUR-19200-EN. Luxembourg: CEC.

Nilsen, T. & Bøhmer, N. (1994). Sources to radioactive contamination in Russian counties of Murmansk and Arkhangelsk. Bellona Report No. 1. Oslo: Bellona Foundation.

North Atlantic Treaty Organization (1995). NATO/CCMS/NACC Pilot Study, Cross-Border Environmental Problems Emanating from Defence-Related Installations and Activities, Vol. 1: Radioactive Contamination. Final Report Phase 1 1993–95. Report 204. Brussels: NATO.

Norwegian Radiation Protection Authority (1997). Joint Norwegian–Russian Expert Group for the investigation of the radioactive contamination in the Northern Areas, Sources contributing to the radioactive contamination of the Techa River and areas surrounding the 'Mayak' Production Association, Urals, Russia. Programme of investigations of possible impacts of the 'Mayak' PA activities on radioactive contamination of the Barents and Kara Seas. Østerås, Norway: NRPA.

Nuclear Energy Agency of the OECD (1974). *Guidelines for Sea Disposal Packages of Radioactive Waste*. Paris: OECD/NEA.

Nuclear Energy Agency of the OECD (1979). *Recommended Operational Procedure for Sea Dumping of Radioactive Waste*. Paris: OECD/NEA.

Nuclear Energy Agency of the OECD (1983). *Interim Oceanographic Description of the North-East Atlantic Site for the Disposal of Low-Level Radioactive Waste*. Paris: OECD/NEA.

Nuclear Energy Agency of the OECD (1985). Review of the continued suitability of '*The Dumping Site for Radioactive Waste in the North-East Atlantic*'. Paris: OECD/NEA.

Nuclear Energy Agency of the OECD (1990). Emergency preparedness for nuclear-powered satellites. Report of an Ad Hoc Meeting. Stockholm, Sweden, 24–26 April 1989. Paris: OECD/NEA.

Office of the President of the Russian Federation (1993). "*White Book*". *Facts and Problems Related to Radioactive Waste Disposal in Seas Adjacent to the Territory of the Russian Federation* (Materials for a Report by the Government Commission on Matters Related to Radioactive Waste Disposal at Sea, Created by Decree No. 613 of the Russian Federation President, October 24, 1992). Moscow: Office of the President of the Russian Federation. English translation by the International Maritime Organization, LC/16/INF.2.

Office of Technology Assessment (1996). An analysis of Arctic and other regional impacts from Soviet nuclear contamination. OTA-ENV-623. Washington, DC: U.S. Government Printing Office.

Panteleyev, G. P., Livingston, H. D., Sayles, F. L. & Medkova, O. N. (1995). Deposition of plutonium isotopes and Cs-137 in sediments of the Ob delta from the beginning of the nuclear age. In P. Strand & A. Cooke (Eds), *Environmental Radioactivity in the Arctic*. Østerås, Norway: NRPA.

Papucci, C. et al. (1996). Time evolution and levels of man-made radioactivity in the Mediterranean sea. In P. Guéguéniat, P. Germain & H. Métivier (Eds), *Radionuclides in the Oceans* (pp. 177–197). Les Ulis, France: Les Éditions de Physique.

Petrov, O. (1995). Radioactive waste and spent nuclear fuel in the navy of Russia. In P. Strand & A. Cooke (Eds), *Proc. of the Second Int. Conf. on Environmental Radioactivity in the Arctic*. Oslo, 21–25 August 1995. Østerås, Norway: Norwegian Radiation Protection Authority.

Petrov, Y. V. & Schlayhter, A. I. (1992). Possible ecological consequences of nuclear powered satellite fall, physics and technology of nuclear reactors (pp. 121–128). St. Petersburg (in Russian).

Petterson, H. B. L. et al. (1995). Radionuclide analysis of samples from the 1994 Japanese–Korean–Russian Expedition to the Sea of Japan. Monaco: IAEA-MEL.

Pettersson, H. B. L. et al. (1996). Radionuclide analysis of samples from the 1995 Japanese–Korean–Russian Expedition to the far Eastern Seas. Monaco: IAEA-MEL.

Ringot, C., Devillers, C. & Warniez, P. (1987). The Mont Louis accident and nuclear safety. In *Proc. Int. Symp. on the Packaging and Transportation of Radioactive Materials*. Davos, 16–20 June 1986. Vienna: IAEA (in French).

Robertson, D. E., Silker, W. B., Langford, J. C., Petersen, M. R. & Perkins, R. W. (1973). Transport and depletion of radionuclides in the Columbia River. In *Proceedings of the Conference on Radioactive Contamination of the Marine Environment* (Seattle, 10–14 July 1972). STI/PUB/313. Vienna: IAEA.

SANDIA National Laboratories (1964). SNAP-9A Incident Summary. SANDIA Report No. Sc-T-64-947. Albuquerque, NM.

Sharpe, R. (Ed.) (1994). *Jane's Fighting Ships, 1994–95* (97th edn). Surrey, UK: Jane's Information Group.

Shirokorad, A. B. (1997). *Soviet Submarines of Postwar Construction*. Moscow: Arsenal Press (in Russian).

Sivintsev, Y. & Kiknadze, O. E. (1998). Inventory of radionuclides in the ship nuclear reactors dumped in the sea of Japan. In *Proceedings of part 2 of an International Symposium on Radionuclides in the Oceans* (RADOC 96-97). April 7–11, 1997. Norwich/Lowestoft.

Sivintsev, Y. & Kiknadze, O. E. (1999). Assessment of potential impact to the environment from radionuclide thermoelectric generator lost in the sea of Okhotsk. In *Proc. Int. Symp. on Marine Pollution*. Monaco, 5–9 October 1998. IAEA-TECDOC-1094. Vienna: IAEA.

Sivintsev, Y. & Vysotskii, V. L. (1994). Radiological consequences of radiation accident on the nuclear submarine at Chazhma Bay, *Atomic Energy*, *76* (2), 158–160.

Smith, J. N. et al. (1994). Sediment mixing and burial of the plutonium-239/240 pulse from the 1968 Thule Greenland nuclear weapon accident. *Journal Environmental Radioactivity*, *25*, 135–159.

Stockholm International Peace Research Institute (2001). *SIPRI Year Book 2000, Armaments, Disarmament and International Security*. Oxford: Oxford University Press.

Tsaturov, Y. & Polikarpov, G. G. (1993). *Sources to Environmental Contamination in the Former USSR*. A. Aarkrog (Ed.). Denmark: RisøNational Laboratory.

United Nations (1977). Sources and effects of ionizing radiation. Report to the General Assembly with Annexes, Scientific Committee on the Effects of Ionizing Radiation (UNSCEAR), UN, New York.

United Nations (1980). Questions relating to the use of nuclear power sources in outer space: United States of America: Working paper. Studies on technical aspects and safety measures of nuclear power sources in outer space, United Nations Committee On the Peaceful Uses of Outer Space (COPUOS). New York: UN.

United Nations (1982). Sources and effects of ionizing radiation. Report to the General Assembly with Annexes, Scientific Committee on the Effects of Ionizing Radiation (UNSCEAR). New York: UN.

United Nations (1988). Sources and effects of ionizing radiation. Report to the General Assembly with Annexes, Scientific Committee on the Effects of Ionizing Radiation (UNSCEAR). New York: UN.

United Nations (1993). Sources and effects of ionizing radiation. Report to the General Assembly with Annexes, Scientific Committee on the Effects of Ionizing Radiation (UNSCEAR). New York: UN.

United Nations (1997a). Report submitted by the Russian delegation to the Thirty-Fourth Session of the Scientific and Technical Subcommittee of the United Nations Committee on the Peaceful Uses of Outer Spaces on the Circumstances Surrounding the Unsuccessful Launch of the Automatic Interplanetary Station Mars' 96, 97-10291 (53). Vienna, February 1997.

United Nations (1997b). Report by the delegation of the Russian Federation to the Thirty-Fourth Session of the Scientific and Technical Subcommittee of the United Nations Committee on the Peaceful Uses of Outer Spaces, Radiation Safety of Radionuclide Power Sources from the Mars' 96 Space Probe, 97-10292 (53). Vienna, February 1997.

United Nations (2000). Sources and effects of ionizing radiation. Report to the General Assembly with Annexes, Scientific Committee on the Effects of Ionizing Radiation (UNSCEAR). New York: UN.

United States Department of the Navy (1984). *Final Environmental Impact Statement on the Disposal of Decommissioned Defueled Naval Submarine Reactor Plants*, Vol. 1. Washington, DC: Dept. of the Navy.

United States Environmental Protection Agency (1992). Letter dated 15 October 1992 from Robert S. Dyer, Deputy Director, Radiation Studies Division, Office of Radiation and Indoor Air, Washington, DC.

United States National Oceanic and Atmospheric Administration (1995). Communication from United States National Oceanic and Atmospheric Administration. USNOAA.

MARINE RADIOACTIVITY
Hugh D. Livingston (Editor)

Chapter 5

Reactive radionuclides as tracers of oceanic particle flux

Michael P. Bacon

Woods Hole Oceanographic Institution, Woods Hole, MA 02543, USA

1. Introduction

Radionuclides and other substances that are strongly bound to marine particulate matter are commonly referred to as 'particle-reactive' or, simply, 'reactive'. Because of this property, their distribution and transport in the ocean are strongly influenced by particle flux, and hence, a number of reactive radionuclides have found important uses as tracers of particle transport. The best examples are cases of radioactive disequilibrium within the natural radioactive decay series brought about and maintained by removal of reactive decay products by uptake on particles and subsequent sedimentation. This chapter will draw primarily upon those examples but will also note important recent work on the cosmogenic nuclides. The fate of anthropogenic radionuclides, particularly the transuranic nuclides, can also be strongly influenced by particle transport, but this group of nuclides is well covered in other chapters of this book and will not receive attention here.

For a thorough coverage of the earlier literature on this topic and other aspects of the natural decay series nuclides in the ocean, the reader is referred to the review by Cochran (1992) and other chapters in the compendium on uranium-series disequilibrium edited by Ivanovich & Harmon (1992). I do not purport here to give a comprehensive literature review, but I have attempted to provide a representative coverage of some of the most important work published over approximately the past decade.

2. Chemical scavenging in the sea

2.1. *The vertical dimension*

The term 'chemical scavenging' or, simply, 'scavenging' refers to the uptake of reactive substances by particulate matter and their subsequent removal from the water column by sedimentation. Knowledge of this process is the basis for all of the oceanographic applications of

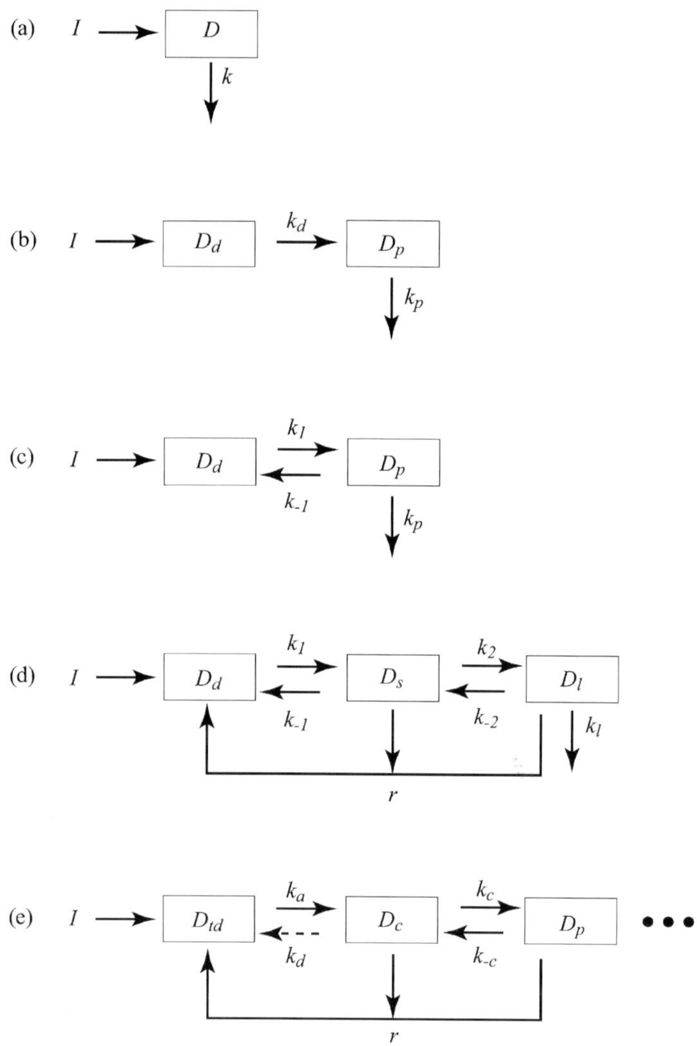

Fig. 1. Models, in order of increasing complexity, that have been used to describe the process of chemical scavenging in the oceanic water column. See text for explanation and definitions of symbols.

particle-reactive radionuclides that have been developed. The clearest and most quantifiable demonstrations of scavenging in the ocean are provided by various radioactive disequilibrium within the natural decay series, in each of which a reactive daughter nuclide is deficient in seawater relative to the concentration it would have in radioactive equilibrium with its parent. The process can be described and quantified by reservoir models of varying complexity, as illustrated in Fig. 1. In each of the models, supply of tracer to a 'dissolved' reservoir is indicated by I and, in the case of a decay-series nuclide produced by in situ decay, is equal to λP, where λ is the radioactive decay constant of the *daughter* and P is the concentration of the parent nuclide in seawater in *activity* units. Although it is not shown explicitly in the figure, a loss term applies to each reservoir and is of the form λD, where D is the concentration of the daughter nuclide in activity units.

The most rudimentary model (Fig. 1a) does not include particles explicitly but can be used to determine the turnover rate k (or its inverse, the residence time τ) of a scavenged daughter radionuclide in the water column from observation of the degree of disequilibrium in a sample of seawater. At steady state, supply by radioactive decay of the parent must be balanced by scavenging and radioactive decay of the daughter, so we can write

$$\lambda P = \lambda D + kD, \tag{1}$$

which can be solved for k:

$$k = \lambda\left(\frac{P}{D} - 1\right). \tag{2}$$

This is the turnover rate of the daughter element by the pathway of chemical scavenging alone and is the same for all isotopes of an element. The corresponding residence time τ is simply $1/k$. The residence time is a useful quantity, for it indicates the characteristic time for response of a tracer concentration to a perturbation. It is important to recognize that the total turnover rate of a scavenged radionuclide, $\lambda + k$, is faster than this, depending on the radioactive decay rate, and the corresponding turnover time, $1/(\lambda + k)$, is shorter. Thus a very long-lived radionuclide, such as ^{230}Th, with a half-life of 75,000 years, has a turnover time in the oceanic water column that is very much shorter (about 20 years) because of its rapid scavenging and can be influenced by processes such as thermohaline ventilation on decadal time scales (see below).

In studies where filtration is used to determine the distribution of reactive radionuclides between dissolved and particulate forms, a slightly more sophisticated model (Fig. 1b) can be applied and used to quantify the particle sinking rates and turnover times (Krishnaswami et al., 1976; Bacon et al., 1976; Coale & Bruland, 1985). We write the steady-state mass balance equations for both dissolved and particulate forms of the daughter:

$$\lambda P = \lambda D_\mathrm{d} + k_\mathrm{d} D_\mathrm{d} \tag{3}$$

and

$$k_\mathrm{d} D_\mathrm{d} = \lambda D_\mathrm{p} + k_\mathrm{p} D_\mathrm{p}, \tag{4}$$

where the subscripts “d” and “p” refer to the dissolved and particulate forms of the daughter nuclide, k_d is the turnover rate of the dissolved form of the daughter, and k_p is the turnover rate of the particulate form. It is important to keep in mind that, in all of these models, the reservoirs are highly idealized. In practice D_d and D_p must be defined operationally by the filter that is used on a sample of real seawater (typically with effective pore size in the 0.2–1.0 μm range), which contains a complex mixture of particles with continuous distributions of size and composition.

The solution of equation (3) for k_d has exactly the same form as equation (2). Elimination of the term $k_d D_d$ by addition of equations (3) and (4) yields the solution for k_p:

$$k_p = \frac{\lambda[P - (D_d + D_p)]}{D_p}, \tag{5}$$

which states simply that the disequilibrium between parent and daughter (the difference between production and decay of the daughter in both dissolved and particulate forms) is maintained entirely by removal of the reactive daughter on particles ($k_p D_p$). Measurement of a vertical profile allows particle sinking velocity to be estimated. Production of the tracer at any depth z must be balanced by the sum of radioactive decay and the vertical flux gradient. For the case of particle sinking velocity S constant with depth

$$S\frac{\partial D_p}{\partial z} = \lambda\big[P - (D_d + D_p)\big]. \tag{6}$$

When D is produced throughout the water column, as in the case of U decay products, D_p increases with depth. For the long-lived decay products ^{230}Th and ^{231}Pa, the decay term $\lambda(D_d + D_p)$ in equation (6) is negligible, and the solution for D_p becomes a simple linear function of depth:

$$D_p = \frac{\lambda P}{S} z, \tag{7}$$

and S can be determined from the slope of a depth profile of D_p, as shown in Fig. 2.

Observations that not only the particulate but also the dissolved form of ^{230}Th increases with depth in the oceanic water column (Moore, 1981; Nozaki et al., 1981) pointed to the need to treat the uptake by particles as a reversible process (Bacon & Anderson, 1982), and measurements of two or more isotopes of the same element but with differing half-lives (^{234}Th and ^{230}Th, for example) can be used with the model illustrated in Fig. 1c to estimate rates of both uptake and release by the particulate matter. The steady-state mass balance between supply and loss is, for the dissolved form,

$$\lambda P + k_{-1} D_p = \lambda D_d + k_1 D_p, \tag{8}$$

and for the particulate form,

$$k_1 D_d = \lambda D_p + k_{-1} D_p + S\frac{\partial D_p}{\partial z}. \tag{9}$$

Equations of the form of equation (8) can be written for different isotopes of the same element, assumed to have the same uptake and release rates but having different values for λ, and solved simultaneously for the two unknowns k_1 and k_{-1}. Results based on Th isotope distributions in the deep ocean indicate reaction times that are short (a few months) compared to the residence time of suspended particles (a few years), implying an approximate state of equilibrium with respect to the exchange between dissolved and particulate forms. In terms of equation (9), this

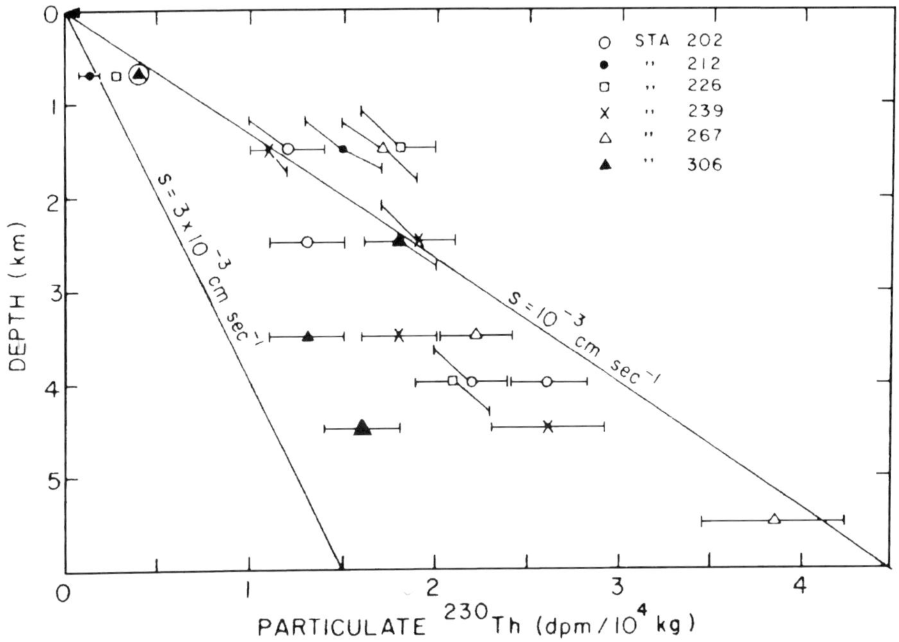

Fig. 2. Determination of the effective sinking velocity of filtered particles based on ^{230}Th measurements made during the Geochemical Ocean Sections (GEOSECS) program. 10^{-3} cm sec^{-1} = 315 m a^{-1}. (From Krishnaswami et al., 1976.)

means that the last term, the particle sinking term, is relatively small and can be neglected for purposes of estimating the distribution of D between dissolved and particulate forms. Thus we can obtain

$$\frac{D_p}{D_d} = \frac{k_1}{\lambda + k_{-1}}. \tag{10}$$

For long-lived radionuclides ($\lambda \ll k_{-1}$), the distribution is simply k_1/k_{-1}, which, when divided by the concentration of suspended particles, gives a distribution coefficient K_d, the quantity most commonly used to compare the particle-reactivities of different elements (Anonymous, 1985).

Particle sinking speeds based on ^{230}Th in filtered particles and equation (6) are of order hundreds of meters per *year* (Fig. 2), contrary to much evidence showing that most of the downward transport of particles in the ocean is at much greater speeds, of order tens or hundreds of meters per *day*, and it became clear that at least two classes of particles must be considered for an adequate description of radionuclide scavenging (Bacon et al., 1985). Fig. 1d is derived from ideas developed by McCave (1975). Small particles are considered to make up the bulk of the particle mass but sink very slowly and contribute little to the downward transport. Large particles, on the other hand, are considered to be responsible for most of the downward transport but are rare and contribute little to the standing stock of particulate matter in the water column. The two particle reservoirs correspond roughly with what is sampled by filtration (small particles) and by sediment traps (large particles) and are thereby defined operationally. (See also Clegg & Whitfield, 1990, for further discussion, in the context of a chemical scavenging model, of the justification for the separation into a sinking and a non-sinking class of particles.) Here the use of the two transfer coefficients k_2 and k_{-2} suggests a reversible exchange between the two particle size classes by processes of aggregation and disaggregation, which may be biologically mediated. As with the reversible uptake and release by small particles, measurement of different isotopes of Th (^{234}Th, ^{230}Th, ^{228}Th) in the two size classes of particle allow both the forward and reverse process to be quantified, because of the different decay constants (Nozaki et al., 1987; Clegg & Whitfield, 1991; Clegg et al., 1991; Cochran et al., 1993, 2000; Murnane et al., 1994, 1996). Also indicated in Fig. 1d (and Fig. 1e) is the possible return of the daughter nuclide to the dissolved reservoir by remineralization (r) of the particles as well as by desorption (k_{-1}).

The uptake and release of reactive tracers by suspended particles, parameterized by k_1 and k_{-1} in Figs 1c and d, have often been loosely referred to as 'adsorption' and 'desorption', as though the arbitrary size cutoff imposed by the filter used to separate D_d and D_p matched a discontinuity that actually existed in nature. A significant aim of many studies over the past 10–15 years has been to develop a better understanding and description of the scavenging process at the sub-micron level. These studies, again based mostly on Th, have favored the view that the uptake k_1 is better described as a process involving coagulation of colloids, to which Th may be irreversibly bound, a view that was first expounded by Tsunogai & Minagawa (1978). The more recent work has received its impetus from two important observations: Santschi et al. (1986) pointed out that the numerical values being reported for k_1 were much too slow to be explained by adsorption of a truly dissolved species at particle surfaces but were consistent with coagulation of colloids being the mechanism for moving scavenged metals up the size spectrum into particles that are retained by filters; and Honeyman et al. (1988) showed that distribution coefficients (K_d) of Th based on oceanic field data have a significant dependence on particle concentration, also inconsistent with a simple adsorption process but consistent with coagulation of colloids. Honeyman & Santschi (1989) developed a theory, based on what they called 'Brownian pumping', to explain these observations. Measurements of ^{234}Th in colloidal matter collected by cross-flow ultrafiltration were first reported by Baskaran et al. (1992) and by Moran & Buesseler (1992). These and subsequent studies (Moran & Buesseler, 1993; Huh & Prahl, 1995; Niven et al., 1995; Greenamoyer & Moran, 1997; Dai & Benitez-Nelson, 2001) showed that a significant fraction of the total ^{234}Th can occur in the colloidal fraction and indicated the need for explicit inclusion of a third, colloidal, class of particles in scavenging models, as shown in Fig. 1e, from which colloid turnover times can be estimated. An especially thorough and helpful review of these studies is provided by Guo & Santschi (1997).

2.2. *Horizontal transport*

The foregoing discussion has depicted scavenging in the sea as a one-dimensional process in which the downward flux of a radionuclide could be inferred simply from its degree of disequilibrium with its parent in the water column. A number of lines of evidence, however, have shown that such a simple vertical mass balance does not always hold and that horizontal fluxes must often be invoked to explain observed distributions and fluxes of scavenged nuclides.

The earliest studies of ^{234}Th/^{238}U and ^{228}Th/^{228}Ra disequilibrium in surface seawater showed much stronger depletions of the Th isotopes in coastal waters than in open-ocean surface waters (Bhat et al., 1969; Broecker et al., 1973), which can be attributed to the higher particle concentrations and fluxes over the continental shelf. The observed concentration gradients (higher concentrations offshore) indicated that net shoreward fluxes of the Th isotopes must occur. A similar phenomenon has been shown for ^{210}Pb (e.g. Nozaki et al., 1976). The trend toward higher rates of scavenging, leading to shoreward accumulation of reactive radionuclides, extends even into estuaries. Olsen et al. (1989) found ^{210}Pb inventories in the Savannah River Estuary far exceeding what could be supplied from the watershed and arrived at the perhaps counter-intuitive conclusion that there is a net supply of ^{210}Pb from the ocean to the estuary.

Studies of ^{210}Pb, following the initial discovery of its disequilibrium relative to ^{226}Ra (Craig et al., 1973), showed that significant horizontal gradients of a reactive nuclide are also maintained in the deep sea as a result of higher particle fluxes and scavenging rates at the ocean margins (Bacon et al., 1976). The term 'boundary scavenging' was introduced to refer to processes that result in the removal of reactive chemical substances at ocean margins at rates much greater than within the ocean interior (Spencer et al., 1981). The importance of boundary scavenging for ^{210}Pb has been confirmed by a number of subsequent studies, for example, in the Bay of Bengal (Cochran et al., 1983; Sarin et al., 2000). The strength of boundary scavenging will presumably be dependent on productivity and particle flux, and eastern boundaries may in general be stronger sinks than western boundaries (Cochran, 1992). Study of sediment inventories of ^{210}Pb over large parts of the North Pacific and North Atlantic tended to confirm the inference from the water-column distributions by showing a correlation between scavenging efficiency and productivity (Cochran et al., 1990).

Significant advances in our appreciation of the importance of boundary scavenging have resulted from studies of ^{231}Pa and the contrast in its behavior with that of ^{230}Th (Anderson et al., 1983; Nozaki & Nakanishi, 1985; Nozaki & Yamada, 1987). The two nuclides have a common source, being produced in seawater from the decay of isotopes of U, which is a conservative element and thus uniformly dispersed throughout the oceans (Chapter 1). The ratio of ^{231}Pa production to ^{230}Th production is rigidly fixed at a value of 0.093 (activity ratio) by the ^{235}U/^{234}U ratio in seawater. Early attempts to use the ^{231}Pa/^{230}Th ratio as a tool for absolute age dating of sediments (see Ku, 1976, for a review), however, revealed that pelagic surface sediments usually had ratios lower than this, often by a factor of 2 or 3. The discovery of the boundary scavenging effect for ^{210}Pb suggested that ^{231}Pa might by similarly affected (Bacon et al., 1976), and the work of Anderson et al. (1983) confirmed this by showing low ^{231}Pa in the vertical flux of particles collected by sediment traps at mid-ocean sites and by showing significant horizontal ^{231}Pa concentration gradients indicating a net transport to the margins. Measurements in ocean margin sediments showed surface values of the ^{231}Pa/^{230}Th ratio substantially higher than the production-rate ratio, in contrast with the abyssal sediments underlying the ocean interior, an effect that is most strongly manifested in the Pacific Ocean (Yang et al., 1986; Walter et al., 1999), as illustrated in Fig. 3.

This pattern is also reflected very clearly in data from deep-sea sediment traps, as shown in Fig. 4: fluxes of ^{230}Th measured in traps are close to those expected from a simple vertical flux

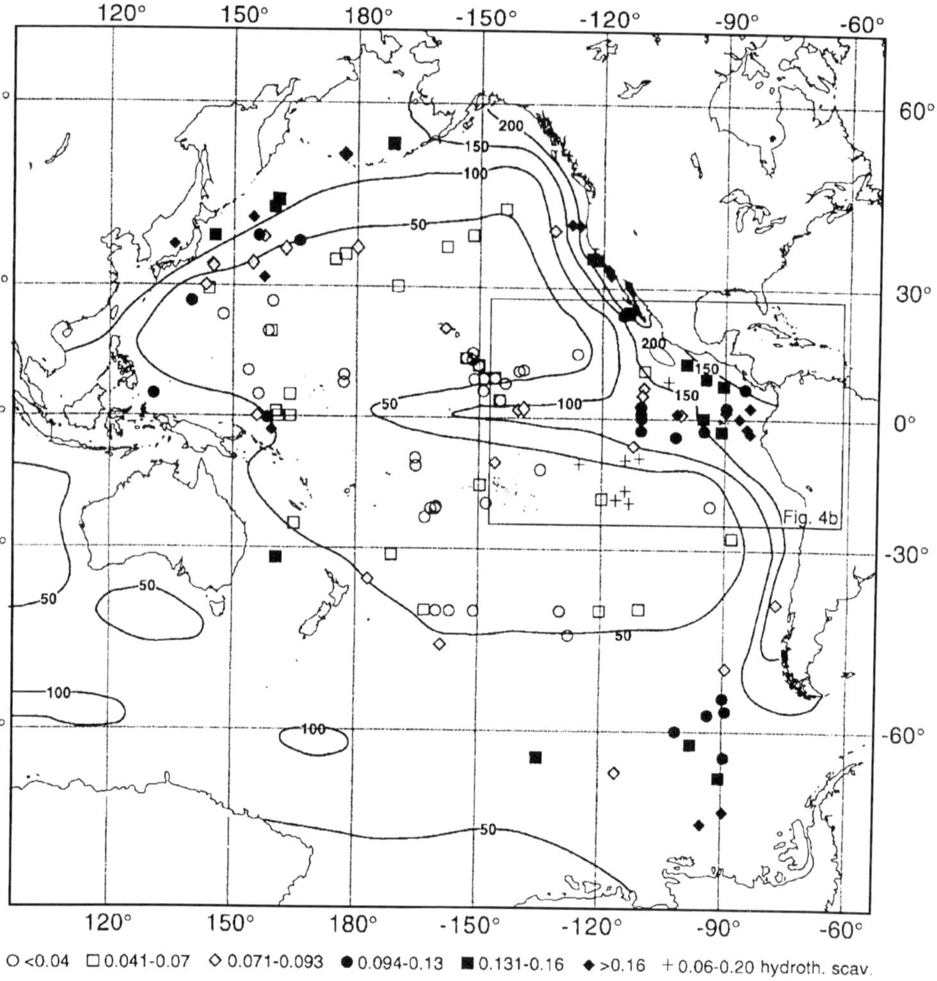

Fig. 3. Map of $xs^{231}Pa/xs^{230}Th$ activity ratios measured in surface sediments of the Pacific Ocean. The expected ratio, based on the $^{235}U/^{234}U$ ratio in seawater is 0.093. The higher ratios at the ocean margins and lower ratios in the ocean interior are due to boundary scavenging of ^{231}Pa, though the extremely low ratios found in the central Pacific, where sediment accumulation rates are very low, may also be influenced by older sediment brought to the surface by bioturbation. Superimposed are contours of surface primary productivity in g C m^{-2} a^{-1}. (From Walter et al., 1999.)

balance and are only weakly dependent on particle flux; fluxes of ^{231}Pa, on the other hand, are significantly lower than expected in areas of low particle flux and higher than expected in areas of high particle flux.

In considering the large-scale fractionation between ^{230}Th and ^{231}Pa in the ocean we can envision two transport pathways for removal by scavenging, one with a strong vertical component due to the particle flux and the other with a strong horizontal component due to the intensified uptake at the margins (boundary scavenging). Differential partitioning between the two pathways explains the observed fractionation between ^{230}Th and ^{231}Pa: ^{230}Th, because

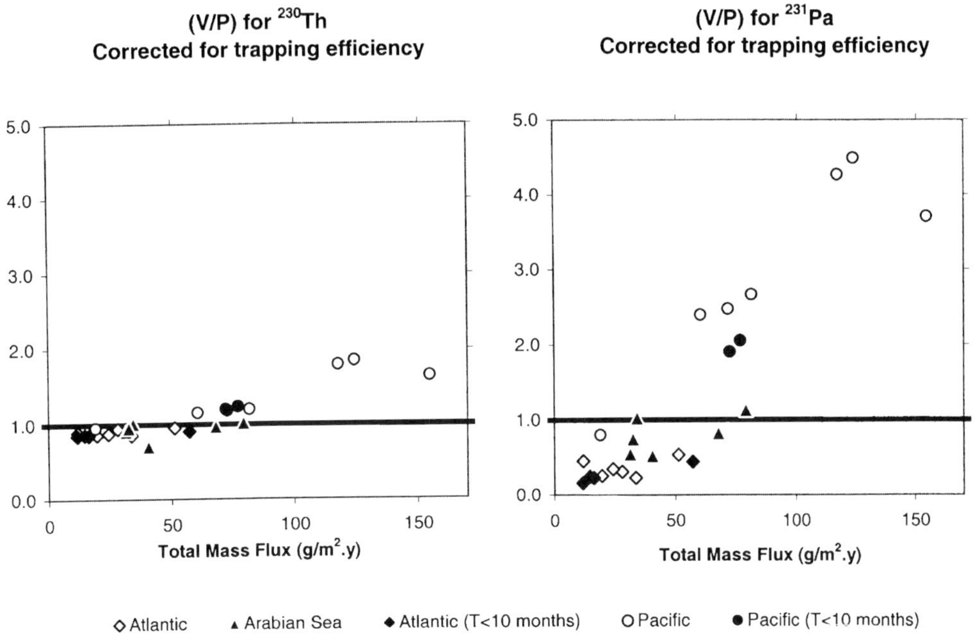

Fig. 4. Ratios of measured (V) to expected (P) vertical fluxes of ^{230}Th and ^{231}Pa in deep-sea sediment traps. Measured fluxes are corrected for trapping efficiency according to Yu et al. (2001a). (From Yu et al., 2001a, but excluding Southern Ocean data, for which trapping efficiency could not be determined.)

of its very strong particle-reactivity, tends to be dominated by the vertical pathway; ^{231}Pa, on the other hand, because it is somewhat less reactive and has a longer oceanic residence time, is dominated more by the horizontal pathway and boundary scavenging (Bacon, 1988). A similar kind of fractionation may occur with the cosmogenic radionuclide pair ^{10}Be/^{26}Al, with ^{26}Al being the more reactive and dominated by the vertical pathway, like ^{230}Th, and ^{10}Be being less reactive and more subject to boundary scavenging, like ^{231}Pa (Bacon, 1988; Anderson et al., 1990; Lao et al., 1992b, 1993; Ku et al., 1995; Kumar et al., 1995; Rutsch et al., 1995; Wang et al., 1996; Luo et al., 2001).

Ventilation of sub-surface levels in the water column by the sinking and lateral advection of water that was recently at the sea surface is another process that can have a significant impact on the distribution and flux of reactive radionuclides. A parcel of water having an initial tracer concentration that is characteristic of surface water will evolve at depth toward a concentration that is governed by the rate of supply by in-situ production and the rate of loss by scavenging and radioactive decay. Since a number of reactive radionuclides (^{210}Pb, ^{230}Th, ^{231}Pa) have deep-water residence times on the order of decades to a century or two, their distributions can be affected by ventilation on those time scales.

Cochran et al. (1987) observed that ^{230}Th profiles in the northwest Atlantic departed from the straight linear increase with depth predicted by the simple vertical scavenging model of Fig. 1c and attributed the departure to the effects of ventilation just described. There have been several subsequent observations of this effect on vertical profiles (Scholten et al.,

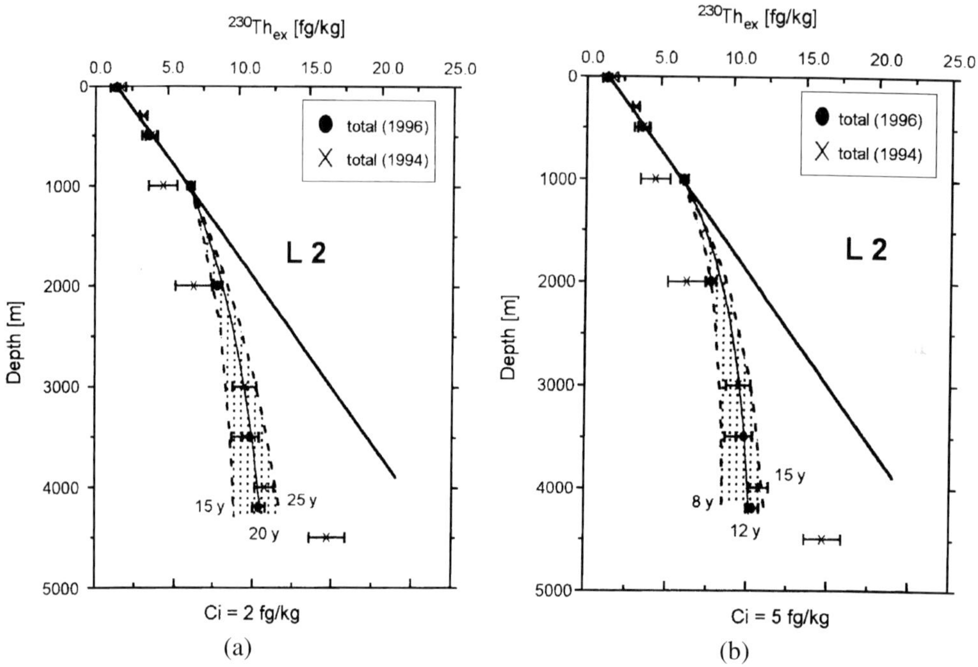

Fig. 5. ^{230}Th profile measured in the eastern North Atlantic compared with predictions of a scavenging-mixing model for two different assumed values of the initial ^{230}Th concentration C_i. The straight line is the expected profile in the absence of ventilation. (From Vogler et al., 1998.)

1995; Moran et al., 1997a; Vogler et al., 1998). A scavenging-mixing model first proposed by Rutgers van der Loeff & Berger (1993), which includes the effect of ventilation, has been shown successfully to explain the observed profiles of ^{230}Th (Fig. 5). On the other hand, ^{230}Th and ^{231}Pa profiles have been interpreted as indicating slow ventilation of Canada Basin Deep Water in the Arctic Ocean (Edmonds et al., 1998). On a larger scale, it has been shown that the thermohaline circulation of the Atlantic Ocean leads to a net export of about 10–20% of the ^{230}Th and ~45% of the ^{231}Pa production (Yu et al., 1996).

An important development of the last few years is the inclusion of metal scavenging in two-dimensional and three-dimensional ocean models to study the effects of lateral processes on the transport of reactive radionuclides and their fluxes to the sediments (Henderson et al., 1999; Igel & von Blanckenburg, 1999; von Blanckenburg & Igel, 1999; Marchal et al., 2000). Henderson et al. (1999) used an ocean general circulation model to study the extent to which horizontal processes can cause departures from the balance that would be predicted between ^{230}Th production in the water column and its deposition in the underlying sediments. Their model-derived maps indicate that about 70% of the ocean floor receives a ^{230}Th flux within 30% of that expected from production. However, the ^{230}Th flux can be as little as 40% of the expected in areas of extremely low particle production and as much as 160% in highly productive regions.

3. Applications

3.1. *Export production*

The flux of particles out of the surface layer of the sea is of fundamental ecological importance. It causes nutrients and carbon to be lost from the euphotic zone and sequestered at depth. Direct measurement of particle flux with sediment traps is difficult in the upper ocean, and significant efforts have been made over the past 10 years or so to develop indirect methods of estimating particle flux based on mass balances of reactive radionuclides. Most of the effort has been devoted to a single nuclide, ^{234}Th, a choice based on its strong reactivity and its suitable half-life (24 days).

The basis of the ^{234}Th method is the disequilibrium that exists between ^{234}Th and ^{238}U in the euphotic zone (Fig. 6), first discovered by Bhat et al. (1969), who attributed it to scavenging by sinking particles. Their observation was confirmed by more extensive measurements of Matsumoto (1975). Later Coale & Bruland (1985) measured ^{234}Th in both dissolved (in the sense of Fig. 1b) and particulate forms and showed that Th uptake rate and particle removal rate, calculated from equations (3) and (4), are both correlated with biological productivity. Eppley (1989) first suggested explicitly that a ^{234}Th method could be developed as a tool for

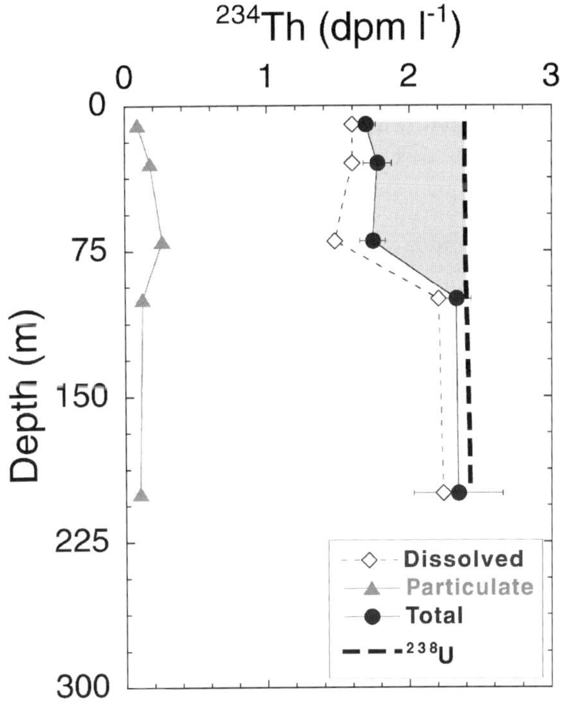

Fig. 6. Profile of ^{234}Th concentrations during an iron fertilization experiment at a station in the Southern Ocean showing the typical deficiency relative to ^{238}U in the euphotic zone (upper 100 m). (From Charette & Buesseler, 2000.)

measuring export production (actually new production, which he equated with export production), reasoning that the particulate Th removal rates (or residence times) could be applied to the standing stock of particulate organic carbon (POC) in the euphotic zone to yield a POC export flux. Murray et al. (1989), however, tested this idea by comparing POC and particulate Th residence times in the equatorial Pacific and found significant differences, which they attributed to differential recycling efficiencies for C and Th, seemingly invalidating the Eppley approach. However, a modification was suggested by Buesseler et al. (1992), who reasoned that the difficulty could be minimized by focusing on the large, rapidly sinking particles at the base of the euphotic zone. They outlined a procedure based on measurement of the C/^{234}Th ratio in sinking particles combined with the integrated deficit of ^{234}Th in the euphotic zone that would yield an estimate of particulate C flux (see also Bacon et al., 1996).

At any depth z, the particulate flux F_D of ^{234}Th must, at steady state, account for the integrated deficit in the overlying water column:

$$F_D = \lambda \int_0^z (P - D_t)\,dz, \tag{11}$$

where D_t is the total (dissolved plus particulate) concentration of ^{234}Th. Thus we can quantify the export of ^{234}Th, and from this the export of carbon, if we can determine the C/^{234}Th ratio in the sinking particles. This method was first applied by Buesseler et al. (1992) during the North Atlantic Bloom Experiment (NABE) of the Joint Global Ocean Flux Study (JGOFS). They used material collected in sediment traps (Martin et al., 1993) to represent the sinking material, in effect ‘calibrating’ the traps with ^{234}Th. The C and N fluxes based on the ^{234}Th method were significantly higher than those based on the uncorrected trap measurements, suggesting significant undertrapping had occurred. The ^{234}Th method was used extensively during JGOFS and in other studies, and there are now published results from several regions (Buesseler et al., 1995, 1998; Cochran et al., 1995; Shimmield et al., 1995; Murray et al., 1996; Moran et al., 1997b; Rutgers van der Loeff et al., 1997; Charette & Moran, 1999; Charette et al., 1999; Hall et al., 2000; Benitez-Nelson et al., 2001; Buesseler et al., 2001a). A compilation of export fluxes based on ^{234}Th was published by Buesseler (1998), who compared them with primary production data. The results indicated that over much of the ocean export/production ratios are low, $<10\%$, indicating efficient recycling of organic matter. Exceptions were associated with blooms, most often where large phytoplankton, particularly diatoms, dominated.

Although most of the attention has been given to ^{234}Th, other reactive nuclides can in principle be used in the same way to constrain estimates of export flux. Indeed Fisher et al. (1988) demonstrated the converse, that the fluxes of several natural and anthropogenic radionuclides could be estimated with reasonable accuracy from knowledge of new (export) production and the concentration factors of the radionuclides in marine plankton. Use of disequilibrium involving the longer lived Th isotopes, ^{230}Th/^{234}U and ^{228}Th/^{228}Ra, has been explored by Luo et al. (1995). An important advantage of the longer-lived isotopes is that the estimates of particle flux may be extended to depths below the euphotic zone and permit rates of remineralization to be determined, but the complicating effects of advection and diffusion may also be more significant than they are for ^{234}Th. The ^{210}Po/^{210}Pb disequilibrium has been the subject of numerous studies since it was first reported (Shannon et al., 1970), but only a few studies have attempted to derive estimates of particle flux from it (Shimmield et al., 1995). A recent study by Friedrich & Rutgers van der Loeff (2002) using both the ^{234}Th/^{238}U and the ^{210}Po/^{210}Pb disequilibrium illustrates the potential power of combining different tracers with different chemical properties. They found that ^{210}Po in the Antarctic Circumpolar Current was preferentially associated with organic particles but that ^{210}Pb and ^{234}Th were less selectively distributed over both organic and siliceous particles. This difference is consistent with evidence that ^{210}Po is taken up by active biological transport as well as by passive adsorption, in contrast with ^{210}Pb and ^{234}Th, for which active transport is much less significant (Fisher et al., 1983). Simultaneous solution of the transport equations for both ^{210}Po and ^{234}Th allowed fluxes of organic and siliceous matter to be separately determined (Friedrich & Rutgers van der Loeff, 2002).

The ^{234}Th method has proven to be successful in avoiding the uncertainties associated with the use of sediment traps in the upper ocean, but it is important to realize that the method has uncertainties of its own. Some of the uncertainties relate to the simplifying assumptions underlying equation (11), which are

(1) that a simple vertical flux balance holds and
(2) that the system is at steady state.

Neither assumption is strictly necessary if measurements are made on appropriate time and space scales. Time-series measurements can be made and a non-steady-state term $(\partial D_t/\partial t)$ added to equation (11), and a grid of stations can be taken for the estimation of contributions due to horizontal transport (see Buesseler et al., 1994, 1995). The number of measurements and additional information required can multiply rather quickly, though advances in sampling and measurement technique (Buesseler et al., 1995, 2001b) have helped to compensate for this.

Perhaps the most difficult uncertainty to deal with, however, is the large variability over the particle size spectrum in the C/^{234}Th ratio, which raises the question whether any sampling method based on filtration can capture a sample of particulate matter that truly reflects the C/^{234}Th ratio in the sinking flux. Recent studies have begun to address this important issue (Burd et al., 2000), but much remains to be done. An especially striking observation relating to this problem was recently reported by Buesseler et al. (2000). They compared the performance of a newly designed neutrally buoyant sediment trap (NBST; Valdes & Price, 2000) with that of the more traditional particle interceptor trap (PIT; Knauer et al., 1979) at a station off Bermuda. The NBST is designed to move with the parcel of water and thus to eliminate flow relative to the trap, in principle also eliminating hydrodynamic bias. Their results for a deployment at 150 m gave total mass, carbon, and nitrogen fluxes that were similar between the two types of trap but fecal pellet and ^{234}Th fluxes that were factors of 2 and 3 lower in the NBST than in the PIT. The C_{org}/^{234}Th ratio in the PIT was a factor of 3.8 lower than it was in the NBST.

Applications of the ^{234}Th method have been focused primarily on the export of particulate organic carbon, but in principle the method can be applied to any chemical substance Y through use of the measured Y/^{234}Th ratio and the calculated ^{234}Th flux. Fluxes of calcium carbonate (Bacon et al., 1996) and opal (Buesseler et al., 2001a) in the upper ocean have been determined in this way. Gustafsson et al. (1997a, b) have determined particulate organic pollutant fluxes based on ^{234}Th. In general, however, the ^{234}Th method as applied to other chemical substances remains underutilized and there remain many unrealized opportunities to exploit it.

3.2. *Sediment trap calibration*

Sediment traps (particle interceptor traps) are widely used for the direct measurement of sinking particle flux in the ocean. Because of the flow of water relative to a trap, it is possible for collection bias to occur (Butman et al., 1986; Butman, 1986; Knauer & Asper, 1989; Gardner, 2000). Several reactive radionuclides have been recognized as potential tracers to evaluate this bias and provide an in situ calibration of traps (Knauer et al., 1979; Brewer et al., 1980; Lorenzen et al., 1981; Moore et al., 1981; Bacon et al., 1985; Bacon, 1996; Gardner, 2000). They include the Th isotopes (^{234}Th, ^{228}Th and ^{230}Th), ^{210}Pb, ^{210}Po and ^{231}Pa. All of these tracers are supplied to the oceanic water column by decay of parent nuclides at rates that can be exactly known, and an expected flux can be determined by measurement of water column profiles and equation (11). Comparison of measured with expected fluxes can indicate whether a significant collection bias has occurred and give a measure of the trapping efficiency. I will discuss first the application to moored sediment traps in the deep ocean and then to the problem of flux measurements in the upper ocean.

Bottom-tethered sediment traps are frequently used to measure the supply of sinking particles to the seafloor, and for evaluating their performance the longer-lived daughter nuclides ^{210}Pb, ^{230}Th and ^{231}Pa, which show deficiencies throughout the entire water column, are the appropriate tracers to use. (In the case of ^{210}Pb, an additional supply from the atmosphere would have to be considered.) Of these the most important is ^{230}Th, because, as we have seen, it is the least affected by boundary scavenging and horizontal redistribution, so that the vertical flux balance (equation (11)) is most likely to hold. In the case of ^{230}Th or ^{231}Pa, P is nearly constant throughout the water column, and the decay term λD_t is negligible, so equation (11) simplifies to

$$F_D = \lambda P z. \tag{12}$$

Thus the expected flux of ^{230}Th or ^{231}Pa is simply proportional to depth in the water column.

It is important to recognize that large short-term departures from this simple balance can occur because of the variability of particle flux on seasonal and other time scales, and as a minimum a year-long record of measured ^{230}Th flux is needed for an appropriate application of equation (12). It is also necessary to recognize that the vertical mass balance does not hold exactly even for ^{230}Th and to consider the effects of horizontal transport on the ^{230}Th budget, although they are usually fairly small. As we have seen, the horizontal transport due to boundary scavenging is more significant for ^{231}Pa, and Anderson et al. (1983) suggested a method of correcting for the effect of it by simultaneously satisfying material balances for both ^{230}Th and ^{231}Pa. In a recent study employing this method, Yu et al. (2001b) determined annual average fluxes of ^{230}Th and ^{231}Pa collected by sediment traps deployed in a variety of locations throughout the oceans. Results from 10 traps deployed in the bathypelagic zone (>1200 m depth) in open ocean regions of the Atlantic, Pacific and Indian Oceans gave an average trapping efficiency of 98% with a standard deviation of 14%, thus validating the general use of baffled, conical sediment traps in such regions, where current velocity is usually low. At shallower depths, within the mesopelagic zone, trapping efficiency tended to be lower and more erratic, even in areas of low current velocity ($<10\ \text{cm s}^{-1}$), and it was suggested that the depth-dependent difference in trapping efficiency reflects changes in the hydrodynamic properties of particles undergoing multiple cycles of aggregation and disaggregation as they move downward through the water column. Similar results were found in a separate study by Scholten et al. (2001).

For determining fluxes in the upper ocean, where currents are usually swifter than in the deep ocean, sediment traps are often deployed in a drifting mode. This can reduce the flow relative to the trap that would cause hydrodynamic collection bias, but it does not eliminate it completely. The shorter-lived radionuclides ^{234}Th, ^{228}Th and ^{210}Po are the most appropriate tracers for validating traps deployed in the upper ocean, and the one used most extensively is ^{234}Th, which is produced by decay of the ^{238}U in seawater. As we have already seen, a deficit of ^{234}Th relative to ^{238}U is usually observed in the euphotic zone (Fig. 6), and from this it is a simple matter to determine the export of ^{234}Th on sinking particles at the base of the euphotic zone for comparison with the flux measured in a sediment trap from equation (11). Buesseler (1991) reviewed all of the published data that allowed this comparison to be made and found that measured fluxes of ^{234}Th differed substantially from the expected fluxes, often by more than a factor of 3, suggesting large hydrodynamic biases. Both positive and negative biases (over- and under-trapping) were observed but tended always to be in the same direction for multiple deployments within the same study.

Use of ^{234}Th in the surface ocean is subject to the same considerations discussed for the use of ^{230}Th in the deep ocean: temporal and spatial variations in particle flux can cause estimates based on a simple steady-state vertical flux balance to be in error. These factors, which could not be quantified, might explain some of the discrepancies noted by Buesseler (1991). Recognizing this, Buesseler et al. (1994) carried out a three-dimensional time-series study of ^{234}Th distributions and fluxes in the Sargasso Sea off Bermuda. Results showed consistent overtrapping during the 4-day period of the study. Evidence from long-term measurements at the same site, on the other hand, suggested that the traps underestimate the annual flux (Michaels et al., 1994), and Buesseler et al. (1994) suggest that the traps overestimate flux during low flux periods and underestimate it during high flux periods.

3.3. *Nepheloid layers*

In many parts of the ocean, especially in regions of strong western boundary currents, layers of elevated turbidity known as bottom nepheloid layers (BNL) are found, where increased quantities of particulate matter are kept in suspension to heights of as much as 1000 m above the seafloor (McCave, 1986). Processes within the BNL are clearly of importance for the transport of sediment and the patterns of sedimentation on the seafloor, and the BNL may also be a site where there are increased rates of a variety of biogeochemical processes, including chemical scavenging (Rutgers van der Loeff & Boudreau, 1997).

There are now several reports showing that a ^{234}Th/^{238}U disequilibrium exists in deep-sea bottom waters where a BNL is present (Bacon & Rutgers van der Loeff, 1989; DeMaster et al., 1991; Turnewitsch & Springer, 2001) but not where a BNL is absent (Amin et al., 1974; Bacon & Rutgers van der Loeff, 1989). Elevated concentrations of ^{234}Th in the suspended particles in the BNL compared with concentrations in surface sediment, together with the disequilibrium, make it possible to estimate the residence time of resuspended particles in the BNL in much the same way that particle residence times in the surface ocean are estimated. For the total ^{234}Th D_t in the BNL we have:

$$\frac{\partial D_t}{\partial t} = \lambda(P - D_t) - k_p C_p (X - X_s), \quad (13)$$

where:

C_p is the total suspended matter (TSM) concentration in the BNL,
X is the concentration of ^{234}Th in the TSM (activity per unit mass),
X_s is the concentration of ^{234}Th in the particles that are resuspended from the sediment surface,

and the other symbols are as defined previously.

At a station in the western North Atlantic with a well-developed BNL, Bacon & Rutgers van der Loeff (1989) found depletions of ^{234}Th in samples from 25 and 64 m above the seafloor amounting to 19% of the equilibrium value and concluded that local resuspension of particles must be a significant factor in maintaining the observed disequilibrium. Particles in the nepheloid layer were greatly enriched in ^{234}Th compared to values observed in surface sediments, and application of equation (13) gave an estimated residence time of resuspended particles in the BNL of 25 days. DeMaster et al. (1991) found much larger depletions of ^{234}Th (60% and more of the equilibrium value) in the bottom 100 m of the water column at the site of the High Energy Benthic Boundary Layer Experiment (HEBBLE) in the North Atlantic, a location where benthic storms have been observed. The most detailed study reported to date is that of Turnewitsch & Springer (2001) for the Porcupine Abyssal Plain in the eastern North Atlantic (Fig. 7). They compared ^{234}Th fluxes based on the integrated ^{234}Th deficit in the bottom water with those based on sediment traps and the integrated ^{234}Th excess in the surface sediments. A large imbalance was observed: the loss of ^{234}Th from the water column could not be accounted for by deposition in the underlying sediment. They accounted for the difference by lateral transport to sites of scavenging at surrounding topographic highs. From this study the inadequacy of a simple one-dimensional steady-state model of such a dynamic system becomes very apparent. Further advances will require carefully designed time-series observations at a grid of stations. The advantages of including other appropriate tracers, such as ^{228}Th and ^{210}Po (Rutgers van der Loeff & Boudreau, 1997) should also be explored.

3.4. *Hydrothermal scavenging*

The hydrothermal vent fluids that discharge from the seafloor at mid-ocean ridges differ markedly in their chemical properties from ambient seawater: they are acidic, anoxic, sulfide-bearing and trace-element-enriched. Upon venting, they mix with cold, well-oxygenated bottom waters, causing precipitation of various metal sulfide and oxide phases. Most of the precipitation and oxidation reactions occur in buoyant hydrothermal plumes that rise hundreds of meters above the seafloor until they reach a level at which neutral buoyancy is attained. Anomalies in temperature, chemical composition, and particle concentration due to hydrothermal plumes can be detected many kilometers away from a ridge crest. Because

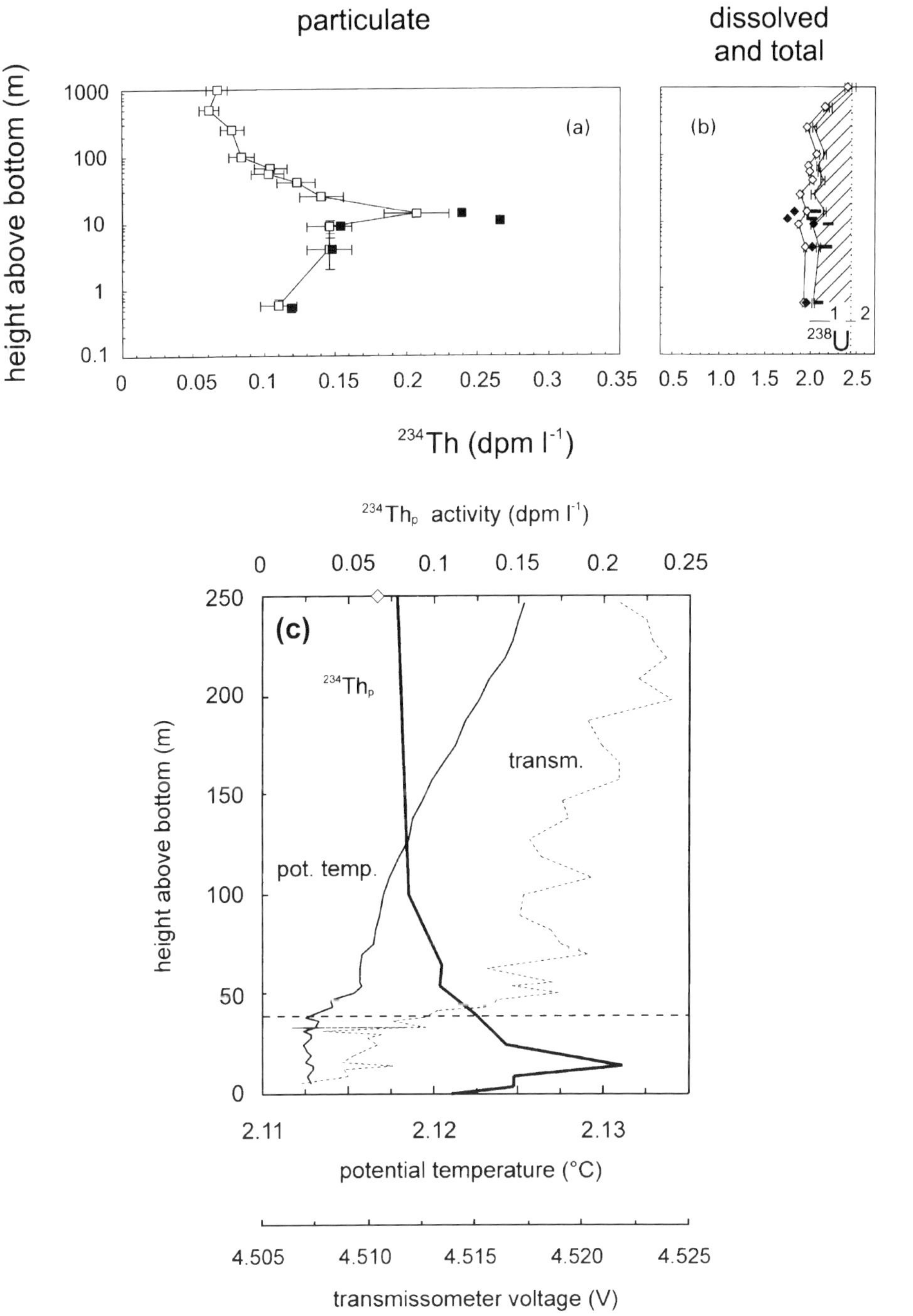

Fig. 7. Profiles showing high concentrations of particulate ^{234}Th (a), a depletion of dissolved and total ^{234}Th (b) and the relationship to profiles of light transmission and potential temperature (c) in the bottom nepheloid layer at a station in the eastern North Atlantic. (From Turnewitsch et al., 2001.)

of the particle-rich environment and the formation of fresh particle surfaces, hydrothermal plumes are likely sites of intensified scavenging of reactive elements from seawater.

Following earlier hints of intensified scavenging of ^{210}Pb near mid-ocean ridge systems (Bacon et al., 1976; Cochran et al., 1983), Kadko et al. (1986–1987) undertook deliberate sampling at a site on the East Pacific Rise to observe the effects of a hydrothermal plume on the distribution of ^{210}Pb and ^{210}Po in the water column. Samples from the plume were found to have ^{210}Pb/^{226}Ra activity ratios ranging from 0.09 to 0.35, among the lowest ever observed in the oceanic water column and indicative of intensified scavenging of ^{210}Pb from the plume. Unusually rapid scavenging of ^{210}Po in the plume was also indicated. These observations were extended by Kadko (1993) at sites on the Endeavour Ridge and the Juan de Fuca Ridge, and subsequent studies of plumes and metalliferous deposits have yielded evidence of intensified scavenging of other reactive radionuclides, including ^{234}Th, ^{230}Th, ^{228}Th, ^{231}Pa and ^{10}Be (German et al., 1991, 1997). Kadko (1993) used ^{210}Pb results from the Endeavor Ridge site to estimate the global effect of scavenging in hydrothermal plumes.

3.5. *Paleoflux and paleoproductivity*

The particle-reactive radionuclides have had a long history of application to the geochronology of marine sediments and the determination of sediment accumulation rates, and several good reviews of the established methods based on uranium-series and other nuclides are available (Goldberg & Bruland, 1974; Ku, 1976; Turekian & Cochran, 1978; Ivanovich et al., 1992; Huh & Kadko, 1992). More recent work has been concerned less with chronology per se and more with the interpretation of radionuclide profiles in sediment cores to determine past variability in the flux of particles to the seafloor.

As we have seen, the evidence from deep-sea sediment traps (Fig. 4) and from the results of global ocean modeling (Henderson et al., 1999) show that the flux of particulate ^{230}Th to the seafloor is nearly in balance with its rate of production in the water column over large parts of the ocean, so that equation (12) can be assumed to hold to a good approximation in many places. Furthermore, because of the long residence time of uranium in the ocean (Cochran, 1992), it is not likely that the production rate of ^{230}Th has varied significantly over the past few hundred thousand years. These considerations have led to the development of ^{230}Th as a constant-flux reference tracer against which variations in the fluxes of other sedimentary components can be measured (Bacon, 1984; Suman & Bacon, 1989; Francois et al., 1990; Thomson et al., 1995a). In practice, a depth profile of excess ^{230}Th (i.e. the ^{230}Th in the sediment that is not supported by ^{234}U) in a sediment core is measured, and each data point is corrected for radioactive decay to the time of deposition, which is based on an independent chronology established, for example, through radiocarbon dating or oxygen-isotope stratigraphy. The mass flux at any depth in a core is then inversely proportional to the age-corrected excess ^{230}Th ($\mathrm{xs}^{230}\mathrm{Th}_0$), or in general for any sedimentary component i the flux is given by

$$F_i = \lambda P z \frac{C_i}{\mathrm{xs}^{230}\mathrm{Th}_0}. \tag{14}$$

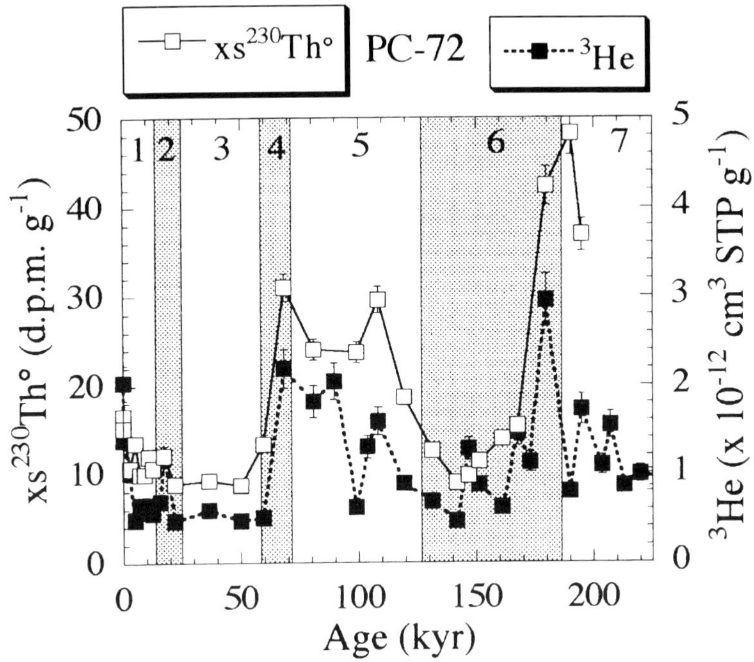

Fig. 8. Profiles of $xs^{230}Th_0$ and ^3He versus age in a core from the equatorial Pacific Ocean. The strong correlation suggests that the variations in ^3He concentration are due mainly to variable dilution and not to variations in the rate of ^3He supply. (From Marcantonio et al., 1996.)

An important consequence of normalizing to $xs^{230}Th_0$ is that variations in sediment accumulation due to variable focusing by redistribution due to bottom currents are effectively canceled out.

The past ten years have seen increasing application of the ^{230}Th profiling method to studies of carbonate, opal, and clay sedimentation (Yang et al., 1990; Francois & Bacon, 1991; Francois et al., 1993; McManus et al., 1998; Thomson et al., 1999; Hall & McCave, 2000), pulsed inputs of ice-rafted debris known as Heinrich events (Francois & Bacon, 1994; Thomson et al., 1995b, 1999; McManus et al., 1998) the flux of cosmogenic nuclides such as ^{10}Be (Anderson et al., 1990; Frank et al., 1995), and the flux of interplanetary dust particles to Earth as recorded by the ^3He content in deep-sea sediments (Marcantonio et al., 1995, 1996, 1998, 1999, 2001). Figure 8 shows the strong correlation that exists between the ^3He and $xs^{230}Th_0$ in two cores from the equatorial Pacific, which suggests that the varying concentrations of both are determined by variable dilution. From these and other profiles, it has been concluded that the accretion rate of extraterrestrial ^3He is fairly constant and that previously variations in ^3He burial rate reflect primarily the variations in sediment focusing due to redistribution by bottom currents and do not reflect systematic changes in the flux of ^3He (Marcantonio et al., 2001).

The ^{230}Th profiling method provides estimates of past fluxes only to the extent that a component is preserved in the sediment. For example, a calcium carbonate paleoflux determined by this method is the net flux (carbonate rain to the seafloor minus dissolution after deposition). The correlation between the ^{231}Pa flux and particle flux (Fig. 4), which is one of the main factors involved in the observed boundary scavenging of ^{231}Pa, has led to the use of the ^{231}Pa/^{230}Th ratio in sediments as an indicator of past changes in biological productivity, or particle export flux, of surface waters. The theory is that the ratio is preserved in the sediment even if the biogenic phases (organic matter, carbonate, opal) are remineralized (Anderson et al., 1983; Bacon, 1988; Lao et al., 1992a, b; Francois et al., 1993; Kumar et al., 1993, 1995; Yang et al., 1995). However, the ^{231}Pa/^{230}Th ratio in the particle flux depends on other variables such as horizontal transport (Bacon, 1988; Rutgers van der Loeff & Berger, 1993; Yu et al., 1996; Walter et al., 1997) and particle composition (Anderson et al., 1983; Shimmield & Price, 1988; Rutgers van der Loeff & Berger, 1993; Walter et al., 1997; Luo & Ku, 1999). Walter et al. (1999) have reviewed the limitations of the ^{231}Pa/^{230}Th method and concluded that the conditions under which it applies are best met in the Pacific Ocean. It has been argued that the ^{10}Be/^{26}Al ratio may prove to be a more reliable paleoproductivity indicator (Luo et al., 2001), though many of the same considerations apply to its use.

4. Conclusion

The value of the particle-reactive radionuclides as ocean tracers for the study of oceanic particle flux, both present and past, has been well demonstrated, and there has been impressive growth in their use over the past decade. Until recently, their measurement has been laborious, and progress in the development of applications has been data-limited. Recent advances in measurement technique, such as inductively-coupled plasma mass spectrometry (ICP-MS) for the longer-lived U- and Th-series nuclides or accelerator mass spectrometry (AMS) for the cosmogenic nuclides, are making the accumulation of larger data sets more feasible, and the incorporation of chemical scavenging processes in large-scale ocean models is allowing the extrapolation of results beyond the immediate observations. It is likely that the current phase of rapid growth in this area of research will extend well into the future.

Acknowledgments

My research on reactive radionuclides over the years has been made possible by the generous financial support of the U.S. Department of Energy (most recently by its Ocean Carbon Sequestration Research Program, Biological and Environmental Research (BER), grant DE-FG02-00ER63020) and the U.S. National Science Foundation (most recently by its Chemical Oceanography Program, grant OCE-0117922).

References

Amin, B. S., Krishnaswami, S. & Somayajulu, B. L. K. (1974). ^{234}Th/^{238}U activity ratios in Pacific Ocean bottom waters. *Earth and Planetary Science Letters*, *21*, 342–344.

Anderson, R. F., Bacon, M. P. & Brewer, P. G. (1983). Removal of ^{230}Th and ^{231}Pa at ocean margins. *Earth and Planetary Science Letters*, *66*, 73–90.

Anderson, R. F., Lao, Y., Broecker, W. S., Trumbore, S. E., Hofmann, H. J. & Wolfli, W. (1990). Boundary scavenging in the Pacific Ocean: A comparison of ^{10}Be and ^{231}Pa. *Earth and Planetary Science Letters*, *96*, 287–304.

Anonymous (1985). *Sediment K_ds and Concentration Factors for Radionuclides in the Marine Environment* (73 pp.). Vienna: IAEA.

Bacon, M. P. (1984). Glacial to interglacial changes in carbonate and clay sedimentation in the Atlantic Ocean estimated from ^{230}Th measurements. *Isotope Geoscience*, *2*, 97–111.

Bacon, M. P. (1988). Tracers of chemical scavenging in the ocean: Boundary effects and large-scale chemical fractionation. *Philosophical Transactions of the Royal Society of London A*, *325*, 147–160.

Bacon, M. P. (1996). Evaluation of sediment traps with naturally occurring radionuclides. In V. Ittekkot, P. Schäfer, S. Honjo & P. J. Depetris (Eds), *Particle Flux on the Ocean* (pp. 85–90). Chichester: Wiley.

Bacon, M. P. & Anderson, R. F. (1982). Distribution of thorium isotopes between dissolved and particulate forms in the deep sea. *Journal of Geophysical Research*, *87*, 2045–2056.

Bacon, M. P. & Rutgers van der Loeff, M. M. (1989). Removal of thorium-234 by scavenging in the bottom nepheloid layer of the ocean. *Earth and Planetary Science Letters*, *92*, 157–164.

Bacon, M. P., Spencer, D. W. & Brewer, P. G. (1976). ^{210}Pb/^{226}Ra and ^{210}Po/^{210}Pb disequilibrium in seawater and suspended particulate matter. *Earth and Planetary Science Letters*, *32*, 277–296.

Bacon, M. P., Huh, C.-A., Fleer, A. P. & Deuser, W. G. (1985). Seasonality in the flux of natural radionuclides and plutonium in the deep Sargasso Sea. *Deep-Sea Research*, *32*, 273–286.

Bacon, M. P., Cochran, J. K., Hirschberg, D., Hammar, T. R. & Fleer, A. P. (1996). Export flux of carbon at the equator during the EqPac time-series cruises estimated from ^{234}Th measurements. *Deep-Sea Research II*, *43*, 1133–1153.

Baskaran, M., Santschi, P. H., Benoit, G. & Honeyman, B. D. (1992). Scavenging of thorium isotopes by colloids in seawater of the Gulf of Mexico. *Geochimica et Cosmochimica Acta*, *56*, 3375–3388.

Benitez-Nelson, C., Buesseler, K. O., Karl, D. M. & Andrews, J. (2001). A time-series study of particulate matter export in the North Pacific Subtropical Gyre based on ^{234}Th:^{238}U disequilibrium. *Deep-Sea Research I*, *48*, 2595–2611.

Bhat, S. G., Krishnaswamy, S., Lal, D., Rama & Moore, W. S. (1969). ^{234}Th/^{238}U ratios in the ocean. *Earth and Planetary Science Letters*, *5*, 483–491.

Brewer, P. G., Nozaki, Y., Spencer, D. W. & Fleer, A. P. (1980). Sediment trap experiments in the deep North Atlantic: Isotopic and elemental fluxes. *Journal of Marine Research*, *38*, 703–728.

Broecker, W. S., Kaufman, A. & Trier, R. M. (1973). The residence time of thorium in surface sea water and its implications regarding the rate of reactive pollutants. *Earth and Planetary Science Letters*, *20*, 35–44.

Buesseler, K. O. (1991). Do upper-ocean sediment traps provide an accurate record of particle flux? *Nature*, *353*, 420–423.

Buesseler, K. O. (1998). The decoupling of production and particulate export in the surface ocean. *Global Biogeochemical Cycles*, *12*, 297–310.

Buesseler, K. O., Bacon, M. P., Cochran, J. K. & Livingston, H. D. (1992). Carbon and nitrogen export during the JGOFS North Atlantic Bloom Experiment estimated from ^{234}Th:^{238}U disequilibrium. *Deep-Sea Research*, *39*, 1115–1137.

Buesseler, K. O., Michaels, A. F., Siegel, D. A. & Knap, A. H. (1994). A three dimensional time-dependent approach to calibrating sediment trap fluxes. *Global Biogeochemical Cycles*, *8*, 179–193.

Buesseler, K. O., Andrews, J. A., Hartman, M. C., Belastock, R. & Chai, F. (1995). Regional estimates of the export flux of particulate organic carbon derived from thorium-234 during the JGOFS EqPac program. *Deep-Sea Research II*, *42*, 777–804.

Buesseler, K. O., Ball, L., Andrews, J., Benitez-Nelson, C., Belastock, R., Chai, F. & Chao, Y. (1998). Upper ocean export of particulate organic carbon in the Arabian Sea derived from thorium-234. *Deep-Sea Research II*, *45*, 2461–2487.

Buesseler, K. O., Steinberg, D. K., Michaels, A. F., Johnson, R. J., Andrews, J. E., Valdes, J. R. & Price, J. F. (2000). A comparison of the quantity and composition of material caught in a neutrally buoyant versus surface-tethered sediment trap. *Deep-Sea Research I*, *47*, 277–294.

Buesseler, K. O., Ball, L., Andrews, J., Cochran, J. K., Hirschberg, D. J., Bacon, M. P., Fleer, A. & Brzezinski, M. (2001a). Upper ocean export of particulate organic carbon and biogenic silica in the Southern Ocean along 170°W. *Deep-Sea Research II*, *48*, 4275–4297.

Buesseler, K. O., Benitez-Nelson, C., Rutgers van der Loeff, M., Andrews, J., Ball, L., Crossin, G. & Charette, M. A. (2001b). An intercomparison of small- and large-volume techniques for thorium-234 in seawater. *Marine Chemistry*, *74*, 15–28.

Burd, A. B., Moran, S. B. & Jackson, G. A. (2000). A coupled adsorption-aggregation model of the POC/^{234}Th ratio of marine particles. *Deep-Sea Research I*, *47*, 103–120.

Butman, C. A. (1986). Sediment trap biases in turbulent flows: Results from a laboratory flume study. *Journal of Marine Research*, *44*, 645–693.

Butman, C. A., Grant, W. D. & Stolzenbach, K. D. (1986). Predictions of sediment trap biases in turbulent flows: A theoretical analysis based on observations from the literature. *Journal of Marine Research*, *44*, 601–644.

Charette, M. A. & Buesseler, K. O. (2000). Does iron fertilization lead to rapid carbon export in the Southern Ocean? *Geochemistry, Geophysics, Geosystems*, *1*. Paper number 2000GC000069.

Charette, M. A. & Moran, S. B. (1999). Rates of particle scavenging and particulate organic carbon export estimated using ^{234}Th as a tracer in the subtropical and equatorial Atlantic Ocean. *Deep-Sea Research II*, *46*, 885–906.

Charette, M. A., Moran, S. B. & Bishop, J. K. B. (1999). ^{234}Th as a tracer of particulate organic carbon export in the subarctic Northeast Pacific Ocean. *Deep-Sea Research II*, *46*, 2833–2861.

Clegg, S. L. & Whitfield, M. (1990). A generalized model for the scavenging of trace metals in the open ocean – I. Particle cycling. *Deep-Sea Research*, *37*, 809–832.

Clegg, S. L. & Whitfield, M. (1991). A generalized model for the scavenging of trace metals in the open ocean – II. Thorium scavenging. *Deep-Sea Research*, *38*, 91–120.

Clegg, S. L., Bacon, M. P. & Whitfield, M. (1991). Application of a generalized scavenging model to thorium isotope and particle data at equatorial and high-latitude sites. *Journal of Geophysical Research*, *96*, 20655–20670.

Coale, K. H. & Bruland, K. W. (1985). ^{234}Th:^{238}U disequilibrium within the California Current. *Limnology and Oceanography*, *30*, 22–33.

Cochran, J. K. (1992). The oceanic chemistry of the uranium- and thorium-series nuclides. In M. Ivanovich & R. S. Harmon (Eds), *Uranium-Series Disequilibrium: Applications to Earth, Marine, and Environmental Sciences* (pp. 334–395). Oxford: Clarendon Press.

Cochran, J. K., Bacon, M. P., Krishnaswami, S. & Turekian, K. K. (1983). ^{210}Po and ^{210}Pb distributions in the central and eastern Indian Ocean. *Earth and Planetary Science Letters*, *65*, 433–452.

Cochran, J. K., Livingston, H. D., Hirschberg, D. J. & Surprenant, L. D. (1987). Natural and anthropogenic radionuclide distributions in the Northwest Atlantic Ocean. *Earth and Planetary Science Letters*, *84*, 135–152.

Cochran, J. K., McKibbin-Vaughan, T., Dornblaser, M. M., Hirschberg, D., Livingston, H. D. & Buesseler, K. O. (1990). ^{210}Pb scavenging in the North Atlantic and North Pacific Oceans. *Earth and Planetary Science Letters*, *97*, 332–352.

Cochran, J. K., Buesseler, K. O., Bacon, M. P. & Livingston, H. D. (1993). Thorium isotopes as indicators of particle dynamics in the upper ocean: Results from the JGOFS North Atlantic Bloom Experiment. *Deep-Sea Research I*, *40*, 1569–1595.

Cochran, J. K., Barnes, C., Achman, D. & Hirschberg, D. J. (1995). Thorium-234/Uranium-238 disequilibrium as an indicator of scavenging rates and particulate organic carbon fluxes in the Northeast Water Polynya, Greenland. *Journal of Geophysical Research*, *100*, 4399–4410.

Cochran, J. K., Buesseler, K. O., Bacon, M. P., Wang, H. W., Hirschberg, D. J., Ball, L., Andrews, J., Crossin, G. & Fleer, A. (2000). Short-lived thorium isotopes (^{234}Th, ^{228}Th) as indicators of POC export and particle cycling in the Ross Sea, Southern Ocean. *Deep-Sea Research II*, *47*, 3451–3490.

Craig, H., Krishnaswami, S. & Somayajulu, B. L. K. (1973). ^{210}Pb-^{226}Ra: Radioactive disequilibrium in the deep sea. *Earth and Planetary Science Letters*, *17*, 295–305.

Dai, M. H. & Benitez-Nelson, C. R. (2001). Colloidal organic carbon and ^{234}Th in the Gulf of Maine. *Marine Chemistry*, *74*, 181–196.

DeMaster, D. J., Brewster, D. C., McKee, B. A. & Nittrouer, C. A. (1991). Rates of particle scavenging, sediment reworking, and longitudinal ripple formation at the HEBBLE site based on measurements of ^{234}Th and ^{210}Pb. *Marine Geology*, *99*, 423–444.

Edmonds, H. N., Moran, S. B., Hoff, J. A., Smith, J. N. & Edwards, R. L. (1998). Protactinium-231 and thorium-230 abundances and high scavenging rates in the western Arctic Ocean. *Nature*, *280*, 405–407.

Eppley, R. W. (1989). New production: History, methods, problems. In W. H. Berger, V. S. Smetacek & G. Wefer (Eds), *Productivity of the Ocean: Present and Past* (pp. 85–97). Chichester: Wiley.

Fisher, N. S., Burns, K. A., Cherry, R. D. & Heyraud, M. (1983). Accumulation and cellular distribution of ^{241}Am, ^{210}Po, and ^{210}Pb in two marine algae. *Marine Ecology Progress Series, 11*, 233–237.

Fisher, N. S., Cochran, J. K., Krishnaswami, S. & Livingston, H. D. (1988). Predicting the oceanic flux of radionuclides on sinking biogenic debris. *Nature, 335*, 622–625.

Francois, R. & Bacon, M. P. (1991). Variations in terrigenous input into the deep equatorial Atlantic during the past 24,000 years. *Science, 251*, 1473–1476.

Francois, R. & Bacon, M. P. (1994). Heinrich events in the North Atlantic: Radiochemical evidence. *Deep-Sea Research I, 41*, 315–334.

Francois, R., Bacon, M. P. & Suman, D. O. (1990). Thorium-230 profiling in deep-sea sediments: High-resolution records of flux and dissolution of carbonate in the equatorial Atlantic during the last 24,000 years. *Paleoceanography, 5*, 761–787.

Francois, R., Bacon, M. P., Altabet, M. A. & Labeyrie, L. D. (1993). Glacial/interglacial changes in sediment rain rate in the SW Indian sector of subantarctic waters as recorded by ^{230}Th, ^{231}Pa, U, and δ^{15}N. *Paleoceanography, 8*, 611–629.

Frank, M., Eisenhauer, A., Bonn, W. J., Walter, P., Grobe, H., Kubik, P. W., Dittrich-Hannen, B. & Mangini, A. (1995). Sediment redistribution versus paleoproductivity change: Weddell Sea margin sediment stratigraphy and biogenic particle flux of the last 250,000 years deduced from $^{230}Th_{ex}$, ^{10}Be and biogenic barium profiles. *Earth and Planetary Science Letters, 136*, 559–573.

Friedrich, J. & Rutgers van der Loeff, M. M. (2002). A two-tracer (^{210}Po–^{234}Th) approach to distinguish organic carbon and biogenic silica export flux in the Antarctic Circumpolar Current. *Deep-Sea Research I, 49*, 101–120.

Gardner, W. D. (2000). Sediment trap sampling in surface waters. In R. B. Hanson, H. W. Ducklow & J. G. Field (Eds), *The Changing Ocean Carbon Cycle: A Midterm Synthesis of the Joint Global Ocean Flux Study* (pp. 240–281). Cambridge: Cambridge University Press.

German, C. R., Fleer, A. P., Bacon, M. P. & Edmond, J. M. (1991). Hydrothermal scavenging at the Mid-Atlantic Ridge: Radionuclide distributions. *Earth and Planetary Science Letters, 105*, 170–181.

German, C. R., Bourlès, D. L., Brown, E. T., Hergt, J., Colley, S., Higgs, N. C., Ludford, E. M., Nelsen, T. A., Feely, R. A., Raisbeck, G. & Yiou, F. (1997). Hydrothermal scavenging on the Juan de Fuca Ridge: $^{230}Th_{xs}$, ^{10}Be, and REEs in ridge-flank sediments. *Geochimica et Cosmochimica Acta, 61*, 4067–4078.

Goldberg, E. D. & Bruland, K. W. (1974). Radioactive geochronologies. In E. D. Goldberg (Ed.), *The Sea* (pp. 451–489). New York: Wiley.

Greenamoyer, J. M. & Moran, S. B. (1997). Investigation of Cd, Cu, Ni and ^{234}Th in the colloidal size range in the Gulf of Maine. *Marine Chemistry, 57*, 217–226.

Guo, L. & Santschi, P. H. (1997). Composition and cycling of colloids in marine environments. *Reviews of Geophysics, 35*, 17–40.

Gustafsson, Ö., Gschwend, P. M. & Buesseler, K. O. (1997a). Settling removal rates of PCBs into the northwestern Atlantic derived from ^{238}U–^{234}Th disequilibrium. *Environmental Science and Technology, 31*, 3544–3550.

Gustafsson, Ö., Gschwend, P. M. & Buesseler, K. O. (1997b). Using ^{234}Th disequilibrium to estimate the vertical removal rates of polycyclic aromatic hydrocarbons from the surface ocean. *Marine Chemistry, 57*, 11–23.

Hall, I. R. & McCave, I. N. (2000). Palaeocurrent reconstruction, sediment and thorium focusing on the Iberian margin over the last 140 ka. *Earth and Planetary Science Letters, 178*, 151–164.

Hall, I. R., Schmidt, S., McCave, I. N. & Reyss, J. L. (2000). Particulate matter distribution and ^{234}Th/^{238}U disequilibrium along the Northern Iberian Margin: Implications for particulate organic carbon export. *Deep-Sea Research I, 47*, 557–582.

Henderson, G., Heinze, C., Anderson, R. F. & Winguth, A. M. E. (1999). Global distribution of the ^{230}Th flux to ocean sediments constrained by GCM modelling. *Deep-Sea Research I, 46*, 1861–1893.

Honeyman, B. D. & Santschi, P. H. (1989). A Brownian-pumping model for oceanic trace metal scavenging: Evidence from Th isotopes. *Journal of Marine Research, 47*, 951–992.

Honeyman, B. D., Balistrieri, L. S. & Murray, J. W. (1988). Oceanic trace metal scavenging: The importance of particle concentration. *Deep-Sea Research, 35*, 227–246.

Honeyman, B. D. & Santschi, P. H. (1992). The role of particles and colloids in the transport of radionuclides and trace metals in the oceans. In J. Buffle & H. P. van Leeuwen (Eds), *Environmental Particles* (pp. 379–423). Boca Raton: Lewis Publishers.

Huh, C.-A. & Kadko, D. C. (1992). Marine sediments and sedimentation processes. In M. Ivanovich & R. S. Harmon (Eds), *Uranium-Series Disequilibrium: Applications to Earth, Marine, and Environmental Sciences* (pp. 460–486). Oxford: Clarendon Press.

Huh, C.-A. & Prahl, F. G. (1995). Role of colloids in upper ocean biogeochemistry in the northeast Pacific Ocean elucidated from ^{238}U–^{234}Th disequilibrium. *Limnology and Oceanography*, *40*, 528–532.

Igel, H. & von Blanckenburg, F. (1999). Lateral mixing and advection of reactive isotope tracers in ocean basins: Numerical modeling. *Geochemistry, Geophysics, Geosystems*, *1*, Paper number 1999GC000003.

Ivanovich, M. & Harmon, R. S. (1992). *Uranium-Series Disequilibrium: Applications to Earth, Marine, and Environmental Sciences* (910 pp.). Oxford: Clarendon Press.

Ivanovich, M., Latham, A. G., & Ku, T.-L. (1992). Uranium-series disequilibrium applications in geochronology. In M. Ivanovich & R. S. Harmon (Eds), *Uranium-Series Disequilibrium: Applications to Earth, Marine, and Environmental Sciences* (pp. 62–94). Oxford: Clarendon Press.

Kadko, D. (1993). An assessment of the effect of chemical scavenging within submarine hydrothermal plumes upon ocean geochemistry. *Earth and Planetary Science Letters*, *120*, 361–374.

Kadko, D., Bacon, M. P. & Hudson, A. (1986–1987). Enhanced scavenging of ^{210}Pb and ^{210}Po by processes associated with the East Pacific Rise near 8°45′N. *Earth and Planetary Science Letters*, *81*, 349–357.

Knauer, G. & Asper, V. (1989). *Sediment Trap Technology and Sampling* (94 pp.). Woods Hole, MA: US JGOFS Planning and Coordination Office.

Knauer, G. A., Martin, J. H. & Bruland, K. W. (1979). Fluxes of particulate carbon, nitrogen, and phosphorus in the upper water column of the Northeast Pacific. *Deep-Sea Research*, *26*, 97–108.

Krishnaswami, S., Lal, D., Somayajulu, B. L. K., Weiss, R. & Craig, H. (1976). Large-volume in-situ filtration of deep Pacific waters: Mineralogical and radioisotope studies. *Earth and Planetary Science Letters*, *32*, 420–429.

Ku, T.-L. (1976). The uranium-series methods of age determination. *Annual Review of Earth and Planetary Sciences*, *4*, 347–379.

Ku, T.-L., Wang, L., Luo, S. & Southon, J. R. (1995). ^{26}Al in seawater and $^{26}Al/^{10}Be$ as paleo-flux tracer. *Geophysical Research Letters*, *22*, 2163–2166.

Kumar, N., Gwiazda, R., Anderson, R. F. & Froelich, P. N. (1993). $^{231}Pa/^{230}Th$ ratios in sediments as a proxy for past changes in Southern Ocean productivity. *Nature*, *362*, 45–48.

Kumar, N., Anderson, R. F., Mortlock, R. A., Froelich, P. N., Kubik, P., Dittrich-Hannen, B. & Suter, M. (1995). Increased biological productivity and export production in the glacial Southern Ocean. *Nature*, *378*, 675–680.

Lao, Y., Anderson, R. F. & Broecker, W. S. (1992a). Boundary scavenging and deep-sea sediment dating: Constraints from excess ^{230}Th and ^{231}Pa. *Paleoceanography*, *7*, 783–798.

Lao, Y., Anderson, R. F., Broecker, W. S., Trumbore, S. E., Hofmann, H. J. & Wolfli, W. (1992b). Transport and burial rates of ^{10}Be and ^{231}Pa in the Pacific Ocean during the Holocene period. *Earth and Planetary Science Letters*, *113*, 173–189.

Lao, Y., Anderson, R. F., Broecker, W. S., Hofmann, H. J. & Wolfli, W. (1993). Particulate fluxes of ^{230}Th, ^{231}Pa, and ^{10}Be in the northeastern Pacific Ocean. *Geochimica et Cosmochimica Acta*, *57*, 205–217.

Lorenzen, C. J., Shuman, F. R. & Bennett, J. T. (1981). In situ calibration of a sediment trap. *Limnology and Oceanography*, *26*, 580–585.

Luo, S. & Ku, T.-L. (1999). Oceanic $^{231}Pa/^{230}Th$ ratio influenced by particle composition and remineralization. *Earth and Planetary Science Letters*, *167*, 183–195.

Luo, S., Ku, T.-L., Kusakabe, M., Bishop, J. K. B. & Yang, Y.-L. (1995). Tracing particle cycling in the upper ocean with ^{230}Th and ^{228}Th: An investigation in the equatorial Pacific along 140°W. *Deep-Sea Research II*, *42*, 805–829.

Luo, S., Ku, T.-L., Wang, L., Southon, J. R., Lund, S. P. & Schwartz, M. (2001). ^{26}Al, ^{10}Be and U–Th isotopes in Blake Outer Ridge sediments: Implications for past changes in boundary scavenging. *Earth and Planetary Science Letters*, *185*, 135–147.

Marcantonio, F., Kumar, N., Stute, M., Anderson, R. F., Seidl, M. A., Schlosser, P. & Mix, A. (1995). A comparative study of accumulation rates derived by He and Th isotope analysis of marine sediments. *Earth and Planetary Science Letters*, *133*, 549–555.

Marcantonio, F., Anderson, R. F., Stute, M., Kumar, N., Schlosser, P. & Mix, A. (1996). Extraterrestrial ^{3}He as a tracer of marine sediment transport and accumulation. *Nature*, *383*, 705–707.

Marcantonio, F., Higgins, S., Anderson, R. F., Stute, M., Schlosser, P. & Rasbury, E. T. (1998). Terrigenous helium in deep-sea sediments. *Geochimica et Cosmochimica Acta*, *62*, 1535–1543.

Marcantonio, F., Turekian, K. K., Higgins, S., Anderson, R. F., Stute, M. & Schlosser, P. (1999). The accretion rate of extraterrestrial ^{3}He based on oceanic ^{230}Th flux and the relation to Os isotope variation over the past 200,000 years in an Indian Ocean core. *Earth and Planetary Science Letters, 170*, 157–168.

Marcantonio, F., Anderson, R. F., Higgins, S., Stute, M. & Schlosser, P. (2001). Sediment focusing in the central equatorial Pacific Ocean. *Paleoceanography, 16*, 260–267.

Marchal, O., François, R., Stocker, T. F. & Joos, F. (2000). Ocean thermohaline circulation and sedimentary ^{231}Pa/^{230}Th ratio. *Paleoceanography, 15*, 625–641.

Martin, J. H., Fitzwater, S. E., Gordon, R. M., Hunter, C. N. & Tanner, S. J. (1993). Iron, primary production and carbon-nitrogen flux studies during the JGOFS North Atlantic Bloom Experiment. *Deep-Sea Research II, 40*, 115–134.

Matsumoto, E. (1975). ^{234}Th–^{238}U radioactive disequilibrium in the surface layer of the ocean. *Geochimica et Cosmochimica Acta, 39*, 205–212.

McCave, I. N. (1975). Vertical flux of particles in the ocean. *Deep-Sea Research, 22*, 491–502.

McCave, I. N. (1986). Local and global aspects of the bottom nepheloid layers in the world ocean. *Netherlands Journal of Sea Research, 20*, 167–181.

McManus, J. F., Anderson, R. F., Broecker, W. S., Fleisher, M. Q. & Higgins, S. M. (1998). Radiometrically determined sedimentary fluxes in the sub-polar North Atlantic during the last 140,000 years. *Earth and Planetary Science Letters, 155*, 29–43.

Michaels, A. F., Bates, N. R., Buesseler, K. O., Carlson, C. A. & Knap, A. H. (1994). Carbon-cycle imbalances in the Sargasso Sea. *Nature, 372*, 537–540.

Moore, W. S. (1981). The thorium isotope content of ocean water. *Earth and Planetary Science Letters, 53*, 419–426.

Moore, W. S., Bruland, K. W. & Michel, J. (1981). Fluxes of uranium and thorium series isotopes in the Santa Barbara Basin. *Earth and Planetary Science Letters, 53*, 391–399.

Moran, S. B. & Buesseler, K. O. (1992). Short residence time of colloids in the upper ocean estimated from ^{238}U–^{234}Th disequilibrium. *Nature, 359*, 221–223.

Moran, S. B. & Buesseler, K. O. (1993). Size-fractionated ^{234}Th in continental shelf waters off New England: Implications for the role of colloids in oceanic trace metal scavenging. *Journal of Marine Research, 51*, 893–922.

Moran, S. B., Charette, M. A., Hoff, J. A., Edwards, R. L. & Landing, W. M. (1997a). Distribution of ^{230}Th in the Labrador Sea and its relation to ventilation. *Earth and Planetary Science Letters, 150*, 151–160.

Moran, S. B., Ellis, K. M. & Smith, J. N. (1997b). ^{234}Th/^{238}U disequilibrium in the central Arctic Ocean: Implications for particulate organic carbon export. *Deep-Sea Research II, 44*, 1593–1606.

Murnane, R. J., Cochran, J. K. & Sarmiento, J. L. (1994). Estimates of particle- and thorium-cycling rates in the Northwest Atlantic Ocean. *Journal of Geophysical Research, 99*, 3373–3392.

Murnane, R. J., Cochran, J. K., Buesseler, K. O. & Bacon, M. P. (1996). Least-squares estimates of thorium, particle, and nutrient cycling rate constants from the JGOFS North Atlantic Bloom Experiment. *Deep-Sea Research I, 43*, 239–258.

Murray, J. W., Downs, J. N., Strom, S., Wei, C.-L. & Jannasch, H. W. (1989). Nutrient assimilation, export production and ^{234}Th scavenging in the eastern equatorial Pacific. *Deep-Sea Research, 36*, 1471–1489

Murray, J. W., Young, J., Newton, J., Dunne, J., Chapin, T., Paul, B. & McCarthy, J. (1996). Export flux of particulate organic carbon from the central equatorial Pacific determined using a combined drifting trap-^{234}Th approach. *Deep-Sea Research II, 43*, 1095–1132.

Niven, S. E. H., Kepkay, P. E. & Boraie, A. (1995). Colloidal organic carbon and colloidal ^{234}Th dynamics during a coastal phytoplankton bloom. *Deep-Sea Research II, 42*, 257–273.

Nozaki, Y. & Nakanishi, T. (1985). ^{231}Pa and ^{230}Th profiles in the open ocean water column. *Deep-Sea Research, 32*, 1209–1220.

Nozaki, Y. & Yamada, M. (1987). Thorium and protactinium isotope distributions in waters of the Japan Sea. *Deep-Sea Research, 34*, 1417–1430.

Nozaki, Y., Thomson, J. & Turekian, K. K. (1976). The distribution of ^{210}Pb and ^{210}Po in the surface waters of the Pacific Ocean. *Earth and Planetary Science Letters, 32*, 304–312.

Nozaki, Y., Horibe, Y. & Tsubota, H. (1981). The water column distributions of thorium in the western North Pacific. *Earth and Planetary Science Letters, 54*, 203–216.

Nozaki, Y., Yang, H.-S. & Yamada, M. (1987). Scavenging of thorium in the ocean. *Journal of Geophysical Research, 92*, 772–778.

Olsen, C. R., Thein, M., Larsen, I. L., Lowry, P. D., Mulholland, P. J., Cutshall, N. H., Byrd, J. T. & Windom, H. L. (1989). Plutonium, lead-210, and carbon isotopes in the Savannah estuary: Riverborne versus marine sources. *Environmental Science and Technology*, *23*, 1475–1481.

Rutgers van der Loeff, M. M. & Boudreau, B. P. (1997). The effect of resuspension on chemical exchanges at the sediment–water interface in the deep sea – a modelling and natural radiotracer approach. *Journal of Marine Systems*, *11*, 305–342.

Rutgers van der Loeff, M. M. & Berger, G. W. (1993). Scavenging of ^{230}Th and ^{231}Pa near the Antarctic Polar Front in the South Atlantic. *Deep-Sea Research I*, *40*, 339–357.

Rutgers van der Loeff, M. M., Friedrich, J. & Bathmann, U. V. (1997). Carbon export during the Spring Bloom at the Antarctic Polar Front, determined with the natural tracer ^{234}Th. *Deep-Sea Research II*, *44*, 457–478.

Rutsch, H.-J., Mangini, A., Bonani, G., Dittrich-Hannen, B., Kubik, P. W., Suter, M. & Segl, M. (1995). ^{10}Be and Ba concentrations in West African sediments trace productivity in the past. *Earth and Planetary Science Letters*, *133*, 129–143.

Santschi, P. H., Nyffeler, U. P., Li, Y.-H. & O'Hara, P. (1986). Radionuclide cycling in natural waters: Relevance of scavenging kinetics. In P. Sly (Ed.), *Sediments and Water Interactions* (pp. 183–191). New York: Springer-Verlag.

Sarin, M. M., Krishnaswami, S., Dalai, T. K., Ramaswamy, V. & Ittekkot, V. (2000). Settling fluxes of U- and Th-series nuclides in the Bay of Bengal: Results from time-series sediment trap studies. *Deep-Sea Research I*, *47*, 1961–1985.

Scholten, J. C., Rutgers van der Loeff, M. M. & Michel, A. (1995). Distribution of ^{230}Th and ^{231}Pa in the water column in relation to the ventilation of the deep Arctic basins. *Deep-Sea Research II*, *42*, 1519–1531.

Scholten, J. C., Fietzke, J., Vogler, S., Rutgers van der Loeff, M. M., Mangini, A., Koeve, W., Waniek, J., Stoffers, P., Antia, A. & Kuss, J. (2001). Trapping efficiencies of sediment traps from the deep eastern North Atlantic: The ^{230}Th calibration. *Deep-Sea Research II*, *48*, 2383–2408.

Shannon, L. V., Cherry, R. D. & Orren, M. J. (1970). Polonium-210 and lead-210 in the marine environment. *Geochimica et Cosmochimica Acta*, *34*, 701–711.

Shimmield, G. B. & Price, N. B. (1988). The scavenging of U, of ^{230}Th and ^{231}Pa during pulsed hydrothermal activity at 20°S East Pacific Rise. *Geochimica et Cosmochimica Acta*, *52*, 669–677.

Shimmield, G. B., Ritchie, G. D. & Fileman, T. W. (1995). The impact of marginal ice zone processes on the distribution of ^{210}Pb, ^{210}Po and ^{234}Th and implications for new production in the Bellingshausen Sea, Antarctica. *Deep-Sea Research II*, *42*, 1313–1335.

Spencer, D. W., Bacon, M. P. & Brewer, P. G. (1981). Models of the distribution of ^{210}Pb in a section across the North equatorial Atlantic Ocean. *Journal of Marine Research*, *39*, 119–138.

Suman, D. O. & Bacon, M. P. (1989). Variations in Holocene sedimentation in the North American Basin determined from Th-230 measurements. *Deep-Sea Research*, *36*, 869–878.

Thomson, J., Colley, S., Anderson, R., Cook, G. T. & MacKenzie, A. B. (1995a). A comparison of sediment accumulation chronologies by the radiocarbon and $^{230}Th_{excess}$ methods. *Earth and Planetary Science Letters*, *133*, 59–70.

Thomson, J., Higgs, N. C. & Clayton, T. (1995b). A geochemical criterion for the recognition of Heinrich events and estimation of their depositional fluxes by the $(^{230}Th_{excess})_0$ profiling method. *Earth and Planetary Science Letters*, *135*, 41–56.

Thomson, H., Nixon, S., Summerhayes, C. P., Schönfeld, J., Zahn, R. & Grootes, P. (1999). Implications for sedimentation changes on the Iberian margin over the last two glacial/interglacial transitions from $(^{230}Th_{excess})_0$ systematics. *Earth and Planetary Science Letters*, *165*, 255–270.

Tsunogai, S. & Minagawa, M. (1978). Settling model for the removal of insoluble chemical elements in seawater. *Geochemical Journal*, *12*, 47–56.

Turekian, K. K. & Cochran, J. K. (1978). Determination of marine chronologies using natural radionuclides. In J. P. Riley & R. Chester (Eds), *Chemical Oceanography* (pp. 313–360). London: Academic Press.

Turnewitsch, R. & Springer, B. M. (2001). Do bottom mixed layers influence ^{234}Th dynamics in the abyssal near-bottom water column? *Deep-Sea Research I*, *48*, 1279–1307.

Valdes, J. R. & Price, J. F. (2000). A neutrally buoyant, upper ocean sediment trap. *Journal of Atmospheric and Oceanic Technology*, *17*, 62–68.

Vogler, S., Scholten, J., Rutgers van der Loeff, M. & Mangini, A. (1998). ^{230}Th in the eastern North Atlantic: The importance of water mass ventilation in the balance of ^{230}Th. *Earth and Planetary Science Letters*, *156*, 61–74.

von Blackenburg, F. & Igel, H. (1999). Lateral mixing and advection of reactive isotope tracers in ocean basins: Observations and mechanisms. *Earth and Planetary Science Letters*, *169*, 113–128.

Walter, H.-J., Rutgers van der Loeff, M. M. & Hoeltzen, H. (1997). Enhanced scavenging of ^{231}Pa relative to ^{230}Th in the South Atlantic south of the Polar Front: Implications for the use of the ^{231}Pa/^{230}Th ratio as a paleoproductivity proxy. *Earth and Planetary Science Letters*, *149*, 85–100.

Walter, H.-J., Rutgers van der Loeff, M. M. & François, R. (1999). Reliability of the ^{231}Pa/^{230}Th activity ratio as a tracer for bioproductivity of the ocean. In G. Fischer & G. Wefer (Eds), *Use of Proxies in Paleoceanography: Examples from the South Atlantic* (pp. 393–408). Berlin: Springer-Verlag.

Wang, L., Ku, T.-L., Luo, S., Southon, J. R. & Kusakabe, M. (1996). ^{26}Al–^{10}Be systematics in deep-sea sediments. *Geochimica et Cosmochimica Acta*, *60*, 109–119.

Yang, H.-S., Nozaki, Y., Sakai, H. & Masuda, A. (1986). The distribution of ^{230}Th and ^{231}Pa in the deep-sea surface sediments of the Pacific Ocean. *Geochimica et Cosmochimica Acta*, *50*, 81–99.

Yang, Y.-L., Elderfield, H. & Ivanovich, M. (1990). Glacial to Holocene changes in carbonate and clay sedimentation in the equatorial Pacific Ocean estimated from thorium-230 profiles. *Paleoceanography*, *5*, 789–809.

Yang, Y.-L., Elderfield, H. & Pederson, T. F. (1995). Geochemical record of the Panama basin during the last glacial maximum carbon event shows that the glacial ocean was not suboxic. *Geology*, *23*, 1115–1118.

Yu, E.-F., Francois, R. & Bacon, M. P. (1996). Similar rates of modern and last-glacial ocean thermohaline circulation inferred from radiochemical data. *Nature*, *379*, 689–694.

Yu, E.-F., Francois, R., Bacon, M. P. & Fleer, A. P. (2001a). Fluxes of ^{230}Th and ^{231}Pa to the deep sea: Implications for the interpretation of excess ^{230}Th and ^{231}Pa/^{230}Th profiles in sediments. *Earth and Planetary Science Letters*, *191*, 219–230.

Yu, E.-F., Francois, R., Bacon, M. P., Honjo, S., Fleer, A. P., Manganini, S. J., Rutgers van der Loeff, M. M. & Ittekot, V. (2001b). Trapping efficiency of bottom-tethered sediment traps estimated from the intercepted fluxes of ^{230}Th and ^{231}Pa. *Deep-Sea Research I*, *48*, 865–889.

MARINE RADIOACTIVITY
Hugh D. Livingston (Editor)

Chapter 6

Radionuclides in the biosphere

Scott W. Fowler[a], Nicholas S. Fisher[b]

[a]*IAEA Marine Environment Laboratory, 4 Quai Antoine 1er, MC 98000, Monaco*
[b]*Marine Sciences Research Center, State University of New York, Stony Brook, NY 11794-5000, USA*

Introduction

That radionuclides become associated with marine organisms is well documented, and the overall topic has been the subject of several in-depth reviews (e.g. Lowman et al., 1971; Bowen et al., 1971; Fowler, 1982; Coughtrey & Thorne, 1983; Coughtrey et al., 1984). Nevertheless, one aspect of radionuclide-biota interactions less well understood is how and to what degree organisms affect the distribution and fate of these substances in the marine environment. Movements of radionuclides associated with biological material are subject to physical and geological transport processes but, in addition, are affected by bioaccumulation, retention, and subsequent food chain transfer, horizontal and vertical migration of many species, and passive sinking of biodetritus. It follows that the relative importance of these biological transport mechanisms compared to physical and chemical processes will be a function of the oceanic biomass at any given location, and most particularly the downward flux of material emanating from biological activity.

In the following sections these key biological processes will be discussed with respect to both artificial and natural radionuclides. Artificial radionuclides were some of the first marine contaminants to be monitored and studied scientifically, hence, the information base on their interaction with biota is quite large and the literature abounds with examples of baseline measurements of radionuclide contaminants in different marine organisms from tropical to polar ecosystems. No attempt is made here to compile a listing of artificial radionuclide concentrations in different marine species as such levels are transient, in some cases site specific, and in most areas are changing due to radioactive decay and/or decreasing as inputs from atmospheric fallout slowly decline. On the other hand, certain examples or case studies are cited where they are considered pertinent to the understanding of the processes involved in the transfer, behavior and fate of radionuclides in the marine environment.

1. Association of radionuclides with marine biota

Accumulation from water

Uptake from water occurs either by adsorption of the radionuclide onto cell or organism surfaces, absorption in body surfaces such as cell membranes, gill and gut, or active transport across organism surfaces mediated by enzymatic activity. For heterotrophs the alternative mode of accumulation is through the ingestion and assimilation of contaminated food. The relative ability of organisms to concentrate radionuclides is often expressed as a concentration factor, defined as the ratio of the amount of radionuclide per unit fresh weight of tissue to that dissolved in an equal weight of seawater. Since these ratios take into account only the radionuclide concentration in water and the organism, they give no information on the relative importance of the different pathways of uptake, nor do they reflect the influence of speciation of the radionuclides in the water on bioavailability. Contaminant concentrations in marine species are in a state of dynamic equilibrium, being the net result of both radionuclide uptake and elimination processes. These dynamics are controlled by many factors such as exposure time, the physico-chemical form of the radionuclide, salinity, temperature, competitive effects with other substances, life cycle of the organism, physiology, feeding habits, etc. For this reason, concentration factors are best viewed as general ranges rather than as absolute values.

Depending on the organism and the radionuclide, concentration factors range from roughly 10^0–10^6 (IAEA, 2004). Clearly radionuclides with the highest concentration factors will be those most readily transported by the biota. In many marine species, especially the smaller ones, radionuclides of Pb, Ru, Zr, certain lanthanides and transuranics are normally concentrated more than physiologically important elements such as Zn, Cu and Co. This occurs since many of these nonessential elements are particle-reactive in seawater and are more apt to sorb to surfaces, especially those of small planktonic organisms with high surface area to volume ratios (Lowman et al., 1971; Fowler, 1982). On the other hand, radionuclides which are less particle-reactive and behave more conservatively in seawater such as ^{137}Cs and ^{99}Tc typically display much lower concentration factors, although exceptions have been noted. For example, some experimental studies have demonstrated very low technetium concentration factors ($\sim10^0$–10^2) in phytoplankton, bivalve molluscs and small crustacea when the organisms were exposed to ^{95m}Tc in seawater either as the pertechnetate anion (VII oxidation state) or in a reduced form (IV) (Fowler et al., 1981; Fisher, 1982), whereas similar laboratory experiments as well as field data have indicated that lobsters and brown algae can reach Tc concentration factors as high as 10^3–10^4 and 10^5, respectively (Swift, 1985; Smith et al., 2001). Thus, despite the very conservative behavior of ^{99}Tc in seawater, the uptake from contaminated seawater can be an important vector in certain species.

Phytoplankton, because of its large surface area to volume ratio, quickly takes up radionuclides and reaches extremely high concentration factors (Table 1). The biphasic process involves rapid sorption to the cell surface, perhaps by cation exchange, followed by slower diffusion across the cell membrane and subsequent binding within the cell (Davies, 1978; Fisher et al., 1983a; Fisher & Reinfelder, 1995). Equilibration times are generally short (minutes to hours) and there is evidence in the case of transuranic and other particle-reactive nuclides that uptake is a passive process (ibid.). Controlled laboratory experiments have shown that uptake

Table 1
Selected trace element and radionuclide concentration factors (element g^{-1} wet animal divided by element g^{-1} water) for phytoplankton and crustacean zooplankton. Data taken from Fowler (1990) and IAEA (2004)

Element/RN	Phytoplankton	Microzooplankton**	Macrozooplankton^{+}
*Hg	1×10^5	1×10^4	4×10^3
*Ni	8×10^3	3×10^3	6×10^2
*Co	2×10^3	7×10^3	6×10^3
*Cd	1×10^3	6×10^4	2×10^4
*Cu	1×10^4	5×10^4	1×10^5
*Fe	4×10^5	3×10^5	2×10^5
*Zn	1×10^4	1×10^5	1×10^5
Tc	4×10^0	1×10^2	1×10^2
$^{239+240}Pu$	9×10^4–1×10^5	4×10^3	1×10^2
^{241}Am	2×10^4–1×10^5	4×10^3	1×10^3
^{144}Ce	9×10^4	6×10^3	–
^{106}Ru	2×10^5	3×10^4	–
^{238}U	2×10^1	3×10^1	–
^{232}Th	2×10^4	2×10^4	–
^{230}Th	8×10^3	4×10^3	–
^{228}Th	2×10^4	6×10^3	–
^{226}Ra	2×10^3	1×10^2	–
^{210}Po	7×10^4	3×10^4	1×10^4

*Computed using recent values for trace element concentration in seawater.
$^{+}$Euphausiids.
**Mainly copepods.

of many radionuclides and corresponding stable metals depends on the element concentration in seawater, the length of exposure, and the algal species. However, it is possible to generalize that interspecific differences in bioconcentration factors are relatively small compared to inter-radionuclide differences (Fisher & Reinfelder, 1995).

Despite the wide variety of different species in the zooplankton community, most of the experimental studies on radionuclide bioaccumulation by zooplankton have been carried out with micro- or macrocrustaceans such as copepods (Fisher & Reinfelder, 1995) and euphausiids (Fowler, 1982). Like plants and many larger invertebrates, heterotrophic zooplankton obtain elements directly from seawater but, in addition, accumulate them through assimilation of ingested food (Fowler, 1982; Wang et al., 1996b). Direct uptake from seawater occurs both by adsorption onto body surfaces and absorption of the elements across surfaces, such as gills or gut linings. Once across the cellular boundaries the elements are translocated to other organs and tissues by either active or passive processes where they are stored or eventually eliminated (Mason & Jenkins, 1995). Uptake rates strongly depend on the element, with reported equilibration times ranging from several hours to several days (Fowler, 1982, for review; Wang & Fisher, 1998a). Radionuclide concentrations in marine plankton are in a state of dynamic equilibrium and are the net result of both uptake and elimination processes occurring simultaneously. The rates of these processes are controlled by exposure time, the physico-chemical form of the element, salinity, temperature, competitive effects with other substances, life cycle of the organism, physiology and feeding habits. Hence, concentration factors in zooplankton vary greatly ranging over several orders of magnitude (Table 1). In general, the highest concentration factors are noted for those radionuclides that are 'particle reactive' in seawater, whereas low concentration factors are typical for the radionuclides which behave more conservatively in seawater.

Amongst the plankton, smaller organisms generally attain higher concentration factors than larger species because of the former's greater relative surface area for adsorption. Furthermore, for certain particle-reactive radionuclides such as plutonium and americium, concentration factors are directly related to the surface to volume ratio of zooplankton and phytoplankton species (Fisher, 1985; Fisher & Fowler, 1987). For these small planktonic organisms concentration factors on the order of 10^4–10^5 are not uncommon for several elements (Table 1). Unfortunately, because of difficulties in sampling, comparable data for the smaller bacterio-plankton do not exist; however, there is some evidence from laboratory experiments using ^{241}Am that marine bacteria may reach volume concentration factors in the range 10^6–10^7 (Fowler, 1990). Such high enrichment factors alone make small species potentially important in affecting subsequent radionuclide redistribution throughout the water column. Moreover, their rapid life cycles, migratory behavior, physiology and feeding strategies further enhance their importance in the biological transport of elements in the sea.

Macroalgae, macroinvertebrates, and fish also absorb radionuclides from water, although the degree of relative uptake is usually much less than that of smaller organisms since the role of surface area in total accumulation is of far lesser importance in larger species. Examples of the accumulation of radionuclides by a myriad of species under a wide variety of laboratory and field conditions can be found in several comprehensive reviews and will not be reiterated here (Pentreath, 1977; Fowler, 1982; Phillips, 1980; Pentreath & Fowler, 1979). In brief, uptake is generally non-linear and often biphasic with an initial rapid component representing surface adsorption followed by a slower rate of radionuclide bioaccumulation into internal tissues. The uptake rate generally decreases until a steady state is reached between the radionuclide in the water and the organism's tissues. Because in larger organisms internal tissues are often isolated from the surrounding seawater, equilibration times based on radionuclide absorption from water are normally much longer (days to weeks) than those observed in small species such as plankton.

The importance of the initial component of uptake depends to some extent on the surface characteristics of the organism. Hard-shelled, calcareous animals may deposit much of the radionuclide in the shell during growth. Indeed, substantial concentrations of radionuclides are present in mollusc shells and exoskeletons of crustaceans and echinoderms (Fowler et al., 1975; Bowen et al., 1976; Higgo et al., 1980; Pentreath, 1981; Guary et al., 1982; Koide et al., 1982). Soft-bodied organisms with no hard, external covering are able to equilibrate their internal tissues more rapidly. Mucus coating the surface of many of these species, including fish, is important in the initial complexing of the radionuclide (Fowler et al., 1975). When the radionuclide has diffused through the epithelium, the blood or haemolymph circulation in invertebrates is the principal vector for radionuclide transfer to the various tissues. The degree of radionuclide accumulation in these tissues depends on the chemistry of the radionuclides, the number of binding sites, retention time in a tissue, and general physiology of the organism. Often liver and kidney of both invertebrates and vertebrates contain the highest concentrations of radionuclides accumulated from water, whereas muscle normally concentrates absorbed radionuclides to a much lesser extent (Pentreath, 1977; Fowler & Guary, 1977; Fowler, 1982).

The degree of radionuclide uptake from water largely depends on the physical and chemical form of the element. It has been shown that radio-iron hydroxide particles readily sorb to the surface of diatoms, a mechanism which may enhance the uptake of other particle-reactive radionuclides since metal hydroxides are known to scavenge other elements in seawater (Davies, 1978). From such tracer studies it is believed that only inorganic ionic species, and usually only the free ions are available for uptake by phytoplankton (Campbell, 1995). The response of other organisms to different chemical forms of radionuclides varies greatly for certain radionuclides. For example, chelation or complexation greatly reduces ^{65}Zn uptake in mussels (Keckes et al., 1969). For other radionuclides like ^{60}Co, the response can be varied. Lowman & Ting (1973) found that juvenile lobsters took up four times more ionic radio-cobalt than when the nuclide was complexed as cobalamine. In contrast, similar studies with phytoplankton and fish indicated a strong preference for radio-cobalamine over the ionic form of radiocobalt (Nolan et al., 1992). In the case of transuranics like plutonium, differences in the uptake response to different oxidation states of the radionuclide appear to be minimal for a variety of marine organisms (Fowler et al., 1975). For fission products like radio-ruthenium and chromium, it has been shown that ^{106}Ru chloride complexes are far more bioavailable than the ^{106}Ru nitrosyl-nitrate forms (Keckes et al., 1972), and hexavalent ^{51}Cr is taken up in preference to the trivalent ion by certain molluscs (Chipman, 1966). Although relatively few studies have addressed the question of exactly how chemical forms of anthropogenic radionuclides in seawater affect uptake processes, it is evident from the existing literature that this single factor may largely govern the initial transfer of the radionuclide from water to organism.

As a general rule, for many marine organisms the uptake of radionuclides from water is proportional to its ambient concentrations in seawater. This holds true particularly for plankton, macroalgae and certain marine invertebrates (e.g. Chipman, 1966; Bryan, 1976; Fowler & Small, 1975). On the other hand, radionuclides of several biologically essential elements (e.g. ^{65}Zn, ^{59}Fe, ^{54}Mn) may be physiologically regulated so that internal concentrations in certain species would show little variation in response to changing levels in their surroundings (Cross et al., 1973; Bryan, 1976).

Environmental factors also affect bioaccumulation of radionuclides from seawater, and temperature and salinity probably exert the strongest effect. Generally speaking, radionuclide uptake rates correlate positively with temperature in a variety of species; however, there are exceptions indicating that temperature has little or no effect. In the case of crustaceans that normally molt more frequently at higher temperatures, radionuclide loss with the molts leads to lower levels than those in animals exposed at lower temperatures (Fowler et al., 1969; Cross et al., 1969). Like many heavy metals, uptake rates of the corresponding radionuclides in marine species generally show an inverse correlation with salinity (Bryan, 1976; Wang et al., 1996a). This effect, attributed to lesser amounts of competing ions in low salinity waters and chloro-complexation of some elements, would be most noticeable in estuaries and coastal waters receiving runoff. One exception appears to be the monovalent ion radio-cesium that concentrates to higher levels in fish living in high-salinity waters (Pentreath, 1975).

Accumulation from food

Like trace metals, the absorption of radionuclides from ingested food takes place in the gut with transport to the various tissues via the circulatory system. Tissue accumulation of radionuclides depends on the assimilation efficiency and the amount retained. Once assimilated from the gut, many of the same factors mentioned above determine the fate of the residual ingested radionuclides. The radionuclides of biologically essential elements such as Zn, Fe and Mn are rapidly absorbed from the gut and assimilated into tissues of many marine species, although quantitative differences may occur in the tissue distribution compared to that following uptake from water (Fowler & Small, 1975; Fowler, 1982). In contrast, radionuclides of non-essential elements, or those that are particle-reactive like Ru, Ce, Pu and Am, are often poorly assimilated and are excreted with feces. Nevertheless, there are exceptions; for example, due to specialized digestive metabolism, assimilation efficiencies are high and retention is strong with elements such as Pu, Am and Tc in crabs, starfish and lobsters (Fowler & Guary, 1977; Pentreath, 1981; Guary et al., 1982; Swift, 1985, 1992). In the case of fish, radionuclides absorbed from food generally accumulate to a high degree in the liver, with elasmobranch livers taking up more of the assimilated radionuclide than those of teleosts (Pentreath, 1977, 1978).

Experimentally determined assimilation efficiencies for selected radionuclides ingested with food are given for several species in Table 2. Although experimental conditions varied considerably, it is evident that radionuclide assimilation into tissues is highly dependent both on species and element. As with bioconcentration factors, assimilation efficiencies vary greatly with the physiological state of the animal, the food type, and diverse environmental conditions; hence it is most appropriate to consider ranges of assimilation efficiencies for any given combination of animal and radionuclide.

The degree to which the food pathway for radionuclide uptake predominates in the natural environment will depend on many parameters, in particular, the length of radionuclide exposure and food availability and density. As a general rule, for radionuclides of elements with high assimilation efficiencies, such as those that form activation products (Reinfelder & Fisher, 1991), the contribution of the food pathway can be significant. However, even for such radionuclides as ^{241}Am and ^{60}Co, where the assimilation efficiencies are very low, modeling has shown that the dietary pathway can account for most of the steady-state body burden of these radionuclides in some invertebrates (Wang et al., 1996a). Certain crustaceans and echinoderms display an amazing ability to assimilate and retain transuranic elements (Table 2). However, despite measured assimilation efficiencies ranging from 50–90% for these elements, direct absorption from water may still play a large role in obtaining equilibrium radionuclide distributions in the organism's tissues in nature. An example where knowledge of uptake pathways is critical in determining the existence of radionuclide 'biomagnification' is plutonium transfer in the mussel–starfish food chain. In earlier studies, the observation of greatly enhanced plutonium concentrations in whole starfish that were feeding on mussels led to the conclusion that plutonium, a radionuclide that is generally poorly assimilated in invertebrates, was biomagnified via the food chain. However, controlled radiotracer experiments showed that when starfish took up ^{237}Pu from water, the resulting radiotracer tissue distribution closely matched the natural distribution of $^{239+240}$Pu in tissues of starfish contaminated by fallout and released nuclear wastes; such was not the case when the plutonium

Table 2
The fraction* (%) of ingested radionuclides of different elements absorbed from contaminated food and assimilated into tissues of marine organisms

Organism	Food	Element														Reference
		Ag	Zn	Cs	Po	Pb	Fe	Cf	Co	Am	Np	Ce	Ru	Pu	Tc	
Zooplankton: Copopods	Phytoplankton	8–19	49–80		29–55		5–35		14–45	0.9–10				0.8		Reinfelder & Fisher (1991), Fisher et al. (1991a), Hutchins et al. (1995), Wang et al. (1996b), Wang & Fisher (1998a), Stewart & Fisher (2003)
Euphausiids	Brine shrimp, phytoplankton		59–71					<5		<3	~1	~0.1–1	~50	~1		Fowler (1982), Fowler et al. (1976), Fowler & Aston (1982), Aston & Fowler (1983), Fisher et al. (1983b)
Ciliates	Phytoplankton	23				<1										Fisher et al. (1995)
Benthic crustaceans: *Nephrops norvegicus*	Shrimp														6	Swift (2001)
Lysmata seticau data	Brine shrimp, worms									10				10–15	26–79	Fowler (1982), Guary & Fowler (1990)
Crangon crangon	Mussel		~60						~20							Fowler (1982)
Carcinus maenas	Worms, clams, brine shrimp		58				35	23	~99	29	30			20–60		Fowler (1982), Fowler et al. (1986), Guary & Fowler (1990)
Cancer pagurus	Mussel, worms, shrimp									10–20				30–40		Guary & Fowler (1990)
Semibalanus balanoides (larvae)	Phytoplankton	10							40							Anastasia et al. (1998)

Table 2
(Continued.)

Organism	Food	Element														Reference
		Ag	Zn	Cs	Po	Pb	Fe	Cf	Co	Am	Np	Ce	Ru	Pu	Tc	
Dyspanopeus sayi (larvae)	Phytoplankton											0				Anastasia et al. (1998)
Benthic molluscs:																
Mytilus edulis (includes *Mytilus galloprovincialis*)	Phytoplankton	3–34	15–48		17	56–64			8–43	0.6–6				0.9		Bjerregaard et al. (1985), Wang et al. (1995, 1996a), Wang & Fisher (1996), Fisher et al. (1996), Wildgust et al. (2000)
Perna viridis	Phytoplankton		21–36	3–4												Chong & Wang (2000), Wang et al. (2000)
Macoma balthica	Phytoplankton	38–49	73						30–53	10–26						Reinfelder et al. (1997), Hutchins et al. (1998)
Ruditapes philippinarum	Phytoplankton		30–59													Chong & Wang (2000)
Mercenaria mercenaria	Phytoplankton	22–35	86						29–34	38						Reinfelder et al. (1997)
Mercenaria mercenaria (larvae)	Phytoplankton	16	41						27	5						Reinfelder & Fisher (1994a)
Crassostrea virginica	Phytoplankton	44	73						34	11						Reinfelder et al. (1997)
Crassostrea virginica (larvae)	Phytoplankton	33	79						19	8						Reinfelder & Fisher (1994a)
Scrobicularia plana	Phytoplankton								83							Fowler (1982)
Elminius modestus	Phytoplankton		37–92													Rainbow & Wang (2001)

Table 2
(Continued.)

Organism	Food	Element														Reference
		Ag	Zn	Cs	Po	Pb	Fe	Cf	Co	Am	Np	Ce	Ru	Pu	Tc	
Octopus vulgaris	Worm									33						Guary & Fowler (1982)
Starfish:																
Coscinasterias tenuispina	Mussel													~70–90		Fowler (1982)
Asterias rubens	Mussel					42										Boisson et al. (2002)
Marthasterias glacialis	Mussel	69	78	82					73	~100						Fowler & Teyssié (1997)
Asterias forbesi	Clam								22–39	26–57						Hutchins et al. (1996)
Brittlestar:																
Ophiura texturata	Mussel							97		87						Fowler & Carvalho (1985)
Fish:																
Raja clavata	Worm, crab hepatopancreas	49												~0.1–1		Fowler (1982)
Pleuronectes platessa	Worm, crab hepatopancreas, shrimp	4	36				32		4		~0.2			<0.1–1		Fowler (1982), Pentreath (1981), Swift (1993)
Serranus scriba	Worms								5	0.7–1.7						Carvalho et al. (1983), Nolan et al. (1992)
Lutjanus argentri maculates	Zooplankton, clam, herbivorous fish			78–88												Zhao et al. (2001)
Menidia menidia	Brine shrimp		6						2							Reinfelder & Fisher (1994b)
Morone saxatilis	Brine shrimp	18	23							6						Baines et al. (2002)
Periophthalmus cantonensis	Zooplankton		4–30													Ni et al. (2000)

*Certain values have been estimated from the original data. In several cases mean values have been computed.

tracer was incorporated in starfish fed contaminated mussels (Guary et al., 1982). The tracer study clearly demonstrated that the high concentrations of plutonium noted in starfish were principally a result of direct uptake from water and were not biomagnified through the food chain.

The general public often holds the view that biomagnification of contaminants in aquatic organisms is a common occurrence in nature. For radionuclides, the phenomenon is rare and at present has only been demonstrated conclusively for one anthropogenic radionuclide. Biomagnification of ^{137}Cs has been observed in both freshwater and marine fish food chains and is thought to result from a very high assimilation efficiency as well as the high percentage of body weight (>50%) represented by fish muscle coupled with its high cesium concentration (Pentreath, 1977; Kasamatsu & Ishikawa, 1997; Zhao et al., 2001). In top marine mammalian predators such as seals and porpoises, ^{137}Cs accumulates mainly in muscle and the concentration tends to increase with body weight and/or age of the predator. Nevertheless the degree to which biomagnification occurs in these animals at the top of the food chain is still unclear. For example, seals from coastal waters of the Irish Sea were reported to contain ^{137}Cs concentrations 3–4 times higher than those in the local fish, their main source of food (Watson et al., 1999), whereas seals sampled in the high Arctic showed very little difference between their ^{137}Cs bioconcentration factors and those in their prey (Carroll et al., 2002). Clearly, more data for complete, well-defined food chains with top mammalian predators would help clarify this issue.

Although to date radiocesium is the only anthropogenic radionuclide that has been demonstrated to biomagnify, radionuclide contaminants of certain other elements should be examined in specific food chains. For example, high assimilation and subsequent strong retention of certain activation products (e.g. ^{65}Zn, ^{60}Co, ^{110m}Ag) are known to occur in tissues of starfish (Fowler & Teyssié, 1997). Hence, there is a clear potential for these radionuclides to biomagnify in these echinoderms, and a more rigorous test of the biomagnification hypothesis in their food chain would be of interest. Overall, when assimilation efficiencies are high relative to efflux rates, there is the potential for biomagnification with any contaminant (Reinfelder et al., 1998), and other specific examples for trace metals in certain marine food chains have been discussed (Wang, 2002). With regard to natural radionuclides, ^{210}Po which originates from the decay of ^{238}U, has also been shown to be more concentrated in zooplankton than in the phytoplankton food that they consume; in this case the apparent biomagnification is tied both to high assimilation efficiencies of ingested ^{210}Po in marine copepods and very slow rates of loss from the animals (Stewart & Fisher, 2003). Because of the high alpha dose delivered by incorporated ^{210}Po (Cherry & Heyraud, 1982) as well as its potential use as a tracer of feeding relationships in marine foodwebs (Heyraud & Cherry, 1979), this radionuclide has garnered much interest in terms of its transfer, metabolism and distribution in tissues of marine organisms (Wildgust et al., 2000; Bustamante et al., 2002; Durand et al., 2002).

Accumulation from sediments

The ultimate marine sink for radionuclides is usually the sediments, and correlations between radionuclide concentrations in marine species and in their surrounding sediments indicate that sediments can act as source of radionuclides for benthic organisms (Bowen et al., 1976; Aarkrog, 1977). The accumulation process can occur either by sediment or suspension feeding organisms ingesting contaminated sediment and organic particles therein, or by direct uptake of the radionuclide from the pore water where it is in equilibrium with that adsorbed to sediment grains (cf. Kd value). Depending upon the source term, subsequent radionuclide assimilation and metabolism occur by the same processes as they do following uptake from water or from food (see above). Furthermore, epifauna and benthopelagic organisms living in close proximity to sediments can also accumulate radionuclides released from the sediments to the overlying waters (Osterberg et al., 1963; Pearcy & Vanderploeg, 1973).

While field studies clearly demonstrate contaminated sediments as a source of radionuclides in marine biota, the exact mechanism of radionuclide uptake has remained obscure; however, controlled laboratory radiotracer experiments have proven useful in delineating specific uptake pathways. For example, Beasley & Fowler (1976) exposed polychaetes to sediments naturally contaminated with plutonium and americium and found that transfer factors based on uptake from total sediment concentrations were very low (TF $\sim 10^{-3}$) whereas concentration factors based on direct uptake from water were much higher (CF ~ 200). When these concentration factors were applied to the actual partitioning of these radionuclides between sediment and pore water, it was concluded that direct uptake from the pore water was most likely the predominant pathway. The same study showed that whether the sediments had been contaminated via nuclear testing activities or by liquid waste effluent releases, plutonium was more bioavailable than americium by a factor of roughly two to three.

Accumulation of radionuclides from sediments by infauna is also a function of sediment type. For example, radiotracer bioaccumulation experiments carried out by Vangenechten et al. (1983) in which both Atlantic and Pacific deep sea sediments were labeled with americium showed that both worms and clams in siliceous Pacific sediments accumulated two to five times more americium than those exposed to the carbonate-rich Atlantic sediments (Fig. 1). Despite the fact that the Kd values for Am in the two sediments were nearly the same, subsequent geochemical leaching techniques indicated that in the Atlantic sediments far more Am (62%) was present in a highly resistant form than in the labeled Pacific sediments (12%), a fact which could explain the similar relative differences observed in bioavailability. Such results indicate that geochemical associations of certain radionuclides may be a far more sensitive indicator of relative bioavailability from sediments than Kd values alone.

The bioavailability of radionuclides to benthic invertebrates from sediments can also differ between oxic and anoxic sediments. Table 3 presents data on the assimilation efficiencies of radionuclides from oxic and anoxic sediment in the tissues of polychaetes and bivalves feeding upon these sediments. While no consistent pattern emerges for all radionuclides and animal species, generally it appears that these radionuclides are somewhat more bioavailable from oxic than from anoxic sediment. These assimilation efficiencies, analogous to the assimilation efficiencies of radionuclides from biological food, suggest that radionuclides are indeed available to benthic animals even from anoxic sediment. Thus, while sediments may be considered as a final repository for particle-reactive radionuclides, they can also serve as an enriched source of radionuclides for some benthic food chains. It is noteworthy that the bioavailable fraction of radionuclides in sediment is that fraction which is adsorbed to sediment grains rather than part of the matrix of the sediment grains themselves (Griscom et al., 2002b). Further, ingestion of contaminated sediment in certain cases can account for a greater fraction of the total radionuclide in invertebrate tissues than that obtained from the dissolved phase – either pore water or overlying water (Griscom et al., 2002b; Wang et al., 1997).

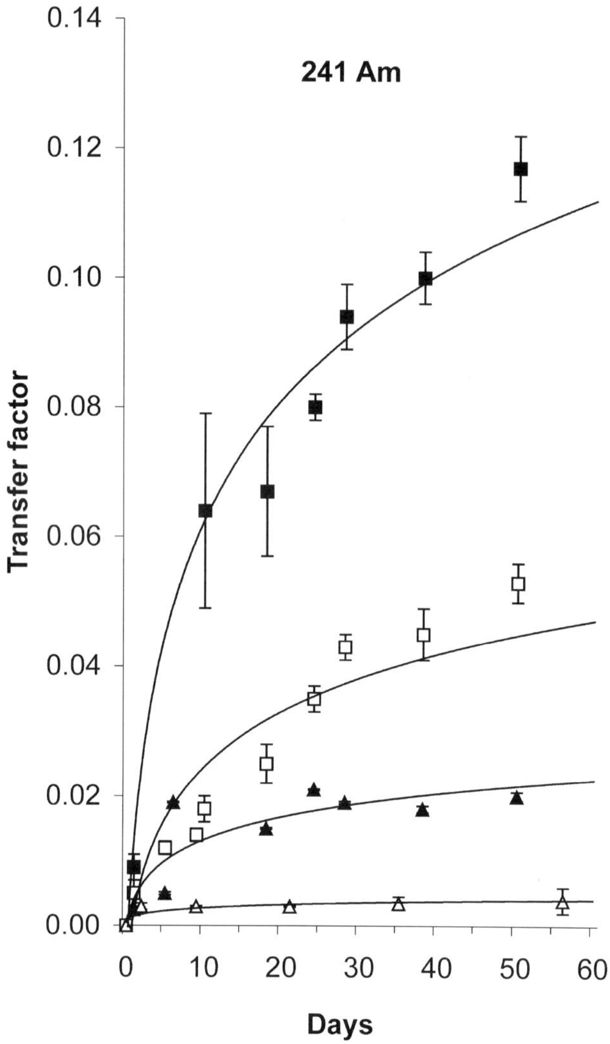

Fig. 1. Transfer of ^{241}Am to polychaetes (*Hermione hystrix*) and clams (*Venerupis decussata*) from contaminated Atlantic and Pacific sediments. Polychaetes in Atlantic (□) and Pacific (■) sediments; clams in Atlantic (△) and Pacific (▲) sediments (after Vangenechten et al., 1983).

Despite an overall low degree of radionuclide transfer from sediments to the benthos, the question arises as to what extent, over the long term, can contaminated sediments act as a source of radionuclides to biota that can mobilize and transport them further from the source. An interesting data set to test this hypothesis exists for transuranic elements in Thule, Greenland, where some 1.1 TBq of plutonium were accidentally deposited as a point source in coastal sediments (Aarkrog, 1977; Aarkrog et al., 1984). Six years after the accident, the contamination was measurable at a distance of approximately 20 km from the site. At that time

Table 3
Assimilation efficiencies (%) of ingested radionuclides in marine invertebrates feeding upon contaminated sediments

Species	Type of sediment	^{51}Cr	^{60}Co	^{65}Zn	^{110m}Ag	Reference
Polychaete:						
Nereis succinea	Oxic		89	46	27	Wang et al. (1999)
	Anoxic		39	29	12	Wang et al. (1999)
Bivalve:						
Mytilus edulis	Oxic	1				Wang et al. (1997)
	Anoxic	5				Griscom et al. (2000)
Mytilus edulis	Oxic				4	Lee et al. (2000)
	Anoxic				3	Lee et al. (2000)
Mytilus edulis	Oxic		9	22	13	Griscom et al. (2000)
	Anoxic		16	32	5	Griscom et al. (2000)
Macoma balthica	Oxic		28		20	Griscom et al. (2002a)
	Anoxic		15		10	Griscom et al. (2002a)

the half-distance for a decrease in Pu concentration from the point source was 3 km in sediments but 5–6 km for certain marine organisms indicating some mobility of Pu. Furthermore, contaminated starfish and shrimp were found more distant from the site than sedentary molluscs underscoring the greater potential of the former species to transport the contamination horizontally. The most significant finding of these temporal studies was that even in this high energy environment off Greenland, greater than 99% of the computed Pu inventory still remained bound to the sediments. In addition, no Pu from the accident was measured in the overlying surface seawater, zooplankton, fish, seaweeds, birds or marine mammals. Clearly the Pu was confined to the benthic habitat, and in fact 11 years after the accident, the entire benthos at the site contained less than 1% of the plutonium in the sediments (Aarkrog et al., 1984).

In summary, both field and experimental data indicate that radionuclide bioavailability from contaminated sediments is typically low, with transfer factors being generally less than 1 (Table 4). However, it is also clear that for some radionuclides and some organisms, the sediments can serve as an enriched source of radionuclides for benthic food chains. Besides serving as a potential source of radioactivity to organisms, sediment bound radionuclides can be redistributed by the bioturbative action of marine biota, particularly infaunal species. For example, the large echiuroid *Mamulleria lankesteri* which lives in the Irish Sea can bioturbate the sediment to a depth of about 40 cm. Analyses of its burrow lining and adjacent sediments showed that $^{239+240}Pu$ and ^{241}Am recently deposited at the surface could be redistributed to depth principally by the action of sediment ingestion and subsequent fecal pellet deposition in the burrow (Kershaw et al., 1983). Such bioturbation activities of these and other infaunal organisms will likely affect estimates of the quantities and rates of sediment plutonium and americium remobilization back to the water column, but to date there have been few quantitative studies to test this hypothesis.

Table 4
Radionuclide transfer factors from contaminated sediments

Infaunal species	Days of exposure	$^{239+240}Pu$	^{241}Am	^{252}Cf	^{99}Tc	^{55}Fe	^{95}Zr–^{95}Nb	^{137}Cs	^{106}Ru	^{60}Co	Reference
Polychaete:											
Nereis diversicolor	40 (15)	0.0014 (200)	0.0005								Beasley & Fowler (1976)
Nereis diversicolor	50	0.0019	0.0013								Hamilton et al. (1991)
Nereis diversicolor	88					0.019					Jennings & Fowler (1980)
Nereis japonica	11 (11)						0.01 (4)	0.2 (6)	0.006 (6)	0.06 (6)	Ueda et al. (1977)
Hermione hystrix	4–50	0.05 (∼300)	0.05–0.12 (1000)								Vangenechten et al. (1983), Grillo et al. (1981), Aston & Fowler (1984)
Hermine hystrix	133			0.046							Fowler et al. (1986)
Arenicola marina	14	0.002 (10)	0.003 (20)								Miramand et al. (1982)
Bivalve:											
Venerupis decussata	40–50	0.006 (∼70)	0.004–0.02 (330)								Grillo et al. (1981), Aston & Fowler (1984), Vangenechten et al. (1983)
Venerupis decussata	147			0.0061							Fowler et al. (1986)
Venerupis decussata	36				0.04–0.6						Fowler et al. (1983a)
Scrobicularia plana	14	0.01 (200)	0.008 (200)								Miramand et al. (1982)
Isopod:											
Cirolana borealis	40–50		0.006–0.032								Vangenechten et al. (1983)
Amphipod:											
Corophium volutator	14	0.10 (780)	0.11 (1000)								Miramand et al. (1982)

TF = dpm g^{-1} animal ÷ dpm g^{-1} sediment.
( ) = Numbers in parentheses are CF from water.

2. Biocycling of radionuclides

Biomass and organism size

It is now well established that the plankton community plays a fundamental role in affecting the transfer and transport of radionuclides through the marine environment. The term 'plankton' is applied to a group of small animals and plants which live freely in the water column, have limited or no powers of locomotion, and thus are more or less passively carried in a given water mass. Nearly every major group of organisms is represented in the plankton and they range in size from microscopic submicron-size viruses and bacteria to jellyfish several centimeters in diameter. Because of their ubiquity and vast number, together phytoplankton and zooplankton make up globally the largest fraction of the marine biomass and are therefore potentially important in the biogeochemical cycling of radionuclides in the oceans.

Because plankton is comprised of so many different groups of organisms, it is not practical to examine radionuclide bioaccumulation processes in each of the different species. For this reason plankton is often treated as a homogeneous community of organisms and most studies on plankton biogeochemistry have focused on plankton as a functional group or on the dominant species therein. Although the plankton may range in size over some 7–8 orders of magnitude, nearly all the species are relatively small ($<$ few cm's) which results in a high surface area to volume ratio for each entity. The general feature of having an enormous surface area in contact with seawater in many cases leads to enhanced surface adsorption of radionuclides, and results in the very high levels and concentration factors of elements observed in plankton (Lowman et al., 1971; Fowler, 1977, 1990; Fisher, 1986; IAEA, 2004). It therefore follows that bioaccumulation and excretion processes in the plankton may have a profound effect on the fluxes and residence times of radionuclides and many marine contaminants.

Radionuclide distributions in the sea strongly depend upon current and water mass movements, eddy diffusion and sedimentation processes. Movements of elements associated with plankton are also subject to physical and geological transport processes but, in addition, are affected by bioaccumulation, retention, and subsequent food chain transfer, horizontal and vertical migration, and passive sinking of biodetritus produced by these organisms. Clearly, the relative importance of these biological transport mechanisms compared to physical and chemical processes is a function of the oceanic biomass at any given location. Various aspects of radionuclide bioaccumulation, bioavailability, food chain transfer and metabolism in marine plankton have been the subject of in-depth reviews (Lowman et al., 1971; Davies, 1978; Fowler, 1982; Fisher & Reinfelder, 1995). In this section, we have only attempted to highlight the status of our current knowledge about some of the more important biologically-mediated uptake, transfer and transport processes which occur from the time radionuclides enter and leave the surface layers of the water column until they ultimately reach depth. Emphasis is placed on geochemical implications of radionuclide redistribution by sinking biogenic particulates, a process which is currently receiving much attention in many oceanographic studies (Fowler & Knauer, 1986; Fisher & Fowler, 1987; Fisher et al., 1988, 1991b, c).

Biodetrital transport processes

Bioaccumulation is not the only process affecting the flux and cycling of elements by the planktonic community; food chain dynamics, excretion, molting and death are other key factors that result in the redistribution and cycling of radionuclides. With respect to food chain transfer, many recent studies have been focused on the fractions of elements assimilated at each step of the plankton food chain (Fowler, 1982; Reinfelder & Fisher, 1991; Fisher & Reinfelder, 1995; Wang & Fisher, 1999a). Assimilation efficiencies for ingested radionuclides for the most part have been shown to be very low. In fact, most evidence from both laboratory and field studies demonstrates that radionuclide concentrations in marine micro-organisms tend to decrease with each increasing step in the food chain, i.e. radionuclide biomagnification does not occur, although exceptions have been noted, as discussed above.

Regardless of the mode of radionuclide bioaccumulation in plankton, the subsequent elimination of elements through excretion is important in maintaining elemental balance within the organism as well as affecting the marine biogeochemical cycles of many radionuclides. Elimination from intact organisms occurs by passive desorption or ion exchange, active excretion of the soluble element, and particulate loss via zooplankton feces, molts and reproductive products. In addition, when organisms die and decompose, the radionuclides associated with their tissues can be released into the ambient seawater. Most often trace elements are lost from plankton more slowly than they are accumulated. Loss rates are rarely constant; hence, there are biological half-times characteristic of the various individual radionuclide pools within the organism. Biological half-times for radionuclide loss vary from several hours to a few days for phytoplankton and zooplankton species (Fowler, 1982; Fisher et al., 1983a, b; Fisher & Reinfelder, 1995). As one example, earlier studies measured the soluble excretion of radionuclides from zooplankton that had been contaminated in situ by nuclear testing in the North Pacific (Keunzler, 1969a, b). Rapid loss rates of 1–22% h^{-1} were noted for radioactive I, Zn, Co and Fe, reflecting both the short exposure time and the fact that loss was only measured for a few hours. However, more recent evidence indicates that long-term depuration from zooplankton does not follow a single exponential rate but takes place from both fast and slowly exchanging compartments (Fowler, 1982; Wang & Fisher, 1998a). Therefore, the rates measured for the naturally contaminated zooplankton probably reflect loss from a rapidly exchanging pool that may represent only a small fraction of the organism's total radionuclide load. Various biokinetic models which have been developed to assess the flux of radionuclides and metals through zooplankton depend heavily on soluble excretion rates to determine total excretion (Small et al., 1973; Wang & Fisher, 1998b; Fisher et al., 2000). Therefore, it is important that soluble loss from all labeled pools within the plankton are taken into account when deriving the soluble excretion parameter for such models.

Other factors that can affect radionuclide eliminationare temperature and length of time that zooplankton are exposed to the radionuclide. Furthermore, elimination rates vary greatly with species; for example, at frequent intervals planktonic crustaceans lose a large fraction of their element body burden when they molt. Molting rates increase with temperature, therefore radionuclide elimination via molting is likely to be more important in the warmer, upper layers of the water column.

Often the most important route of loss from zooplankton is defecation. Especially for non-biologically essential elements or those that are poorly assimilated, loss via feces becomes increasingly important. Element assimilation from food is therefore an important aspect of radionuclide bioaccumulation in zooplankton and consequently it is currently an active area of research in the field of trace element biogeochemistry (Fowler, 1982; Fisher et al., 1991a;

Reinfelder & Fisher, 1991; Wang et al., 1996b; Anastasia et al., 1998; Wang & Fisher, 1999a; Stewart and Fisher, 2003). Using accurate knowledge of assimilation rates and the varied parameters that control them, reasonable estimates can be deduced of the fractions of radionuclides that will be released in fecal pellets for eventual cycling in or removal from the water column.

The production of zooplankton fecal pellets is therefore one of the key vectors controlling the vertical flux and resultant residence times of elements in the sea. In the case of radionuclides, initial enrichment occurs on the smallest planktonic forms and, primarily through grazing activities of zooplankton, these small particles (e.g. phytoplankton cells) and organic aggregates are 'packaged' into larger detrital particles which rapidly sediment due to their increased sinking speeds (Higgo et al., 1977). Alternatively, plankton can transport incorporated radionuclides through their horizontal and vertical movements. Phytoplankton migrates very little but many species of larger zooplankton exhibit diel vertical migrations over several hundred meters. Nevertheless, model studies indicate that diel vertical migration is restricted to roughly the top 1000 meters for these species and conclude that sinking detrital products (e.g. fecal pellets, molts, shells, carcasses, phytodetritus, etc.) from the plankton are quantitatively more important than vertical migration for the downward transport of elements (Lowman et al., 1971; Small & Fowler, 1973).

Specialized field collection techniques have been used to sample zooplankton excreta for specific radionuclide analyses. The data indicate that biogenic detrital particles originating from zooplankton are rich in many radionuclides (Table 5). Furthermore, element concentrations are often higher in these types of particles than in the organisms producing them or their precursor prey (Fowler, 1977; Higgo et al., 1977; Krishnaswami et al., 1985). Element enrichment in pellets occurs when organic rich food particles are ingested and subsequently stripped of nutritive material (loss of dry weight) with the resultant residual fecal pellet, composed mainly of non-digestible hard parts, displaying further radionuclide enrichment. Since the bulk of biogenic debris, like fecal pellets, is produced in the surface layers, it is not surprising that these particles are primary vectors for removing trace elements and radionuclides from the upper water column. Furthermore, fecal pellets may continue to scavenge elements and radionuclides as they sink through the water column (Fisher et al., 1991c). In fact, using elemental concentrations in biogenic particles and mass balance considerations of river input and sediment output of the same elements, model studies have concluded that the settling of fecal pellets and fecal aggregates is the most important mechanism affecting the vertical transport, and thus residence time, of many trace elements (Cherry et al., 1978; Li, 1981). However, the data used for these models, like those in Table 5, are limited and far more information on radionuclides in a variety of fecal pellets and other biogenic detrital particles is needed in order to confirm the relative importance of zooplankton feces in controlling radionuclide residence times.

Testing the radionuclide vertical transport hypothesis in the field is difficult; however, the use of sediment traps and large volume filtration systems has allowed direct examination of certain aspects of the processes by which plankton accumulate radionuclides, repackage them and transport them to depth (Fowler et al., 1983b, 1987, 1990, 1991; Bacon et al., 1985; Fowler & Knauer, 1986; Fisher et al., 1988). Such data coupled with knowledge of the radionuclide concentrations in the dissolved and suspended particulate phases have given new

Table 5
Environmental levels of anthropogenic and natural radionuclides in zooplankton and their particulate products

Element	Zooplankton	Fecal pellets ($Bq\ kg^{-1}$)	Molts (%)*	Reference
	Phytoplankton			Fowler et al. (1983b), Fowler (1987)
$^{239+240}Pu$	1.0, 0.35			
^{241}Am	0.22, 0.052			
	Copepods			Fowler et al. (1983b, 1987, 1991), Higgo et al. (1977)
$^{239+240}Pu$	0.48, 0.046, 0.18	1.2, 3.3		
^{241}Am	0.12, 0.022	1.0, 2.7		
^{210}Po	126			
^{232}Th	0.63			
^{238}U	12.6			
	Euphausiids			Higgo et al. (1977, 1980)
$^{239+240}Pu$	0.015	3.6	0.18 (90)	
^{210}Po	40.7	907	13.3 (2.5)	
^{232}Th	0.013	9.25	0.096 (57)	
^{238}U	0.78	19.2	9.07 (90)	
	Salps			Krishnaswami et al. (1985), Fowler et al. (1991)
$^{239+240}Pu$	–	8.9		
^{241}Am	–	5.0		
^{210}Po	260	658		
^{210}Pb	18.3	400		
^{234}Th	780	12,500		
^{232}Th	1.73	16.8		
^{228}Th	3.97	36.0		
^{238}U	12.5	16.2		

*Percent of total body burden contained in molt.

insight into the processes controlling the vertical transport and residence times of radionuclides. As one example, a data set from the central North Pacific gyre is shown in Table 6. Both live and detrital particles from the gyre contained relatively high concentrations of plutonium and americium. Mixed microplankton, primarily composed of copepodites, diatoms, cyanobacteria, radiolarians and dinoflagellates, contained 0.42 $Bq\ kg^{-1}$ and 0.35 $Bq\ kg^{-1}$ dry of $^{239+240}Pu$ and ^{241}Am, respectively. For pure copepods in the size fraction 500–1000 μm and 1000–2000 μm, corresponding transuranic concentrations were approximately 10 times lower. Large amounts of mucilaginous floc ('marine snow'), containing a myriad of detrital and living particles, were netted in the upper 150 m and closely resembled the material collected in the sediment traps. $^{239+240}Pu$ and ^{241}Am levels in the marine snow were 0.688 $Bq\ kg^{-1}$ and 0.385 $Bq\ kg^{-1}$ dry. The highest concentrations (5.33–19.6 $Bq\ kg^{-1}$ dry) for both radionuclides were found in copepod fecal pellets, again underscoring the potential importance of these particular particles in effecting transuranic transport through the water column. In all biological samples, the Am/Pu activity ratios (0.44–0.82) were somewhat higher than that normally found in fallout at these latitudes (~0.3), indicating a relative enrichment of ^{241}Am in the plankton.

In a related seasonal study in the North Pacific, Fowler et al. (1991) showed that the downward flux of particulate-associated transuranic elements was closely related to grazing activity of zooplankton in the overlying water column. Concentrations of $^{239+240}$Pu and ^{241}Am in sinking particles were in good agreement with those measured in zooplankton fecal pellets. Concentrations of Pu and Am in the pellets of salps (Table 5), which were several times higher than transuranic concentrations in copepod fecal pellets, could therefore contribute significantly to transuranic flux in areas where salp swarming occurs. Coupling these biologically driven vertical fluxes of $^{239+240}$Pu with concentrations of $^{239+240}$Pu in the upper water column results in average computed residence times of plutonium of approximately four years in this region of the North Pacific.

Material collected in sediment traps at different depths showed a marked increase in transuranic concentrations with depth (at least to 1000 m), suggesting scavenging of ^{241}Am and $^{239+240}$Pu by plankton detritus as it settled through the water column (Table 6). The concomitant increase in Am/Pu ratio in these particles with depth indicates that ^{241}Am was being adsorbed and scavenged by the particles to a greater extent than $^{239+240}$Pu. The question remains as to the fate of these high transuranic concentrations in biogenic particles as they sink into deeper waters and eventually deposit in the sediments. There is some indication from the trap samples (Table 6) that Pu may be remineralized from particles sinking between 1000 m and 1500 m. Furthermore, from measurements of surface sediments (top 1 cm) taken from five mid-Pacific gyre cores in the vicinity of this station, it was found that $^{239+240}$Pu and ^{241}Am levels were quite low, approximately 0.07 Bq kg^{-1} dry (Table 6). This suggests that large particles forming these sediments have lost a substantial portion of their associated radionuclide content by the time they reach the sediment. In fact, laboratory radiotracer studies have shown that Pu and Am are not irreversibly bound to fecal pellets and other biodetritus but are leached out with half-times for release on the order of 1–3 weeks (Fowler, 1982; Fisher & Fowler, 1987; Fisher & Reinfelder, 1995). It would therefore appear that much of the transuranic concentration associated with sinking biogenic detritus is recycled in deeper waters.

Another striking example of radionuclide scavenging behavior is the particle flux data from high resolution, time-series sediment traps in the western Mediterranean Sea which clearly demonstrated that Chernobyl fission products arriving at the sea surface essentially as a single pulse on 3 May 1986, were rapidly transported to 200 m in approximately 7 days primarily by zooplankton grazing activities (Table 7). Microscopic examination of the trap samples containing the bulk of the radioactivity indicated they were composed mainly of copepod fecal pellets. Fresh pellets collected from copepods living above the traps were found to contain similar radionuclide concentrations and ratios as those in the trap material, confirming that such pellets were responsible for transporting the surfaced-introduced radioactivity to depth. Furthermore, isotopic ratios in pellets, air, water and copepods indicated that particle reactive radionuclides like ^{144}Ce, ^{141}Ce, ^{106}Ru and ^{103}Ru were scavenged to a far greater extent by sinking fecal pellets than the cesium nuclides, an observation consistent with the radionuclides' chemical behaviour in seawater.

Table 6
Transuranic concentrations (Bq kg^{-1}dry $\pm 1\sigma$) and activity ratios in biological source material and sediment trap samples from the central Northeast Pacific Gyre in 1983 (after Fowler, 1987)

Samples	$^{239+240}$Pu	^{241}Am	^{241}Am/$^{239+240}$Pu
Microplankton (60 μm)	0.422 ± 0.17	0.348 ± 0.10	0.82 ± 0.41
Marine snow	0.688 ± 0.18	0.385 ± 0.13	0.56 ± 0.24
Copepods (500–1000 μm)	0.067 ± 0.01	0.3 ± 0.01	0.44 ± 0.18
Copepods (1000–2000 μm)	0.3 ± 0.01	0.2 ± 0.01	0.75 ± 0.47
Copepod fecal pellets	5.33 ± 4.07	19.6 ± 5.92	3.7 ± 3.0
Sediment trap:			
150 m	0.929 ± 0.096	0.22 ± 0.052	0.24 ± 0.06
250 m	2.15 ± 0.537	0.87 ± 0.22	0.40 ± 0.10
500 m	7.36 ± 1.1	4.66 ± 0.41	0.63 ± 0.09
1000 m	15.0 ± 1.3	10.0 ± 0.81	0.67 ± 0.06
1500 m	11.1 ± 1.7	10.1 ± 0.89	0.91 ± 0.14
Top 1 cm of bottom sediment ($X \pm 1\sigma$, $n = 5$)	0.070 ± 0.01	0.2±0.004	0.3

Table 7
Prominent Chernobyl fallout radionuclides in large particles (Bq g^{-1} dry) collected at 200 m depth in a 2200 m water column 15 miles off Calvi, Corsica (Samples 1–6), and in fecal pellets produced by zooplankton living over the traps (Sample 7). Trap values are decay-corrected for midpoint of sampling period (after Fowler et al., 1987, 1990)

Sample:	1	2	3	4	5	6	7
Date:	13–20 April	20–26 April	26 April–2 May	2–8 May	8–15 May	15–21 May	6 May (pellets)
Dry wt (mg)	167.8	87.5	50.17	51.43	42.06	45.26	42.5
Radionuclide:							
^{95}Zr	<0.07	<0.2	<0.3	<0.2	24.5 ± 1.4	<0.2	1.4 ± 0.8
^{95}Nb	<0.03	<0.1	<0.2	<0.1	31.8 ± 1.1	<0.2	<0.2
^{103}Ru	<0.06	<0.1	<0.2	3.7 ± 0.2	23.6 ± 1.0	14.0 ± 0.4	16.0 ± 1.9
^{106}Ru	<0.2	<0.4	<0.8	1.1 ± 0.5	5.4 ± 1.8	3.5 ± 0.7	5.8 ± 2.9
^{134}Cs	<0.05	<0.05	<0.05	0.41 ± 0.05	2.1 ± 0.2	1.9 ± 0.1	3.4 ± 0.6
^{137}Cs	<0.05	<0.05	0.15 ± 0.08	0.85 ± 0.08	3.8 ± 0.3	4.0 ± 0.1	6.3 ± 1.0
^{141}Ce	<0.2	<0.2	<0.3	1.3 ± 0.7	12.6 ± 0.6	1.1 ± 0.5	0.9 ± 0.4
^{144}Ce	<0.06	<0.3	<0.3	<0.2	13.6 ± 0.7	<0.4	2.5 ± 1.3
$^{239+240}$Pu*	5.43	2.00	3.00	3.22	9.70	4.71	7.4
^{241}Am*	0.87	0.68	1.51	1.05	3.63	2.83	0.63

*Bq kg^{-1}.
Note: Values preceded by "<" indicate limits of detection.

Thus, direct measurements of particles from sediment traps deployed under various, and sometimes fortuitous, conditions have confirmed the importance of plankton debris in rapidly transporting atmospherically derived radionuclides from the euphotic zone to depth. The larger question remains regarding the ultimate fate of these radionuclide-enriched biogenic particles, and several studies have been undertaken to examine the radionuclide remineralization processes and rates associated with sinking particles (Fowler & Knauer, 1986; Fisher & Fowler, 1987; Fisher et al., 1988, 1991b, c; Lee & Fisher, 1992a, b, 1994; Fisher & Wente, 1993; Reinfelder et al., 1993). Information to date suggests that, in general, metabolically essential elements and radionuclides that are bound to proteins are remineralized more rapidly than non-essential, particle-reactive elements which are often associated with surfaces of non-labile materials in these particles. Many of these data are derived from laboratory radiotracer studies and it would be a step forward if a method could be devised to test these hypotheses and ground-truth such experimentally-determined remineralization and scavenging rates in nature.

3. Bioindicators of radionuclides

In radioecology different species have been selected to give information on ambient radionuclide contamination levels in specific environments. Early studies focused on sites where inputs of radioactivity were known or believed to occur. Thus, different types of organisms were chosen in tropical atolls near nuclear testing sites in the Pacific (Kuenzler, 1969a, b; Lowman et al., 1971) from those that were selected to monitor radionuclide inputs from nuclear processing activities in temperate regions (Frazier & Guary, 1976; Pentreath, 1981; Charmasson et al., 1999). Choices of bioindicators are often made based on whether the monitoring studies are purely scientific or health and safety oriented. For example, in open sea areas, plankton are often the only organisms routinely found, but the plankton is made up of different species which vary over time and thus could furnish ambiguous results on trends in ambient radionuclide levels. In coastal regions, edible species such as fish and crustaceans have been selected primarily because of seafood safety concerns for the human population that consumes them, but both types of organisms present drawbacks when trying to assess radionuclide concentrations in their surrounding waters. Fish are highly mobile and not always representative of a given location. Furthermore, they are often poor integrators of radionuclides (i.e. display low concentration factors), and detailed knowledge of radionuclide distribution and behavior in individual fish tissues and organs is usually a prerequisite for relating radionuclide concentrations in fish to corresponding levels in seawater. In the case of benthic and pelagic crustaceans, frequent molting of the animals' exoskeleton and resultant loss of radionuclide burden can lead to highly variable radionuclide concentrations among a given group of crustaceans. For many of these reasons, interest has focused on sedentary, filter-feeding bivalves that have proven useful bioindicators of many inorganic and organic contaminants (Phillips, 1980).

Because biological factors can greatly influence the extent to which trace metals concentrate in mussel tissues, existing monitoring programs make sure that these are considered in their sampling protocols (NOAA, 1998). For example, mussels that are spawning can mobilize certain metals more than others, and a misleading picture of the bioavailable concentrations of those metals can result if spawning mussels are sampled. A number of studies have experimentally examined the factors that influence the bioconcentration of important components of radioactive wastes (e.g. isotopes of Pu, Am, Tc and others) in marine mussels (Guary & Fowler, 1981; Dahlgaard, 1986; Bjerregaard et al., 1985; Fisher & Teyssié, 1986; Fisher et al., 1996). Clearly, the same processes governing metal uptake and retention in these animals are applicable to the long-lived radioactive wastes. Radionuclide concentrations in mussel tissues have been regularly monitored in the US for decades (Valette-Silver & Lauenstein, 1995; NOAA, 1998).

Modeling radionuclide bioaccumulation in marine organisms

As noted in Section 1, aquatic animals can accumulate radionuclides from both the dissolved phase directly and from ingested food. The relative importance of the different uptake pathways has received considerable attention in recent years, in part because contaminants, including radionuclides, accumulated from food deposit in different tissues in animals than those obtained from the dissolved phase, and this may ultimately have consequences for subsequent trophic transfer and toxicity (Bjerregaard et al., 1985; Fisher et al., 1996; Hook & Fisher, 2001). Bioenergetic-based kinetic models to describe the accumulation of contaminants in aquatic animals have been developed relatively recently and have been successfully applied for a variety of organic and inorganic contaminants. These models provide a broad framework for addressing controls on contaminant bioaccumulation for diverse organisms and can be used for studying contaminant bioavailability and determining the relative importance of different routes of contaminant accumulation, including that of radionuclides (Landrum et al., 1992; Wang et al., 1996a). It is noteworthy that this general modeling approach works equally well for environmentally important radionuclides as it does for other inorganic contaminants. The models are flexible enough to incorporate environmental variability in radionuclide sources, radionuclide concentrations, food availability and organism growth rates in their predictions of organism radionuclide levels. One widely used version of these models treats contaminant (or radionuclide) accumulation as a first order function of contaminant concentrations in particles and water.

$$\frac{\mathrm{d}C}{\mathrm{d}t} = (k_\mathrm{u} * C_\mathrm{w}) + (\mathrm{AE} * \mathrm{IR} * C_\mathrm{f}) - (k_\mathrm{e} + g) * C, \qquad (1)$$

where:

C is the contaminant (or radionuclide) concentration in the animals (μg g^{-1} or Bq g^{-1}),
t is the time of exposure (d),
k_u is the uptake rate constant from the dissolved phase (l g^{-1} d^{-1}),
C_w is the contaminant (or radionuclide) concentration in the dissolved phase (μg l^{-1} or Bq l^{-1}),
AE is the assimilation efficiency from ingested particles (%),
IR is the ingestion rate of particles (mg g^{-1} d^{-1}),
C_f is the contaminant (or radionuclide) concentration in ingested particles (μg mg^{-1} or Bq mg^{-1}),
k_e is the efflux rate constant (d^{-1}), and
g is the growth rate constant (d^{-1}).

At steady state, this equation simplifies to:

$$C_\mathrm{ss} = \frac{(k_\mathrm{u} * C_\mathrm{w}) + (\mathrm{AE} * \mathrm{IR} * C_\mathrm{f})}{(k_\mathrm{e} + g)}, \qquad (2)$$

where C_{ss} is the steady-state concentration of contaminant (or radionuclide) in the organism ($\mu g\ g^{-1}$ or $Bq\ g^{-1}$). The efflux parameter, k_e, can be split into solute (k_{ew}) and particle (k_{ef}) components (3).

$$C_{ss} = \frac{(k_u * C_w)}{(k_{ew} + g)} + \frac{(AE * IR * C_f)}{(k_{ef} + g)}. \tag{3}$$

Except for growth, all of the parameters in equation (2) are readily estimated from laboratory radiotracer studies. The importance of growth varies, depending on the species and stage of maturity.

Application of this biokinetic model has shown that laboratory-based measurements of AE, k_u, k_{ew} and k_{ef} are applicable to field situations for populations of marine mussels (Wang et al., 1996a; Fisher et al., 1996); clams (Luoma et al., 1992), copepods (Fisher et al., 2000; Stewart & Fisher, 2003), fish (Baines et al., 2002) and freshwater mussels (Roditi et al., 2000). Site-specific model predictions for metal concentrations in animal tissues are strikingly close to independent field measurements for diverse water bodies, suggesting that it is possible to account for the major processes governing contaminant concentrations in marine animals and that the laboratory-derived kinetic parameters are applicable to natural conditions. A note on the applicability of this approach to modeling radionuclides in marine organisms is appropriate here. With only one exception (that of ^{210}Po: Stewart & Fisher, 2003), no attempt has been made to field verify model predictions of radionuclide concentrations in animal tissues, although there is no reason to expect that radioactive and stable metals should behave any differently from one another with respect to their interactions with biota. Indeed, the laboratory-determined kinetic parameters were determined with appropriate radioisotopes for the metals being considered. Since the field measurements and model predictions match one another so closely, it is reasonable to assume that the kinetic parameters determined with radioisotopes are applicable to understanding the behavior of stable metals in natural waters.

Because some of the parameters in equation (2) may not be available, modifications can be made that would enable useful computations to be made. For example, the estimate of metal uptake from the dissolved phase can be described as:

$$I_w = \alpha_w * FR * C_w = k_u * C_w, \tag{4}$$

where:

I_w is the metal (or radionuclide) influx rate from the dissolved phase (μg [or Bq] $g^{-1}\ d^{-1}$), note that this parameter can be measured experimentally, and

k_u is the dissolved uptake rate constant ($l\ g^{-1}\ d^{-1}$) of the metal/radionuclide, which equals absorption efficiency of dissolved metal/radionuclide across the gill (α_w) times the filtration rate of the animal (FR).

The k_u can be computed from the relationship between I_w and C_w. When C_f is not known, it can be calculated from C_w by applying a Kd value for suspended particulate matter:

$$C_f = C_w * Kd, \tag{5}$$

where Kd is the partition coefficient of a metal or radionuclide on the ingested particles ($l\,kg^{-1}$).

The fraction of total contaminant uptake attributable to uptake from the dissolved phase and particulate ingestion can be calculated as:

$$F_{\mathrm{w}} = \frac{C_{\mathrm{w,ss}}}{C_{\mathrm{ss}}}, \tag{6}$$

$$F_{\mathrm{f}} = \frac{C_{\mathrm{f,ss}}}{C_{\mathrm{ss}}}, \tag{7}$$

where:

F_{w} is the proportion of accumulated metal/radionuclide obtained from the dissolved phase, and

F_{f} is the proportion of accumulated metal/radionuclide obtained from food.

Note that C_{w} (dissolved concentration, μg [or Bq] l^{-1}) and C_{f} (particulate concentration, μg [or Bq] mg^{-1}) are directly dependent on the total suspended solids load (TSS), Kd, and the total concentration (C_{t}, μg [or Bq] l^{-1}) of contaminant in the water column (that is, dissolved plus particulate contaminant concentration).

$$C_{\mathrm{t}} = C_{\mathrm{w}} + (C_{\mathrm{f}} * \mathrm{TSS}), \tag{8}$$

$$C_{\mathrm{t}} = C_{\mathrm{w}} + (C_{\mathrm{w}} * \mathrm{Kd} * \mathrm{TSS}). \tag{9}$$

Thus,

$$C_{\mathrm{w}} = \left[\frac{1}{(1 + \mathrm{TSS} * \mathrm{Kd})}\right] * C_{\mathrm{t}}. \tag{10}$$

It is noteworthy that there is no constant proportionality between C_{ss} and C_{t}; that is, bioconcentration factors vary with biological factors like physiological state and environmental factors like temperature. Changes in any of the kinetic parameters can significantly affect the ratio of C_{ss} to C_{t}, which is a mathematical expression of the widely employed bioaccumulation factor (BAF). The BAF as defined here thus considers both dissolved and particulate metal (or radionuclide) uptake in animals:

$$\mathrm{BAF} = \frac{C_{\mathrm{ss}}}{C_{\mathrm{t}}} = \left[\frac{k_{\mathrm{u}}}{k_{\mathrm{ew}} + g} + \frac{\mathrm{AE} * \mathrm{IR} * \mathrm{Kd}}{k_{\mathrm{ef}} + g}\right] * \left[\frac{1}{1 + \mathrm{TSS} * \mathrm{Kd}}\right]. \tag{11}$$

Recently it has been shown that bioconcentration factors of some metals in some aquatic animals are inversely related to dissolved ambient concentrations (McGeer et al., 2003). This has been shown to be particularly evident for essential metals whose internal tissue concentrations are regulated for physiological homeostasis. It is unlikely that such regulation would occur for those radionuclides (e.g., transuranic elements and certain other fission products) for which there is no biological use, and consequently bioconcentration factors for these radionuclides should be largely independent of ambient concentrations.

While the above model appears to hold relatively well for bivalves and some species of zooplankton and juvenile fish, it should be tested with other bioindicator organisms that display widely different, relative uptake responses to radionuclide inputs from food and water. One group of benthic organisms receiving much attention as contaminant bioindicators are the echinoderms, particularly starfish that are known to accumulate ^{137}Cs, ^{60}Co and transuranium nuclides and retain significant fractions in their tissues for several months (Guary et al., 1982; Hutchins et al., 1996; Fowler & Teyssié, 1997; Warnau et al., 1999). These organisms therefore have the potential to provide an excellent medium to long-term historical record of radionuclide contamination in cases where sampling for monitoring is carried out only infrequently. In such a case, filter-feeding bivalves may purge themselves of the same radionuclides over relatively short time-scales and, hence, the signal for an acute contamination event might be missed. Clearly, when bio-monitoring temporal and spatial trends in radionuclide concentrations, bioindicator species selection should be based not only on the organism's bioconcentration potential but also on the strategies and constraints of the radionuclide monitoring programme.

4. Potential impacts on organisms and ecosystems

In recent years, there has been an increasing awareness of the vulnerability of the marine environment and of the need to protect it against the effects of industrial pollutants including radionuclides. Until now it has been considered that standards of radiation protection that are adequate to protect human health will also ensure that no other species is threatened as a population, even if individuals of the species may be harmed. Environmental protection philosophy as regards radiation has evolved however and, within the overall framework of protecting and enhancing human well-being, increasing emphasis is currently being placed on ensuring the protection of biotic populations other than man (Pentreath, 1999; Stone, 2002).

Natural radioactivity

In the marine environment in order to assess the biological effects of radiation, a sound knowledge of the radiation dose rate regimes experienced by marine biota is an essential prerequisite. Marine organisms have been exposed to low levels of radiation from environmental and cosmic sources to varying degrees throughout geologic time. Only during the last half century or so have the potential biological effects from anthropogenically-introduced radioactivity in the sea been considered against the background of natural radioactivity (Bowen et al., 1971; Templeton et al., 1971; IAEA, 1976, 1979). With the recognition that marine biota receive dose from exposure to radionuclides in their surrounding waters, sediments and those incorporated in their own tissues, it is necessary to have quantitative information on the levels and distributions of natural radionuclides in marine species so that the actual dose received from artificial radionuclide contaminants can be accurately assessed. A summary of typical concentrations of selected natural radionuclides in some principal groups of marine organisms is given in Table 8. It should be noted that whereas such measurements

Table 8
Typical concentrations of selected natural radionuclides (mBq g^{-1} wet)* in marine organisms (data from Cherry & Shannon, 1974; IAEA, 1976; Pentreath, 1977; Higgo et al., 1980; Krishnaswami et al., 1985; Aarkrog et al., 1997)

Radionuclide	Phytoplankton	Zooplankton	Crustacea	Mollusca	Fish (muscle)
^{3}H	0.02–0.1	0.02–0.10	0.02–0.10	0.02–0.10	0.02–0.10
^{14}C	11	11	22	19	15
^{40}K	93	93	93	107	93
^{87}Rb	–	–	1.5	1.9	0.93
^{238}U	0.4–1.9	0.22–0.74	0.10	–	0.003–1.1
^{226}Ra	0.74–3.7	0.22–0.74	0.74	0.74	0.007–0.19
^{210}Pb	3.7–26	0.37–9.3	1.5–2.6	0.19–0.37	0.007–0.09
^{210}Po	3.3–63	1.9–41	6–59	15–41	0.015–5.2
^{228}Th	0.26–2	0.07–0.81	–	0.37	0.037
^{232}Th	0.074	0.04–0.074	0.0017	–	–

*In some cases concentrations were converted to wet weight using a dry/wet weight ratio = 0.1.

have been made in a great diversity of marine biota, complete data sets of natural radionuclides for a single species, or even groups of organisms, are rare or non-existent. Although natural radiation sources, and thus exposures to natural radioactivity, in the marine environment do vary, organisms are exposed to these radionuclides throughout their lives and therefore tissue concentrations probably better reflect equilibrium conditions than do concentrations of artificial radionuclides whose inputs and concentrations vary greatly over space and time.

From Table 8 it can be seen that ^{40}K and ^{210}Po generally display the highest concentrations in most of the organism groups examined. Based on calculations of dose rate emanating from exposure to water, sediments, and internal radionuclide deposition, the only significant source of exposure for these organisms is incorporated radioactivity with a somewhat greater proportion of exposure coming from sediments in the case of benthic crustaceans and molluscs (IAEA, 1976). In fact, of the natural radionuclides in seawater, only ^{40}K contributes significantly to the overall dose rate. With respect to incorporated radioactivity in phytoplankton, zooplankton and pelagic fish, ^{210}Po is the main source of the natural dose with ^{40}K contributing most of the remainder. For benthic fish, crustaceans and molluscs, natural gamma emitters in the surface sediments give a similar radiation dose. The alpha-emitter ^{210}Po is of particular interest because of its non-homogeneous distribution within tissues of many marine species. For example, it is found in very high concentrations in crustacean hepatopancreas and fish viscera (Cherry & Shannon, 1974; Cherry & Heyraud, 1982), a fact which results in extremely high doses of alpha radiation delivered to individual organs or tissues. In the case of pelagic penaeid shrimp, doses in their hepatopancreas on the order of 2 Sv y^{-1} have been recorded, a dose of natural radiation that is roughly three orders of magnitude higher than that received by the human lung (Cherry & Heyraud, 1982). Clearly, an expanded data base on polonium levels in marine organisms and their tissues will help refine dose estimate calculations upon which assessments of the effects of ionizing radiation on marine organisms are strongly dependent. Recent ^{210}Po compilations that have been made at regional levels tend to show that polonium concentrations in various seafood species do not vary significantly from one marine environment to another and that regional differences in concentration, if any, are within the natural variations observed among species (Aarkrog et al., 1997).

Artificial radioactivity and dose

For the purposes of estimating dose rates in marine species, a compilation of fallout radionuclide concentrations in similar groups of marine organisms was made in the mid-1970s (IAEA, 1976). It is evident that such concentrations can only reflect the levels in seawater of a particular environment at a given time and, owing to input fluctuations, decay, dilution and various geochemical processes, they are rarely indicative of an equilibrium situation. Nevertheless, using dose models for the different groups of organisms, it was concluded that in the long term, for the fallout nuclides ^{137}Cs is the major contributor to the dose rate received by all groups of organisms, with $^{239+240}$Pu and ^{90}Sr/^{90}Y being of some significance for plankton and crustaceans, respectively (ibid.). The dose rates from natural radioactivity in the marine environment were estimated to range up to approximately 0.4 μGy h^{-1}. The corresponding dose rates from global fallout were at that time similar to or somewhat higher than those from natural background, but are now declining and are in the same range as the dose rates from natural radiation. Similar calculations have shown that in areas receiving liquid radioactive wastes such as the northeast Irish Sea near the Sellafield reprocessing facility, dose rates were much higher than those from natural background with maximum values reaching 250 μGy h^{-1} at the points of discharge. Nevertheless, in areas where wastes are released, such rates can most likely be considered as an upper limit. For example, since those estimates were derived in the 1970s, Sellafield waste releases have dropped dramatically and corresponding dose rates are certainly much lower at present.

Radiation effects

There is a large body of investigations on the effects of ionizing radiation on marine species that have been thoroughly reviewed and assessed by expert groups over the years (e.g. Templeton et al., 1971; IAEA, 1976; NCRP, 1991). Unfortunately most of the experimental effects studies have employed unusually high doses and dose rates which bear little relevance to those which exist in the marine environment and hence, it is extremely difficult if not impossible to extrapolate those results to low level chronic exposures occurring in the marine ecosystems. Nevertheless an assessment of available relevant radiobiological data indicates that the most radiosensitive marine species presently known are teleost fish, particularly the developing eggs and juveniles of some species (IAEA, 1976). When considering radiation effects on these or any other species at the population or ecosystem level, it is necessary to examine the relative importance of such effects with those caused by conventional pollutants or from mortality under natural conditions. Based on an assessment of the presently existing information from both experimental and field studies, it appears that no deleterious effects on marine populations or ecosystems would be expected at the doses and dose rates that have been observed or estimated to occur in the marine environment (ibid.).

5. Suggestions for future studies

A close examination of the available radioecology literature shows that a disproportionate amount of data and thus our present day knowledge of radionuclide concentrations, distributions, and behavior in marine biota comes from studies carried out in mid-latitude temperate zones or specific tropical islands where nuclear testing has occurred. Until recently, very little was known about radionuclide bioaccumulation and behavior in marine organisms inhabiting polar regions. However, waste dumping activities and nuclear accidents which have taken place in Arctic waters have led to a shift in focus to derive information on existing radioactivity levels in polar marine species as well as how radionuclides behave and at what rates they are accumulated and transferred at low temperatures through Arctic food chains (Hutchins et al., 1996, 1998; Fisher et al., 1999; Carroll et al., 2002; IAEA, 2004). An analogous case can be made for tropical coastal areas that encompass a good part of the world's oceans and supply seafood to a large fraction of the world's population. At present global fallout is the primary source of radioactivity in the low latitudes, but in many tropical countries nuclear power reactors situated in coastal areas are planned for the future, and could result in an additional source of environmental radioactivity. Furthermore, these regions contain some of the most sensitive marine ecosystems, e.g. coral reefs, which are continually under stress from many other sources of contamination and could easily become impacted. Therefore, some effort to obtain radionuclide concentration factor and transfer rate data for coastal tropical species would appear warranted, and it would allow assessing the extent to which numerous corresponding data from temperate zones could in fact be extrapolated to the tropics and applied in present and future radionuclide monitoring and radiation protection programmes.

To adequately address questions related to the transfer of radionuclides through marine food chains and the potential for radionuclide biomagnification, it is imperative to have relevant data for each link in a specific food chain. Marine mammals are considered to be at or near the top of the marine food chain. Furthermore, many species are consumed by indigenous human populations, particularly those in Arctic regions. Therefore, to establish adequate radiation dose models, radionuclide concentrations and transfer factors to this link in the marine food chain leading to man are necessary. It is noteworthy that since marine mammals obtain their radionuclide body burden principally from food and that some mammals feed at very different levels of the food chain, radionuclide levels and hence concentration factors are likely to vary considerably within any one group of mammals. Such variability might occur in some cetaceans; for example, concentrations in whalebone whales, which consume plankton, compared to that in carnivorous toothed whales. Given the importance of diet, transfer mechanisms through the food chain probably control the body burdens in other mammal species as well. Radionuclide data are extremely limited for marine mammals when compared to other marine organisms, and we know little about radionuclide transfer processes in these mammals (IAEA, 2004). Given the difficulties and restrictions related to sampling marine mammals and the near impossibility of controlled experimentation with these animals, making maximum use of samples obtained for other purposes could help enhance the mammalian radioactivity database.

Bioaccumulation processes result in a high enrichment of radionuclides in plankton primarily because of the large relative surface area for adsorption in the different species comprising the plankton community. In terms of trace element and radionuclide concentration data, most field data pertain to zooplankton with lesser amounts available for phytoplankton, and virtually no information about levels in bacterioplankton because of the difficulty in obtaining pure samples in sufficient quantity for analysis. Concerning zooplankton, present information on bioaccumulation potential and the biokinetics involved is largely derived from crustaceans (i.e. copepods and euphausiids) that, while often the dominant forms in the community, are not the only common zooplankton species. Many other crustaceans (amphipods, isopods, mysiids, ostracods, larval decapods, etc.) as well as molluscan species (pteropods and heteropods) are sometimes abundant. Soft-bodied or gelatinous forms such as chaetognaths, polychaetes, salps, medusae and larvaceans are other important members for which comparable information is largely lacking. Moreover, what little we do know from some initial studies indicates that radionuclide bioaccumulation and retention processes and rates in soft-bodied forms may be quite different from those typical of planktonic crustaceans (Gorsky et al., 1984; Krishnaswami et al., 1985; Fisher & Fowler, 1987; Fisher et al., 1991c). Clearly, some effort should be expended to obtain comparable data for these other important plankters.

Through heterotrophic processes, the smallest microorganisms are ingested and aggregated into larger particles that also contain elevated concentrations of many radionuclides. These larger aggregates sink more rapidly than the individual plankton species or suspended detrital particles that form them and, thus, act as rapid conveyors of these elements to depth. During descent, biogenic particles may further adsorb and subsequently scavenge radionuclides from the water column, or release them during remineralization at depth. Recent sediment trap studies have confirmed that large organic aggregates such as zooplankton fecal pellets are instrumental in removing radionuclides from the euphotic zone and rapidly transporting them to depth, yet quantitative field data on element and radionuclide concentrations, scavenging rates, remineralization rates and vertical fluxes are still sparse. What is now needed are properly-designed field experiments using time-series sediment traps in conjunction with real-time sampling of plankton and their freshly-produced particulate products at different depths in order to closely examine radionuclide transport and remineralization processes at higher resolution and in more detail. In this way, greater insight will be obtained about which radionuclides are being scavenged and removed to depth, and which are being remineralized and retained in the water column. Furthermore, through time-series collections of the above-mentioned materials, additional information on the rates at which these biogeochemical processes are proceeding could also be obtained.

Since plankton-derived particles are so important in understanding basic biogeochemical cycles of elements and radionuclides in the sea, some effort should be made to discern the loci of enrichment and micro-distribution of analogue elements of radionuclides within the different types of biogenic particles. Synchrotron-based X-ray fluorescence microprobes are sufficiently sensitive to detect naturally occurring trace element concentrations in individual cells and would appear to be well suited to elucidate elemental concentrations and distributions in both plankton and their detritus, as has recently been shown by Twining et al. (2003).

Wherever artificial radioactivity enters the sea, there will always be interest in knowing whether man himself could suffer health effects from eating contaminated sea food and/or whether organisms and ecosystems will be impacted by these contaminants. Whichever question is addressed, models for calculating the radiation dose received from radioactive contamination must take into account the natural radiation dose in the same organisms. For this, accurate information on the levels of natural radionuclides, in particular ^{210}Po and ^{40}K, is required for all organisms under consideration. At present the existing database for natural radionuclides in marine organisms is relatively limited, and augmenting it would greatly help in refining models used to assess the biological effects of radiation on marine organisms and ecosystems.

Acknowledgments

The IAEA Marine Environment Laboratory operates under an agreement between the International Atomic Energy Agency and the Government of the Principality of Monaco.

References

Aarkrog, A. (1977). Environmental behavior of plutonium accidentally released at Thule, Greenland. *Health Physics*, *32*, 271–284.

Aarkrog, A., Dahlgaard, H., Nilsson, K. & Holm, E. (1984). Further studies of plutonium and americium at Thule, Greenland. *Health Physics*, *46*, 29–44.

Aarkrog, A., Baxter, M. S., Bettencourt, A. O., Bojanowski, R., Bologa, A., Charmasson, S., Cunha, I., Delfanti, R., Duran, E., Holm, E., Jeffree, R., Livingston, H. D., Mahapanyawong, S., Nies, H., Osvath, I., Pingyu, L., Povinec, P. P., Sanchez, A., Smith, J. N. & Swift, D. (1997). A comparison of doses from ^{137}Cs and ^{210}Po in marine food: A major international study. *Journal of Environmental Radioactivity*, *34*, 69–90.

Anastasia, J. R., Morgan, S. G. & Fisher, N. S. (1998). Tagging crustacean larvae: Assimilation and retention of trace elements. *Limnology and Oceanography*, *43*, 362–368.

Aston, S. R. & Fowler, S. W. (1983). Preliminary observations on californium-252 behavior in sea water, sediments and zooplankton. *Health Physics*, *44*, 359–365.

Aston, S. R. & Fowler, S. W. (1984). Experimental studies on the bioaccumulation of plutonium from sea water and a deep-sea sediment by clams and polychaetes. *Journal of Environmental Radioactivity*, *1*, 67–78.

Bacon, M. P., Huh, C.-A., Fleer, A. P. & Deuser, W. G. (1985). Seasonality in the flux of natural radionuclides and plutonium in the deep Sargasso Sea. *Deep-Sea Research*, *32*, 273–286.

Baines, S. B., Fisher, N. S. & Stewart, R. (2002). Assimilation and retention of selenium and other trace elements from crustacean food by juvenile striped bass (*Morone saxatilis*). *Limnology and Oceanography*, *47*, 646–655.

Beasley, T. M. & Fowler, S. W. (1976). Plutonium and americium: Uptake from contaminated sediments by the polychaete *Nereis diversicolor*. *Marine Biology*, *38*, 95–100.

Bjerregaard, P., Topcuoglu, S., Fisher, N. S. & Fowler, S. W. (1985). Biokinetics of americium and plutonium in the mussel *Mytilus edulis*. *Marine Ecology Progress Series*, *21*, 99–111.

Boisson, F., Cotret, O. & Fowler, S. W. (2002). Transfer and distribution of lead in the asteroid *Asterias rubens* following ingestion of contaminated food: A radiotracer study. *Marine Pollution Bulletin*, *44*, 1003–1009.

Bowen, V. T., Olsen, J. S., Osterberg, C. L. & Ravera, J. (1971). Ecological interactions of marine radioactivity. In *Radioactivity in the Marine Environment* (Ch. 8, pp. 200–222). Washington, DC: National Academy of Sciences.

Bowen, V. T., Livingston, H. D. & Burke, J. C. (1976). Distribution of transuranium nuclides in sediment and biota of the North Atlantic Ocean. In *Transuranium Nuclides in the Environment* (pp. 107–120). Vienna: IAEA.

Bryan, G. W. (1976). Heavy metal contamination in the sea. In R. Johnston (Ed.), *Marine Pollution* (pp.185–302). London: Academic Press.

Bustamante, P., Germain, P., Leclerc, G. & Miramand, P. (2002). Concentration and distribution of ^{210}Po in the tissues of the scallop *Chlamys varia* and the mussel *Mytilus edulis* from the coasts of Charente-Maritime (France). *Marine Pollution Bulletin*, *44*, 997–1002.

Campbell, P. G. C. (1995). A critique of the free-ion activity model. In A. Tessier & D. R. Turner (Eds), *Metal Speciation and Bioavailability in Aquatic Systems* (pp. 45–102). Chichester: Wiley.

Carroll, J., Wolkers, H., Andersen, M. & Rissanen, K. (2002). Bioaccumulation of radiocaesium in Arctic seals. *Marine Pollution Bulletin*, *44*, 1366–1371.

Carvalho, F. P., Fowler, S. W. & La Rosa, J. (1983). Assimilation, inter-organ transfer and excretion of americium in two teleost fish. *Marine Biology*, *77*, 59–66.

Charmasson, S., Barker, E., Calmet, D., Pruchon, A.-S. & Thébault, H. (1999). Long-term variations of man-made radionuclide concentrations in a bio-indicator *Mytilus galloprovincialis* from the French Mediterranean coast. *Science of the Total Environment*, *237/238*, 93–103.

Cherry, R. D. & Heyraud, M. (1982). Evidence of high natural radiation doses in certain mid-water oceanic organisms. *Science*, *218*, 54–56.

Cherry, R. D. & Shannon, L. V. (1974). The alpha radioactivity of marine organisms. *Atomic Energy Review*, *12*, 3–45.

Cherry, R. D., Higgo, J. J. W. & Fowler, S. W. (1978). Zooplankton fecal pellets and element residence times in the ocean. *Nature*, *274*, 246–248.

Chipman, W. A. (1966). *Uptake and Accumulation of Chromium-51 by the Clam, Tapes Decussatus, in Relation to Physical and Chemical Form* (pp. 571–582). Vienna: IAEA.

Chong, K. & Wang, W.-X. (2000). Assimilation of Cd, Cr, and Zn by the green mussel *Perna viridis* and the clam *Ruditapes philippinarum*. *Environmental Toxicology and Chemistry*, *19*, 1660–1667.

Coughtrey, P. J. & Thorne, M. C. (1983). *Radionuclide Distribution and Transport in Terrestrial and Aquatic Ecosystems. A Critical Review of Data*, Vol. 1 (496 pp.). Rotterdam: A. A. Balkema.

Coughtrey, P. J., Jackson, D., Jones, C. H., Kane, P. & Thorne, M. C. (1984). *Radionuclide Distribution and Transport in Terrestrial and Aquatic Ecosystems. A Critical Review of Data*, Vol. 4 (565 pp.). Rotterdam–Boston: A. A. Balkema.

Cross, F. A., Dean, J. M. & Osterberg, C. L. (1969). The effect of temperature, sediment, and feeding on the behavior of four radionuclides in a marine benthic amphipod. In D. J. Nelson & F. C. Evans (Eds), *Symposium on Radioecology, Proceedings of the 2nd National Symposium* (pp. 450–461). Ann Arbor: CONF 670503, USAEC-TID-4500.

Cross, F. A., Hardy, L. H., Jones, N. V. & Barber, R. T. (1973). Relation between total body weight and concentrations of manganese, iron, copper, zinc, and mercury in white muscle of bluefish (*Pomatomus saltatrix*) and a bathyl-demersal fish *Antimora rostrata*. *Journal of the Fisheries Research Board of Canada*, *30*, 1287–1291.

Dahlgaard, H. (1986). Effects of season and temperature on long-term in situ loss rates of Pu, Am, Np, Eu, Ce, Ag, Tc, Zn, Co and Mn in a Baltic *Mytilus edulis* population. *Marine Ecology Progress Series*, *33*, 157–165.

Davies, A. G. (1978). Pollution studies with marine plankton, Part II: Heavy metals. *Advances in Marine Biology*, *15*, 381–508.

Durand, J. P., Goudard, F., Barbot, C., Pieri, J., Fowler, S. W. & Cotret, O. (2002). Ferritin and hemocyanin: ^{210}Po molecular traps in marine fish, oyster and lobster. *Marine Ecology Progress Series*, *233*, 199–205.

Fisher, N. S. (1982). Bioaccumulation of technetium by marine phytoplankton. *Environmental Science and Technology*, *16*, 579–581.

Fisher, N. S. (1985). Accumulation of metals by marine picoplankton. *Marine Biology*, *87*, 137–142.

Fisher, N. S. (1986). On the reactivity of metals for marine phytoplankton. *Limnology and Oceanography*, *31*, 443–449.

Fisher, N. S. & Fowler, S. W. (1987). The role of biogenic debris in the vertical transport of transuranic wastes in the sea. In T. P. O'Connor, W. V. Burt & I. W. Duedall (Eds), *Transuranic Processes in Marine Pollution, 2: Physicochemical Processes and Wastes in the Ocean* (Ch. 16, pp. 197–207). Malabar, Florida: R. E. Krieger Publ. Co. Inc.

Fisher, N. S. & Reinfelder, J. R. (1995). The trophic transfer of metals in marine systems. In A. Tessier & D. R. Turner (Eds), *Metal Speciation and Bioavailability in Aquatic Systems* (pp. 363–406). Chichester: Wiley.

Fisher, N. S. & Teyssié, J.-L. (1986). Influence of food composition on the biokinetics and tissue distribution of zinc and americium in mussels. *Marine Ecology Progress Series*, *28*, 197–207.

Fisher, N. S. & Wente, M. (1993). The release of trace elements by dying marine phytoplankton. *Deep-Sea Research*, *40*, 671–694.

Fisher, N. S., Bjerregaard, P. & Fowler, S. W. (1983a). Interactions of marine plankton with transuranic elements. 1. Biokinetics of neptunium, plutonium, americium and californium in phytoplankton. *Limnology and Oceanography*, *28*, 432–447.

Fisher, N. S., Bjerregaard, P. & Fowler, S. W. (1983b). Interactions of marine plankton with transuranic elements. 3. Biokinetics of americium in euphausiids. *Marine Biology*, *75*, 261–268.

Fisher, N. S., Cochran, J. K., Krishnaswami, S. & Livingston, H. D. (1988). Predicting the oceanic flux of radionuclides on sinking biogenic debris. *Nature, 335*, 622–625.

Fisher, N. S., Nolan, C. V. & Fowler, S. W. (1991a). Assimilation of metals in marine copepods and its biogeochemical implications. *Marine Ecology Progress Series, 71*, 37–43.

Fisher, N. S., Nolan, C. V. & Fowler, S. W. (1991b). Scavenging and retention of metals by zooplankton fecal pellets and marine snow. *Deep-Sea Research, 38*, 1261–1275.

Fisher, N. S., Nolan, C. V. & Gorsky, G. (1991c). The retention of cadmium and zinc in appendicularian houses. *Oceanologica Acta, 14*, 427–430.

Fisher, N. S., Breslin, V. T. & Levandowsky, M. (1995). Accumulation of silver and lead in estuarine microzooplankton. *Marine Ecology Progress Series, 116*, 207–215.

Fisher, N. S., Teyssié, J.-L., Fowler, S. W. & Wang, W.-X. (1996). Accumulation and retention of metals in mussels from food and water: A comparison under field and laboratory conditions. *Environmental Science and Technology, 30*, 3232–3242.

Fisher, N. S., Fowler, S. W., Boisson, F., Carroll, J., Rissanen, K., Salbu, B., Sazykina, T. G. & Sjoblom, K.-L. (1999). Radionuclide bioconcentration factors and sediment partition coefficients in Arctic seas subject to contamination from dumped nuclear wastes. *Environmental Science and Technology, 33*, 1979–1982.

Fisher, N. S., Stupakoff, I., Sanudo-Wilhelmy, S. A., Wang, W.-X., Teyssié, J.-L., Fowler, S. W. & Crusius, J. (2000). Trace metals in marine copepods: A field test of a bioaccumulation model coupled to laboratory uptake kinetic data. *Marine Ecology Progress Series, 194*, 211–218.

Fowler, S. W. (1977). Trace elements in zooplankton particulate products. *Nature, 269*, 51–53.

Fowler, S. W. (1982). Biological transfer and transport processes. In *Pollutant Transfer and Transport in the Sea*, Vol. 2 (pp. 1–65). Boca Raton: CRC Press Inc.

Fowler, S. W. (1987). VERTEX: Factors affecting the vertical flux of $^{239+240}$Pu and ^{241}Am in the upper layers of the Northeast Pacific Ocean. In *Colloque Internatinale d'Océanologie, Ecosystème de Marges Continentales* (p. 82). Monaco: CIESM.

Fowler, S. W. (1990). Trace element transfer and transport by marine micro-organisms. In R. Moro & R. Cesareo (Eds), *2nd International Workshop on XRF and PIXE Applications in Life Science* (pp. 65–78). Singapore: World Scientific.

Fowler, S. W. & Aston, S. R. (1982). Application of ^{235}Np in experimental aquatic radioecology: Preliminary observations on neptunium behavior in sea water, sediments and zooplankton. *Health Physics, 42*, 515–520.

Fowler, S. W. & Carvalho, F. P. (1985). Americium biokinetics in benthic organisms as a function of feeding mode. *Bulletin of Environmental Contamination and Toxicology, 35*, 826–834.

Fowler, S. W. & Guary, J. C. (1977). High absorption efficiency for ingested plutonium in crabs. *Nature, 266*, 827–828.

Fowler, S. W. & Knauer, G. A. (1986). Role of large particles in the transport of elements and compounds through the oceanic water column. *Progress in Oceanography, 16*, 147–194.

Fowler, S. W. & Small, L. F. (1975). Procedures involved in radioecological studies with marine zooplankton. In *Design of Radiotracer Experiments in Marine Biological Systems* (pp. 63–84). Technical Report Series No. 167. Vienna: IAEA.

Fowler, S. W. & Teyssié, J.-L. (1997). Assimilation and excretion of selected heavy metals and radionuclides ingested by seastars. *Radioprotection – Colloques, 32* (C2), 317–322.

Fowler, S. W., Small, L. F. & Dean, J. M. (1969). Metabolism of zinc-65 in euphausiids. In D. J. Nelson (Ed.), *Symposium on Radioecology, Proceedings of the 2nd National Symposium* (pp. 399–411). Ann Arbor: CONF 670503, USAEC-TID-4500.

Fowler, S. W., Heyraud, M. & Beasley, T. M. (1975). Experimental studies on plutonium kinetics in marine biota. In *Impacts of Nuclear Releases Into the Aquatic Environment* (pp. 157–177). Vienna: IAEA.

Fowler, S. W., Heyraud, M. & Cherry, R. D. (1976). Accumulation and retention of plutonium by marine zooplankton. In *Activities of the International Laboratory of Marine Radioactivity, 1976 Report* (pp. 42–50). Vienna: IAEA.

Fowler, S. W., Benayoun, G., Parsi, P., Essa, M. W. A. & Schulte, E. H. (1981). Experimental studies on the bioavailability of technetium in selected marine organisms. In *Impacts of Radionuclide Releases into the Marine Environment* (pp. 319–339). Vienna: IAEA.

Fowler, S. W., Aston, S. R., Benayoun, G. & Parsi, P. (1983a). Bioavailability of technetium from artificially labelled North–east Atlantic deep-sea sediments. *Marine Environmental Research, 8*, 87–100.

Fowler, S. W., Ballestra, S., La Rosa, J. & Fukai, R. (1983b). Vertical transport of particulate-associated plutonium and americium in the upper water column of the Northeast Pacific. *Deep-Sea Research*, *30*, 1221–1233.

Fowler, S. W., Carvalho, F. P. & Aston, S. R. (1986). Experimental studies on californium bioavailability to marine benthic invertebrates. *Journal of Environmental Radioactivity*, *3*, 219–243.

Fowler, S. W., Buat-Menard, Y., Yokoyama, S., Ballestra, S., Holm, E. & Nguyen, H. V. (1987). Rapid removal of Chernobyl fallout from Mediterranean surface waters by biological activity. *Nature*, *329*, 56–58.

Fowler, S. W., Ballestra, S., La Rosa, J., Holm, E. & Lopez, J.-J. (1990). Flux of transuranium nuclides in the northwestern Mediterranean following the Chernobyl accident. *Rapp. Comm. Int. Mer Médit.*, *32*, 317.

Fowler, S. W., Small, L. F., La Rosa, J., Lopez, J.-J. & Teyssié, J.-L. (1991). Interannual variation in transuranic flux at the Vertex time-series station in the northeast Pacific and its relationship to biological activity. In P. J. Kershaw & D. S. Woodhead (Eds), *Radionuclides in the Study of Marine Processes* (pp. 286–298). London: Elsevier.

Frazier, A. & Guary, J.-C. (1976). Recherche d'indicateurs biologiques appropriés au contrôle de la contamination du littoral par le plutonium. In *Transuranium Nuclides in the Environment* (pp. 679–689). Vienna: IAEA.

Goldberg, E. D., Bowen, V. T., Farrington, J. W., Harvey, G. R., Martin, J. H., Parker, P. L., Risebrough, R. W., Robertson, W., Schneider, E. & Gamble, E. (1978). The mussel watch. *Environmental Conservation*, *5*, 101–125.

Goldberg, E. D., Koide, M., Hodge, V., Flegal, A. R. & Martin, J. (1983). US Mussel Watch: 1977–1978. Results on trace metals and radionuclides. *Estuarine Coastal and Shelf Science*, *16*, 69–93.

Gorsky, G., Fisher, N. S. & Fowler, S. W. (1984). Biogenic debris from the pelagic tunicate, *Oikopleura dioica*, and its role in the vertical transport of a transuranium element. *Estuarine Coastal and Shelf Science*, *18*, 13–23.

Grillo, M. C., Guary, J. C. & Fowler, S. W. (1981). Comparative studies on transuranium nuclide biokinetics in sediment-dwelling invertebrates. In *Impacts of Radionuclide Releases into the Marine Environment* (pp. 273–291). Vienna: IAEA.

Griscom, S. B. & Fisher, N. S. (2002). Uptake of dissolved Ag, Cd, and Co by the clam, *Macoma balthica*: Relative importance of overlying water, oxic pore water and burrow water. *Environmental Science and Technology*, *36*, 2471–2478.

Griscom, S. B., Fisher, N. S. & Luoma, S. N. (2000). Geochemical influences on assimilation of sediment-bound metals in clams and mussels. *Environmental Science and Technology*, *34*, 91–99.

Griscom, S. B., Fisher, N. S., Aller, R. C. & Lee, B.-G. (2002a). Effects of gut chemistry in marine bivalves on the assimilation of metals from ingested sediment particles. *Journal of Marine Research*, *60*, 101–120.

Griscom, S. B., Fisher, N. S. & Luoma, S. N. (2002b). Kinetic modeling of Ag, Cd and Co bioaccumulation in the clam *Macoma balthica*: Quantifying dietary and dissolved sources. *Marine Ecology Progress Series*, *240*, 127–141.

Guary, J.-C. & Fowler, S. W. (1981). Americium-241 and plutonium-237 turnover in mussels (*Mytilus galloprovincialis*) living in field enclosures. *Estuarine Coastal and Shelf Science*, *12*, 193–203.

Guary, J.-C. &, Fowler, S. W. (1982). Experimental studies on the biokinetics of plutonium and americium in the cephalopod *Octopus vulgaris*. *Marine Ecology Progress Series*, *7*, 327–335.

Guary, J.-C. & Fowler, S. W. (1990). Experimental study of the transfer of transuranium nuclides in marine decapod crustaceans. *Marine Ecology Progress Series*, *60*, 253–270.

Guary, J.-C., Fowler, S. W. & Beasley, T. M. (1982). Routes of plutonium uptake and their relation to biomagnification in starfish. *Marine Pollution Bulletin*, *13*, 99–102.

Hamilton, T. F., Fowler, S. W., La Rosa, J., Holm, E., Smith, J. D., Aarkrog, A. & Dahlgaard, H. (1991). Comparative study of plutonium and americium bioaccumulation from two marine sediments contaminated in the natural environment. *Journal of Environmental Radioactivity*, *14*, 211–223.

Heyraud, M. & Cherry, R. D. (1979). Polonium-210 and lead-210 in marine food chains. *Marine Biology*, *52*, 227–236.

Higgo, J. J. W., Cherry, R. D., Heyraud, M. & Fowler, S. W. (1977). Rapid removal of plutonium from the oceanic surface layer by zooplankton faecal pellets. *Nature*, *266*, 623–624.

Higgo, J. J. W., Cherry, R. D., Heyraud, M., Fowler, S. W. & Beasley, T. M. (1980). Vertical oceanic transport of alpha-radioactive nuclides by zooplankton fecal pellets. In T. F. Gesell & W. M. Lowder (Eds), *Natural Radiation Environment III*, Vol. 1 (pp. 502–513). NTIS, CONF-780422.

Hook, S. E. & Fisher, N. S. (2001). Reproductive toxicity of metals in calanoid copepods. *Marine Biology*, *138*, 1131–1140.

Hutchins, D. A., Wang, W.-X. & Fisher, N. S. (1995). Copepod grazing and the biogeochemical fate of diatom iron. *Limnology and Oceanography*, *40*, 989–994.

Hutchins, D. A., Stupakoff, I. & Fisher, N. S. (1996). Temperature effects on accumulation and retention of radionuclides in the sea star, *Asterias forbesi:* Implications for contaminated northern waters. *Marine Biology, 125*, 701–706.

Hutchins, D. A., Stupakoff, I., Hook, S., Luoma, S. N. & Fisher, N. S. (1998). Effects of Arctic temperatures on distribution and retention of the nuclear waste radionuclides ^{241}Am, ^{57}Co, and ^{137}Cs in the bioindicator bivalve *Macoma balthica. Marine Environmental Research, 45*, 17–28.

IAEA (1976). *Effects of Ionizing Radiation on Aquatic Organisms and Ecosystems.* Technical Report Series No. 172. Vienna: IAEA.

IAEA (1979). *Methodology for Assessing Impacts of Radioactivity on Aquatic Ecosystems.* Technical Report Series No. 190. Vienna: IAEA.

IAEA (2004). *Sediment Distribution Coefficients and Concentration Factors for Biota in the Marine Environment* (95 pp.). Technical Report Series No. 422. Vienna: IAEA.

Jennings, C. D. & Fowler, S. W. (1980). Uptake of ^{55}Fe from contaminated sediments by the polychaete *Nereis diversicolor. Marine Biology, 56*, 277–280.

Kasamatsu, F. & Ishikawa, Y. (1997). Natural variation of radionuclide ^{137}Cs concentration in marine organisms with special reference to the effect of food habits and trophic level. *Marine Ecology Progress Series, 160*, 109–120.

Keckes, S., Ozretic, B. & Krajnovic, M. (1969). Metabolism of Zn-65 in mussels (*Mytilus galloprovincialis* Lam.). Uptake of Zn-65. *Rapp. Comm. Int. Mer Médit., 19*, 949–952.

Keckes, S., Fowler, S. W. & Small, L. F. (1972). Flux of different forms of ^{106}Ru through a marine zooplankter. *Marine Biology, 13*, 94–99.

Kershaw, P. J., Swift, D. J., Pentreath, R. J. & Lovett, M. B. (1983). Plutonium redistribution by biological activity in Irish Sea sediments. *Nature, 306*, 774–775.

Koide, M., Lee, D. S. & Goldberg, E. D. (1982). Metal and transuranic records in mussel shells, byssal threads and tissues. *Estuarine Coastal and Shelf Science, 15*, 679–695.

Krishnaswami, S., Baskaran, M., Fowler, S. W. & Heyraud, M. (1985). Comparative role of salps and other zooplankton in the cycling and transport of selected elements and natural radionuclides in Mediterranean waters. *Biogeochemistry, 1*, 353–360.

Kuenzler, E. J. (1969a). Elimination of iodine, cobalt, iron and zinc by marine zooplankton. In D. J. Nelson & F. C. Evans (Eds), *Symposium on Radioecology, Proceedings of the 2nd National Symposium* (pp. 462–473). Ann Arbor: CONF 670503, USAEC-TID-4500.

Kuenzler, E. J. (1969b). Elimination and transport of cobalt by marine zooplankton. In D. Nelson & F. C. Evans (Eds), *Symposium on Radioecology, Proceedings of the 2nd National Symposium* (pp. 483–492). Ann Arbor: CONF 670503, USAEC-TID-4500.

Landrum, P. F., Lee, H. & Lydy, M. J. (1992). Toxicokinetics in aquatic systems: Model comparisons and use in hazard assessment. *Environmental Toxicology and Chemistry, 11*, 1709–1725.

Lee, B.-G. & Fisher, N. S. (1992a). Degradation and elemental release rates from phytoplankton debris and their geochemical implications. *Limnology and Oceanography, 37*, 1345–1360.

Lee, B.-G. & Fisher, N. S. (1992b). Decomposition and release of elements from zooplankton debris. *Marine Ecology Progress Series, 88*, 117–128.

Lee, B.-G. & Fisher, N. S. (1994). Effects of sinking and zooplankton grazing on the release of elements from planktonic debris. *Marine Ecology Progress Series, 110*, 271–281.

Lee, B.-G., Griscom, S. B., Lee, J.-S., Choi, H. J., Koh, C.-H., Luoma, S. N. & Fisher, N. S. (2000). Influences of dietary uptake and reactive sulfides on metal bioavailability from aquatic sediments. *Science, 287*, 282–284.

Li, Y.-H. (1981). Geochemical cycles of elements and human perturbation. *Geochimica et Cosmochimica Acta, 45*, 2073–2084.

Lowman, F. G. & Ting, R. Y. (1973). The state of cobalt in seawater and its uptake by marine organisms and sediments. In *Radioactive Contamination of the Marine Environment* (pp. 369–384). Vienna: IAEA.

Lowman, F. G., Rice, T. R. & Richards, F. A. (1971). Accumulation and redistribution of radionuclides by marine organisms. In *Radioactivity in the Marine Environment* (Ch. 7, pp. 161–199). Washington, DC: National Academy of Sciences.

Luoma, S. N., Johns, C., Fisher, N. S., Steinberg, N. A., Oremland, R. S. & Reinfelder, J. R. (1992). Determination of selenium bioavailability to a benthic bivalve from particulate and solute pathways. *Environmental Science & Technology, 26*, 485–491.

Mason, A. Z. & Jenkins, K. D. (1995). Metal detoxification in aquatic organisms. In A. Tessier (Ed.), *Metal Speciation and Bioavailability in Aquatic Systems* (pp. 479–608). Chichester: Wiley.

McGeer, J. C., Brix, K. V., Skeaff, J. M., DeForest, D. K., Brigham, S. I., Adams, W. J. & Green, A. (2003). Inverse relationship between bioconcentration factor and exposure concentration for metals: Implications for hazard assessment of metals in the aquatic environment. *Environmental Toxicology and Chemistry, 22*, 1017–1037.

Miramand, P., Germain, P. & Camus, H. (1982). Uptake of americium and plutonium from contaminated sediments by three benthic species: *Arenicola marina, Corophium volutator* and *Scrobicularia plana. Marine Ecology Progress Series, 7*, 59–65.

NCRP (1991). Effects of ionizing radiation on aquatic organisms (115 pp.). NCRP Report No. 109. Bethesda, MD: National Council on Radiation Protection and Measurements.

Ni, I.-H., Wang, W.-X. & Tam, Y. K. (2000). Transfer of Cd, Cr and Zn from zooplankton prey to mudskipper *Periophthalmus cantonensis* and glassy *Ambassis urotaenia* fishes. *Marine Ecology Progress Series, 194*, 203–210.

NOAA (1998). Sampling and analytical methods of the National Status and Trends Program Mussel Watch Project: 1993–1996 update. *Technical Memorandum NOS ORCA 130.* Silver Spring, MD, USA.

Nolan, C. V., Fowler, S. W. & Teyssie, J.-L. (1992). Cobalt speciation and bioavailability in marine organisms. *Marine Ecology Progress Series, 88*, 105–116.

Osterberg, C. L., Carey, A. G. & Curl, H. (1963). Acceleration of sinking rates of radionuclides in the ocean. *Nature, 200*, 176–177.

Pearcy, W. G. & Vanderploeg, H. A. (1973). Radioecology of benthic fishes off Oregon. In *Radioactive Contamination of the Marine Environment* (pp. 245–261). Vienna: IAEA.

Pentreath, R. J. (1975). Radiobiological studies in marine fish. In *Design of Radiotracer Experiments in Marine Biological Systems* (pp. 137–170). Technical Reports Series No. 167. Vienna: IAEA.

Pentreath, R. J. (1977). Radionuclides in marine fish. *Marine Biology Annual Reviews, 15*, 365–460.

Pentreath, R. J. (1978). ^{237}Pu experiments with the thornback ray *Raja clavata. Marine Biology, 48*, 337–342.

Pentreath, R. J. (1981). The biological availability to marine organisms of transuranic and other long-lived radionuclides. In *Impacts of Radionuclide Releases into the Marine Environment* (pp. 241–272). Vienna: IAEA.

Pentreath, R. J. (1999). A system for radiological protection of the environment: Some initial thoughts and ideas. *Journal of Radiological Protection, 19*, 117–128.

Pentreath, R. J. & Fowler, S. W. (1979). Irradiation of aquatic animals by radionuclide incorporation. In *Methodology for Assessing Impacts of Radioactivity on Aquatic Ecosystems* (pp. 129–171). *Technical Reports Series No. 190.* Vienna: IAEA.

Phillips, D. J. H. (1980). *Quantitative Aquatic Biological Indicators* (488 pp.). London, UK: Applied Science.

Rainbow, P. S. & Wang, W.-X. (2001). Comparative assimilation of Cd, Cr, Se, and Zn by the barnacle *Elminius modestus* from phytoplankton and zooplankton diets. *Marine Ecology Progress Series, 218*, 239–248.

Reinfelder, J. R. & Fisher, N. S. (1991). The assimilation of elements ingested by marine copepods. *Science, 251*, 794–796.

Reinfelder, J. R., & Fisher, N. S. (1994a). The assimilation of elements ingested by marine planktonic bivalve larvae. *Limnology and Oceanography, 39*, 12–20.

Reinfelder, J. R. & Fisher, N. S. (1994b). Retention of elements absorbed by juvenile fish (*Menidia menidia, Menidia beryllina*) from zooplankton prey. *Limnology and Oceanography, 39*, 1783–1789.

Reinfelder, J. R., Fisher, N. S., Fowler, S. W. & Teyssié, J. L. (1993). Release rates of trace elements and protein from decomposing planktonic debris. 2. Copepod carcasses and sediment trap particulate matter. *Journal of Marine Research, 51*, 423–442.

Reinfelder, J. R., Wang, W.-X., Luoma, S. N. & Fisher, N. S. (1997). Assimilation efficiencies and turnover rates of trace elements in marine bivalves: A comparison of oysters, clams, and mussels. *Marine Biology, 129*, 443–452.

Reinfelder, J. R., Fisher, N. S., Luoma, S. N., Nichols, J. W. & Wang, W.-X. (1998). Trace element trophic transfer in aquatic organisms: A critique of the kinetic model approach. *Science of the Total Environment, 219*, 117–135.

Roditi, H. A., Fisher, N. S. & Sañudo-Wilhelmy, S. A. (2000). Field testing a metal bioaccumulation model for zebra mussels. *Environmental Science and Technology, 34*, 2817–2825.

Small, L. F. & Fowler, S. W. (1973). Turnover and vertical transport of zinc by the euphausiid *Meganyctiphanes norvegica* in the Ligurian Sea. *Marine Biology, 18*, 284–290.

Small, L. F., Fowler, S. W. & Keckes, S. (1973). Flux of zinc through a macroplanktonic crustacean. In *Radioactive Contamination of the Marine Environment* (pp. 437–452). Vienna: IAEA.

Smith, V., Fegan, M., Pollard, D., Long, S., Hayden, E. & Ryan, T. P. (2001). Technetium-99 in the Irish marine environment. *Journal of Environmental Radioactivity*, *56*, 269–284.

Stewart, G. M. & Fisher, N. S. (2003). Bioaccumulation of polonium-210 in marine copepods. *Limnology and Oceanography*, *48*, 2011–2019.

Stone, R. (2002). Radioecology's coming of age- or its last gasp. *Science*, *297*, 1800–1801.

Swift, D. J. (1985). The accumulation of ^{95m}Tc from sea water by juvenile lobsters (*Homarus gammarus* L.). *Journal of Environmental Radioactivity*, *2*, 229–243.

Swift, D. J. (1992). The accumulation of plutonium by the European lobster (*Homarus gammarus* L.). *Journal of Environmental Radioactivity*, *16*, 1–24.

Swift, D. J. (1993). Laboratory experiments on the retention of ^{235}Np by plaice (*Pleuronectes platessa* L.). *Journal of Environmental Radioactivity*, *21*, 189–201.

Swift, D. J. (2001). The accumulation and retention of ^{95m}Tc by the Norway lobster, *Nephrops norvegicus* L. *Journal of Environmental Radioactivity*, *57*, 139–149.

Templeton, W. L., Nakatani, R. E. & Held, E. E. (1971). Radiation effects. In *Radioactivity in the Marine Environment* (Ch. 9, pp. 223–239). Washington, DC: National Academy of Sciences.

Twining, B. S., Baines, S. B., Fisher, N. S., Maser, J., Vogt, S., Jacobsen, C., Tovar-Sanchez, A. & Sañudo-Wilhelmy, S. A. (2003). Quantifying trace elements in individual aquatic protist cells with a synchrotron X-ray fluorescence microprobe. *Analytical Chemistry*, *75*, 3806–3816.

Ueda, T., Nakamura, R. & Suzuki, Y. (1977). Comparison of influences of sediments and sea water on accumulation of radionuclides by worms. *Journal of Radiation Research*, *18*, 84–92.

Valette-Silver, N. J. & Lauenstein, G. G. (1995). Radionuclide concentrations in bivalves collected along the coastal United States. *Marine Pollution Bulletin*, *30*, 320–331.

Vangenechten, J. H. D., Aston, S. R. & Fowler, S. W. (1983). Uptake of americium-241 from two experimentally labelled deep-sea sediments by three benthic species: A bivalve molluscs, a polychaete and an isopod. *Marine Ecology Progress Series*, *13*, 219–228.

Wang, W.-X. (2002). Interactions of trace metals and different marine food chains. *Marine Ecology Progress Series*, *243*, 295–309.

Wang, W.-X. & Fisher, N. S. (1996). Assimilation of trace elements and carbon by the mussel *Mytilus edulis*: Effects of food composition. *Limnology and Oceanography*, *41*, 197–207.

Wang, W.-X. & Fisher, N. S. (1997). Modeling metal bioavailability for marine mussels. *Reviews of Environmental Contamination and Toxicology*, *151*, 39–65.

Wang, W.-X. & Fisher, N. S. (1998a). Accumulation of trace elements in a marine copepod. *Limnology and Oceanography*, *43*, 273–283.

Wang, W.-X. & Fisher, N. S. (1998b). Excretion of trace element by marine copepods and their bioavailability to diatoms. *Journal of Marine Research*, *56*, 713–729.

Wang, W.-X. & Fisher, N. S. (1999a). Assimilation efficiencies of chemical contaminants in aquatic invertebrates: A synthesis. *Environmental Toxicology and Chemistry*, *18*, 2034–2045.

Wang, W.-X. & Fisher, N. S. (1999b). Delineating metal accumulation pathways for aquatic invertebrates. *Science of the Total Environment*, *237/238*, 459–472.

Wang, W.-X., Fisher, N. S. & Luoma, S. N. (1995). Assimilation of trace elements ingested by the mussel *Mytilus edulis*: Effects of algal food abundance. *Marine Ecology Progress Series*, *129*, 165–176.

Wang, W.-X., Fisher, N. S. & Luoma, S. N. (1996a). Kinetic determinations of trace element bioaccumulation in the mussel *Mytilus edulis*. *Marine Ecology Progress Series*, *140*, 91–113.

Wang, W.-X., Reinfelder, J. R., Lee, B.-G. & Fisher, N. S. (1996b). Assimilation and regeneration of trace elements by marine copepods. *Limnology and Oceanography*, *41*, 70–81.

Wang, W.-X., Griscom, S. B. & Fisher, N. S. (1997). Bioavailability of Cr(III) and Cr(VI) to marine mussels from solute and particulate pathways. *Environmental Science & Technology*, *31*, 603–611.

Wang, W.-X., Stupakoff, I. & Fisher, N. S. (1999). Bioavailability of dissolved and sediment-bound metals to a marine deposit-feeding polychaete. *Marine Ecology Progress Series*, *178*, 281–293.

Wang, W.-X., Ke, C.-H., Yu, K. N., & Lam, P. K. S. (2000). Modeling radiocesium bioaccumulation in a marine food chain. *Marine Ecology Progress Series*, *208*, 41–50.

Warnau, M., Fowler, S. W. & Teyssié, J.-L. (1999). Biokinetics of radiocobalt in the asteroid *Asterias rubens* (Echinodermata): Sea water and food exposures. *Marine Pollution Bulletin*, *39*, 159–164.

Watson, W. S., Sumner, D. J., Baker, J. R., Kennedy, S., Reid, R. & Robinson, I. (1999). Radionuclides in seals and porpoises in the coastal waters around UK. *Science of the Total Environment*, *234*, 1–13.
Wildgust, M. A., McDonald, P. & White, K. N. (2000). Assimilation of ^{210}Po by the mussel *Mytilus edulis* from the alga *Isochrysis galbana*. *Marine Biology*, *136*, 49–53.
Zhao, X., Wang, W.-X., Yu, K. N. & Lam, P. K. S. (2001). Biomagnification of radiocesium in a marine piscivorous fish. *Marine Ecology Progress Series*, *222*, 227–237.

MARINE RADIOACTIVITY
Hugh D. Livingston (Editor)

Chapter 7

Radiological assessment of ocean radioactivity

G. J. Hunt

Centre for Environment, Fisheries and Aquaculture Science, Lowestoft NR33 0HT, England

This chapter begins by setting out the radiological quantities and criteria for protection when assessing ocean radioactivity. The radiological protection standards are those promulgated by the International Commission on Radiological Protection (ICRP). A sophisticated system has evolved to protect man and whilst this has been sufficient in most environments to protect other species, a framework for protection of other species is being considered and this is outlined.

The radiological assessment process is then described using a pathway approach with simple models as examples to describe the consequences of the different marine environmental processes. Ingestion, inhalation and external dose assessments are discussed. For assessment of compliance with dose limits, the selection of an appropriate 'critical group' is a central feature, and this is based on the results of habits surveys. Collective dose is also a consideration in the ICRP methodology and this too is described.

There is then a comparative assessment of sources of ocean radioactivity, looking first at doses due to natural radionuclides, then those due to artificially-enhanced natural radioactivity. Artificial sources due to weapons-test fallout, operations of the nuclear industry, ocean dumping of solid radioactive waste, dumping in the Arctic, and the effect of the Chernobyl accident are all compared in terms of critical group dose and collective dose. Though there are fluctuations near particular sources, generally the highest doses from marine sources derive from natural radionuclides, followed by those from artificially-enhanced natural radionuclides. Weapons-test fallout is the next most significant source of dose in collective terms, but being diffuse, individual doses are very low. Doses via marine pathways due to the nuclear industry, waste dumping operations and the Chernobyl accident have also produced low doses by comparison with natural sources. The differences are illustrated with suitable graphs.

1. Standards for radionuclides in the oceans

1.1. *Introduction*

Other chapters in this book describe inter alia the extent of radioactive labelling of the world oceans. This chapter now examines the methodology for assessment of radiological significance of this labelling. In the course of describing the relevant methodology it is necessary to set out the standards on which the assessment methods are based, and this is done first, setting out the quantities and units used and then examining the relevant radiological protection criteria. The application of these criteria to protection of the oceans is then discussed.

1.2. *Quantities, units and perspectives*

The current generally accepted system of radiological units is part of the Système Internationale (SI), in which the basic quantity of radiation exposure, 'absorbed dose', the energy absorbed per unit mass of material, has the units gray (Gy) (1 Gy = 1 J kg^{-1}). The gray is a large unit, equivalent to 100 rads in the older system of units, thus submultiples mGy and μGy are in common use. In human tissue, the radiological effect of an absorbed dose is obtained by weighting the absorbed dose by a quality factor, Q, which depends on the relative biological effectiveness (RBE) of the type of radiation, and other factors. The result is now termed the 'equivalent dose', for which the unit is the sievert (Sv). Again, this is a large unit (=100 rem in the older system) with submultiples mSv, μSv, etc.

To provide some perspective, indeed for the whole of this chapter, the average natural background radiation dose rate to the world population due to cosmic rays, terrestrial gamma rays, inhalation of radon and from foodstuffs is 2.4 mSv y^{-1} (UNSCEAR, 2000). Table 1 shows this and the main man-made contributions to average doses to the world population in 2000. There can be significant variations between individuals due to location, habits, etc.; the natural background itself would generally be expected to be in the range 1–10 mSv y^{-1}.

1.3. *The basis of radiological protection criteria*

The generally accepted standards for radiological protection, and which are adopted by international organisations, are based on the recommendations of the International Commission on Radiological Protection (ICRP). The ICRP was set up with its current name and form in 1950, but its history dates from 1928. The recommendations of the ICRP have developed significantly over the years. Initially, emphasis was given to occupational exposures and to avoiding doses above prescribed thresholds. However, some harm at low doses was suspected and support grew for an assumption of a linear, no-threshold (LNT) dose-response relationship. Recommendations were consequently based on avoiding threshold effects and keeping risks of cancers and hereditary effects (later termed 'stochastic' effects) to an acceptable level,

Table 1
Annual worldwide average *per caput* effective doses from natural and man-made sources in year 2000 (from UNSCEAR, 2000)

	mSv
Natural background	2.4
Diagnostic X-ray procedures	0.4
Atmospheric nuclear testing	0.005
Chernobyl accident	0.002
Nuclear power production	0.0002
Total	~2.8

based on comparisons of risks in safe industries. In the 1960s, dose limits were prescribed for particular organs of the body. The importance of risks at low doses was emphasised by the recommendation that "all doses shall be kept as low as reasonably achievable, economic and social considerations being taken into account" (ICRP, 1966). The LNT assumption allowed the quantity 'collective dose', i.e. summing doses to individuals over space and time, to describe the overall detriment. Tools such as cost benefit analysis were used to judge the extent of optimisation of protection. Application of dose limitation to the public was achieved by setting limits at one-tenth of those for radiation workers, at the equivalent of 5 mSv y^{-1} for doses other than from natural sources and medical exposures. This limit was to apply to the 'critical group', that small group of those whose habits and customs caused them to receive the highest exposures, thereby ensuring that others were also protected.

In 1977, further recommendations (ICRP, 1977) introduced a more formal system of dose limitation with the principles of justification, optimisation and compliance with dose limits. The differences in organ radiosensitivity were dealt with by introducing a quantity 'effective dose equivalent' with appropriate organ weighting factors and an overall limit to effective dose equivalent. This was set at the same level as before for workers (50 mSv y^{-1}) and the public (5 mSv y^{-1}).

Significant developments took place over the next few years, such that further recommendations ('ICRP-60') were published in 1991 (ICRP, 1991) and these are the basis of the standards in place today. One main development in ICRP-60 was the distinction drawn between 'practices' (which add exposures) and 'interventions' (those intended to reduce exposures, e.g. action in an emergency or remediation of contamination already present). For practices, the principles of justification, optimisation and limitation continue; however, the concept of a 'constraint' to optimisation was introduced to cater inter alia for multiple sources. Constraints are to be applied by radiation protection authorities for particular types of practice. A new quantity 'effective dose' was defined with revised weighting factors from the previous effective dose equivalent. The limit on effective dose to members of the public for practices was set at 1 mSv y^{-1}. For interventions, only the justification and optimisation elements of the system of control are relevant.

Radiation risks have been subject to intensive study and form part of the basis for ICRP standards. The most significant source of data has been the study of the Japanese atomic bomb survivors. Current estimates for stochastic effects, which have not changed a great deal from those used in ICRP-60, suggest a nominal fatal cancer risk of 5×10^{-5} per mSv for a population of all ages. The ICRP-recommended dose limit of 1 mSv y^{-1} for members of the public is consistent with a level of risk between 1 in 10^4 and 1 in 10^5 as the maximum tolerable involuntary risk for a member of the public.

The recommendations of ICRP-60 have been widely adopted by international organisations. The International Atomic Energy Agency (IAEA) have incorporated them into their Basic Safety Standards (IAEA, 1996), adopting also as trivial

(a) a level of dose of 10 μSv (equivalent to a fatal cancer risk of less than 1 in 10^6) and
(b) an annual collective dose level of 1 man-Sv for practices.

Since publication, the ICRP-60 recommendations have been subject to interpretation and clarification by ICRP. With relevance to radioactive waste disposal, ICRP Publication 77 (ICRP, 1997) sets out some interpretive policy principles. A constraint level for a single source of no more than 0.3 mSv y^{-1}, implying a maximum tolerable risk of about 1 in 10^5, is deemed appropriate. Where environmental monitoring is used to assess doses from a combination of sources, it is suggested that control should be based on a dose to the critical group approaching 1 mSv y^{-1}. ICRP Publication 77 also recognises the problems on the use of collective dose, which had been perhaps over-emphasised previously, in summing over long future timescales for which large uncertainties would arise. The suggestion is to divide the assessment into blocks of collective dose received at different rates. It is becoming conventional that meaningful assessments of collective dose would only be available up to 500 years into the future.

ICRP Publications 81 (ICRP, 1998) and 82 (ICRP, 1999) give further guidance on radiation protection as applied to long-lived waste disposal and prolonged exposures, which are relevant to radioactivity in the oceans. The uncertainties in calculating collective doses over long time period are stressed. Earlier recommendations on applying dose limitation to critical groups of members of the public are effectively endorsed.

It is also to be noted that the ICRP-60 recommendations of ICRP are at present under review. Suggested changes have been put forward (Clarke, 1999) which have been under discussion. Revised recommendations are due for promulgation in 2005.

1.4. *Application of standards to radioactivity in the oceans*

Internationally, guidance on application of radiation protection standards to ocean radioactivity has been provided by the IAEA. Major drivers for this role are its relationship to the UN and its designation as the competent authority on radioactivity issues for the Convention of Prevention of Marine Pollution by Dumping of Wastes and Other Matter (The London Convention 1972, previously the London Dumping Convention). The 1961 document "Radioactive Waste Disposal into the Sea" (IAEA, 1961) reviewed the extent of radioactive waste problems and application of then-current radiation protection principles both to sea dumping and pipeline discharges. This document was later updated (IAEA, 1983) to include developments in radiation protection standards. From about that time, and following the cessation of sea dumping of solid radioactive wastes in the 1980s, the needs for international guidance on (a) liquid discharges via pipelines and (b) definitions of 'de minimis' levels of radioactivity in solid wastes developed more separately.

Liquid effluent discharges via pipelines are controlled by appropriate authorisation procedures of each country. Principles for releases of liquid and gaseous effluents and setting of authorisations were reviewed by the IAEA in 1986 (IAEA, 1986) and most recently in 2000 (IAEA, 2000), taking account inter alia of recent ICRP developments, from which were derived the IAEA Basic Safety Standards (IAEA, 1996). Many countries' procedures are subject to regional agreements. One example is the Convention for the Protection of the Marine Environment of the North-east Atlantic (the OSPAR Convention). This Convention, which came into force in 1998 following amalgamation of former Oslo and Paris Conventions, commits contracting parties to take all possible steps to prevent and eliminate pollution of the marine environment of the NE Atlantic by applying the precautionary approach and best available technologies and practices. The Ministerial statement of the contracting parties to OSPAR from the meeting at Sintra, Portugal, in May 1998 included a commitment to progressive and substantial reductions in discharges, emissions and losses of radioactive substances, with the aim of achieving concentrations near background for naturally occurring radioactive substances and close to zero for artificial radioactive substances.

As regards solid radioactive wastes, dumping of high-level waste has always been prohibited by the London Convention. Since 1993, all radioactive wastes have been prohibited from sea dumping by an amendment to the Convention, and this has now been ratified by most countries. However it was recognised that all materials contain some radioactivity, including natural radioactivity, and contamination from pre-existing activities, thus it was necessary to define so-called 'de minimis' levels. The expression 'de minimis' (literally, 'concerning small things') embraces two legal concepts. The first is that of 'exclusion', where sources are not amenable to control, as in the case of unenhanced natural radioactivity. The second is the concept of 'exemption', such that even though sources would otherwise fall within the ambit of control, they are so low as to be trivial and can safely be disregarded.

The IAEA (IAEA, 1999c) has provided guidance for judgements on whether the concepts of exclusion or exemption can be applied to materials for sea dumping, or whether a specific assessment is needed. The basic radiological criteria given earlier are used, i.e. <10 μSv y^{-1} to members of the public or <1 man-Sv y^{-1} collective effective dose from the dumping practice. This IAEA guidance is currently being developed by the London Convention into a stepwise screening process to enable judgements on whether candidate materials may be further considered for sea dumping.

1.5. *Protection of non-human species*

The established system of radiological protection is based on the protection of humankind. In its recommendations of Publication 26 (ICRP, 1977), ICRP stated its belief that "if man is adequately protected then other living things are also likely to be protected". The more recent recommendations of Publication 60 (ICRP, 1991) included the rider "Occasionally, individual members of non-human species might be harmed, but not to the extent of endangering whole species or creating imbalance between species". Dose limits to humans are set at low levels such that the ICRP belief appears to have largely protected the environment from observable harm, but environments exist where this belief is open to challenge. For example, certain ocean environments (Pentreath & Woodhead, 1988), or contaminated areas from which people have been evacuated (UNSCEAR, 1996), involve remote pathways for exposure of people, and the potential exists for harm to the ecosystem, without the ICRP dose limit of 1 mSv y^{-1} for members of the public being exceeded. Consideration of the environment has in recent years been emphasised through the development of the concept of Sustainable Development and the Rio Declaration (UNCED, 1992). Thus there has emerged the need to develop a more explicit framework for radiological protection of the environment.

Discussions to develop such a framework are taking place at international (e.g. IAEA, 1999b; ICRP, 2003) and regional levels (e.g. the European Framework for the Assessment of Environmental Impact (FASSET, 2003). The following key elements of such a framework have been identified (IAEA, 1999b): specification of endpoints of concern; protection criteria; methods of dosimetry; and how to demonstrate compliance.

Consideration of endpoints involves, inter alia, whether harm to the individual or harm at the population level is of primary importance. Whilst for particular protected species harm at the individual level may be of concern, it is more usual to consider harm at the population level. Most radiosensitive stages of the life cycle are those associated with reproduction and effects on gametes and embryos. However, genetic damage can be observed in organisms at dose rates well below those at which reproduction is impaired, and this also needs to be taken in account.

Appropriate protection criteria depend upon data on observable harm. Considerable information is available. A review by IAEA (1992) concluded that there is no convincing evidence that chronic radiation dose rates below 1 mGy d^{-1} will harm annual or plant populations. More recently, the detailed review by UNSCEAR (1996) concluded that detrimental effects on the most sensitive populations would not be expected at dose rates below 1–2 mGy d^{-1} for low-LET radiation. This has been confirmed more recently in the FASSET project (FASSET, 2003) in which it is observed that the dose rate threshold for statistically significant effects in most studies is about 10^2 μGy h^{-1}. By comparison, natural background dose rates in normal marine environments are contributed mainly by alpha-emitting ^{210}Po with absorbed dose rates to gonads of up to several μGy h^{-1}, at least an order of magnitude lower.

Dosimetry for protection of humans involves the use of 'equivalent dose' and 'effective dose' (measured in sieverts) which combine a radiation quality factor with the basic 'absorbed dose' (measured in grays). The extent to which absorbed dose should be modified for non-human species could well vary, making a parallel to equivalent dose uncertain in its application. Therefore, the quantity 'absorbed dose' remains the basis for assessing effects on non-human species.

A number of different approaches have been proposed for the purpose of demonstrating compliance. In particular a 'reference flora and fauna' approach is proposed (ICRP, 2003) based on choices of appropriate reference organisms such that other species in the reference group would also be protected; this approach would complement practice elsewhere in radiological protection. Along the principles of earlier work (e.g. Woodhead, 1979; Pentreath & Woodhead, 2001) a hierarchy of reference models is being developed (FASSET, 2003; ICRP, 2003) to assess absorbed dose rates based on radioactivity concentrations derived either from measurement, or calculated from releases to the environment and appropriate models as discussed later in this chapter. In a radiological assessment the absorbed dose rates so estimated for the critical reference species would be compared with the protection criteria derived as described above.

2. Assessment methodology

2.1. *Introduction*

Assessment of doses to man or the environment involves consideration of the potential pathways by which radioactivity can be transmitted through the environment and lead to exposure. The contributions to exposure from each pathway will be additive, but in many cases a particular pathway (the critical pathway) will dominate. So too will exposures to a particular group of people (the critical group) or type of organism. Dose limits and constraints apply to the doses to the critical group, thus this is the most usual type of assessment. This section concentrates on critical group assessments but adaptation can be made to estimate collective doses to larger populations.

Assessments are often classed as 'prospective' (i.e. predictive of doses due to a proposed release scenario and often carried out for the purposes of setting authorised limits) and 'retrospective' (i.e. looking back at the effects of an existing or former scenario, often done to judge compliance with dose limits). Both types of assessment rely on appropriate models, but the retrospective assessment can make use of measured concentrations of radioactivity in the environment as a result of monitoring programmes provided these levels are detectable.

The complexity of the models chosen for an assessment needs to suit its objectives. Initial calculations are often done assuming pessimistic models and conditions for 'screening' purposes, to identify whether a route of exposure warrants further investigation. Many authorisation assessments particularly for the aquatic environment can assume a steady release rate and an equilibrium situation, suitable for the 'concentration factor' method of calculation (see Section 2.3). However, a more complex dynamic model or 'systems analysis' method (ICRP, 1979) may be needed in assessing time-dependent or short-term releases. In recent years and with advances in computing power, 'probabilistic' codes have been developed in which particular input parameters can be sampled against defined distributions and a distribution of results obtained after iteration. These codes are useful for studying the uncertainties in results of model assessment due to uncertainties and variabilities in input parameters.

The main pathways for exposure of humans due to ocean releases of radioactivity are represented in Fig. 1. The assessment procedure is generally broken down into models representing the separate pathways and processes, usually

(a) dilution and dispersion mechanisms and particulate interaction to estimate water concentrations;
(b) uptake by biota and deposition onto sediments;
(c) dose to man by internal exposure;
(d) dose to man from external pathways.

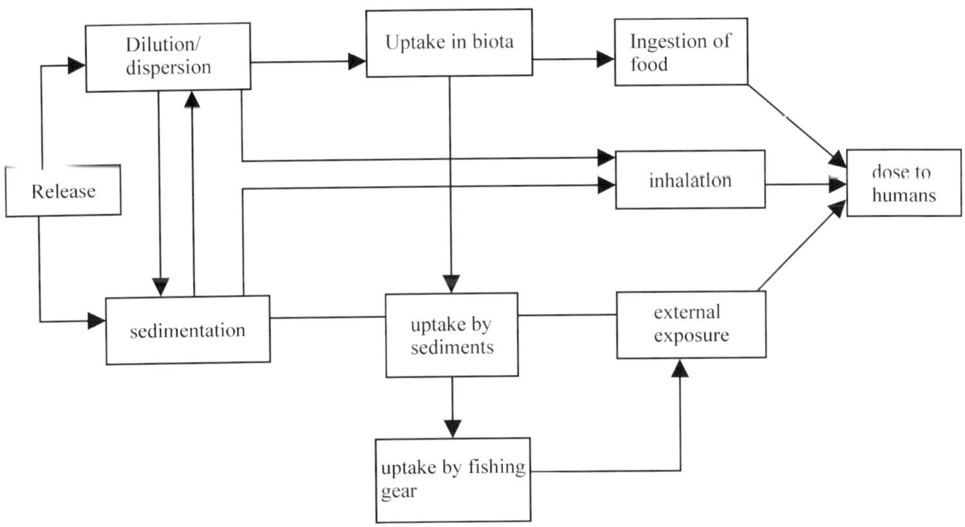

Fig. 1. Pathways for human exposure from ocean releases of radioactivity.

These parts of the overall assessment are described in more detail in the following sections.

2.2. *Dilution, dispersion and particulate interaction*

The objective of this part of the assessment is to derive concentrations of radionuclides in seawater from data on radionuclide releases. Calculations generally rely on mathematical models; there is a range of types and complexities, the choice depending on the objectives of the assessment. The main types are summarised as follows.

Simple screening model
This type is usually used initially with pessimistic parameters, to investigate the need for further study. At its simplest level, sediment interaction may be ignored, and the calculation reduces to:

$$C_i = \frac{\dot{Q}_i}{kV}, \tag{1}$$

where:

C_i is the concentration of radionuclide i and
$\dot{Q}_i$ its release rate into a compartment of volume V with fractional removal rate k (units chosen to be consistent).

k is given by:

$$k = \lambda_i + \frac{\dot{v}}{V}, \tag{2}$$

where:

λ_i is the radioactive decay constant of nuclide i and
$\dot{v}$ is the volume exchange rate of the compartment volume V.

For a coastal site the compartment may be taken to be the immediate tidal area whose dimensions and exchange rate can be obtained from hydrographic charts.

Sediment interaction
Following the most basic screening it is important to consider sediment interaction at an early stage because of the large reductions in concentrations which can be brought about for particle-reactive radionuclides. The interaction is usually described by the distribution coefficient, k_d, which is the ratio between the equilibrium concentration on sediment (usually expressed as dry weight) and the concentration in the water phase. (Units of k_d: $1\ \mathrm{kg}^{-1}$) Values of k_d's for pelagic ocean and coastal sediments have been compiled by IAEA (IAEA, 1985, 2004). The factor f_i by which the water concentration of nuclide i is reduced due to sediment adsorption is given by:

$$f_i = \frac{1}{1 + k_{di} S}, \tag{3}$$

where S is the suspended particulate load in kg l^{-1}.

Advection/diffusion models
These models take account of advection (water movement) and diffusion in the marine environment and are generally used to determine concentrations in the vicinity of a release point up to several tidal ranges away, in which the 'critical group' might be exposed. Simple examples are described by IAEA (IAEA, 2001).

Interlinked compartment ("box") models
A simple model near a release point may consist of one compartment with concentrations calculated using some or all of the methodology given above. For calculations at greater distance, for example to investigate collective doses or effects of discharges on other countries, interlinked box models are often used. These models assume well-mixed behaviour within a given box, which is often both convenient and adequate for radiological calculations; transfer parameters are defined between boxes based on hydrographic data. Sedimentation and other loss (or gain) terms are added. One example of such a box model, for the European coastal environment, is described by the European Commission (EC, 1994). The NE Atlantic part of this model has 44 compartments with associated exchange and sediment parameters. A subset of this model was tested as part of the European MARINA study (CEC, 1990) (see Section 3.4) and found to give broadly good agreement with measured concentrations of ^{99}Tc, ^{137}Cs and $^{239+240}$Pu. The MARINA study has since been updated as MARINA II, with 72 compartments (EC, 2002a). A further example of a large box model is that used by OECD (NEA) to assess exposures due to dumping of solid radioactive wastes in the North Atlantic (NEA, 1985). This is described more fully in Section 3.5.

Finite difference models
These models are generally more complex and based on solving appropriate equations on a grid, either in two or three dimensions. The desired range of physical processes can be represented: hydrodynamic and transport equations with tidal forcing; wind driven flows; sediment transport; seabed deposition and erosion interactions. Recent examples of this type of model are described for the Irish Sea (Aldridge, 1998) and the English Channel (Salomon et al., 1993; Perianez & Reguera, 1999). Both models produce results broadly in line with values derived from observations.

2.3. *Uptake by marine biota and sediments*

The next step in the assessment methodology is to derive radionuclide concentrations in marine biota and on intertidal sediments from concentrations in seawater. In many cases the 'concentration factor' approach may be used, based on equilibrium concentration ratios between the material (Bq kg^{-1}) and filtered seawater (Bq l^{-1}) (ICRP, 1979). The 'systems analysis' approach (ICRP, 1979) is used for time-dependent calculations (e.g. following a short-term release); this requires further information such as characteristic response times in environmental media. Concentration factors (CFs) for the marine environment have been compiled by IAEA (IAEA, 2004) as consensus values derived from ranges found in the literature.

Values necessarily encompass a range of species in each of the classes of fish, crustaceans and molluscs into which it has been found reasonable to group similar behaviour and generally similar CFs. However the following pervading principle in radiological protection applies: "if data more appropriate to the specific situation can be justified, it should be used". One well-known example is the CF for ^{99}Tc in crabs (*Cancer pagurus*) and lobsters (*Homarus gammarus*), where values of 16 and 1200 respectively (e.g. Smith et al., 1998) have been observed; the IAEA (2004) value for crustaceans of 10^3 would seem to be pessimistic in the case of crabs.

For the purposes of calculating external exposure of people occupying intertidal areas it is important to derive concentrations of radionuclides on sediments. Using the 'concentration factor' approach, equilibrium concentrations are derived by multiplying the radionuclide concentration in seawater by the appropriate factor or distribution coefficient (k_d). Compilation of these factors and coefficients may be found in (IAEA, 2004). As for biota, they are consensus values derived from ranges found in the literature, and can err on the side of caution. It is important to consider local conditions. Wide differences can occur depending on sediment composition and particle size. For coarse-grained sands to be found on beaches, a reduction in k_d compared with muddy sediments may be justified for a realistic assessment.

2.4. *Assessment of internal exposure*

Following the recommendations of ICRP (e.g. ICRP, 1991) exposures are calculated as 'effective doses' weighted over the organs of the body (see Section 3). For internally deposited radionuclides, which (depending on half life) can deliver a dose throughout a lifetime, assessment is made of the 'committed effective dose' which integrates the effective dose received due to an intake. For adults, 50 years is assumed to be the average remaining lifespan at the time of intake; for children, a longer integrating time is taken depending on age. It is the committed effective dose received due to one year's intake which requires to be compared with the ICRP-recommended dose limit for the public of 1 mSv y^{-1}; the appropriate consumption or inhalation rate for this comparison is that of an average member of the critical group (see Section 1.3). Thus the annual committed effective dose $E_{i,\text{ing}}$ due to radionuclide i is given for ingestion by:

$$E_{i,\text{ing}} = e_{i,\text{ing}} \sum_j C_{ij} R_j, \tag{4}$$

where:

- $e_{i,\text{ing}}$ is a nuclide-specific factor called the 'dose coefficient' (earlier term: dose per unit intake);
- C_{ij} is the average concentration of radionuclide i in food j; and
- R_j is the amount of food j eaten in a year.

Inhalation of radionuclides in air or sea-spray is usually of lower radiological significance than ingestion, because of concentration mechanisms in biota. However, the annual committed effective dose $E_{i,\text{inh}}$ due to inhalation is similarly given by:

$$E_{i,\text{inh}} = e_{i,\text{inh}} \sum_j a_{ij} B_j, \tag{5}$$

where:

$e_{i,\text{inh}}$ is the dose coefficient for inhalation for radionuclide i in the appropriate form,
a_{ij} is its annual average air concentration in location j, and
B_j the annual volume of air breathed in each location.

Appropriate dose coefficients are calculated on the basis of metabolic models derived by ICRP. The most recent compilation is given in ICRP Publication 72 (ICRP, 1996). Dose coefficients for a wide range of nuclides are given for babies (3 months), infants (1 year), 5, 10 and 15 year-old children and adults. Doses to the embryo and foetus due to intakes by the mother have also recently been calculated (ICRP, 2001) but these are critical for relatively few radionuclides.

Depending on the metabolism of the particular radionuclide in the body, the dose coefficient will be sensitive to assumptions on particular parameters assumed for modelling purposes. One of the potentially most significant parameters is the gut transfer factor, or 'f_1 value'. Many observations of parameters are based on animal studies, and for a particular parameters can vary widely. Much work has been done on the metabolism of plutonium and related elements (ICRP, 1986), and it may be argued that available data do not permit an f_1 value for plutonium to be proposed more precisely than an order of magnitude. In addition, actual parameters are likely to vary significantly between individuals. ICRP practice is to choose parameters to err on the side of caution so as to produce a maximising result in calculating the dose to the critical group.

2.5. *Assessment of external exposure*

There are a number of potential pathways for external exposure; the more radiologically significant pathways usually involve exposure to tide-washed or entrained sediments because of their potential to adsorb radionuclides from seawater. Exposure to the water mass may also be considered, as when swimming or during activities on small boats. Exposure is due to beta particles or gamma rays; alpha particles do not feature because of their small range, being stopped in the upper layers of the skin.

Exposure to gamma rays from sediments

Irradiation due to photons from sediments contaminated by radionuclides is a potential exposure pathway for people who occupy intertidal areas during work or recreational activities. Prospective assessment of this pathway needs to be based on an appropriate model of gamma dose rates in air over sediments, whose radionuclide concentrations can be calculated as described in Section 2.4. Suitable models can be found in the literature (e.g. Beck & de Planque, 1968; Hunt, 1984), and radionuclide tabulations are also given based on photon transport or Monte Carlo calculations (e.g. Kocher & Sjoreen, 1985; ICRU, 1994). It is important to choose models which correctly describe the depth distribution of radioactivity so as to allow for shielding by overlying sediments. For sandy beaches, where sediments are more uniformly contaminated with depth, a simple model based on energy conservation considerations could be adequate. This gives the dose rate D_i in air (μGy h^{-1}) due to radionuclide i as:

$$D_i = 0.288 C_i E_i, \tag{6}$$

where:

C_i is the concentration of the radionuclide (Bq g^{-1}) in sand and
E_i is the photon energy (MeV) per decay (Hunt, 1984).

In the case of muddy sediments there will generally be variation of radionuclide concentration with depth due to discharge history and radioactive decay. Often, assumption of an exponential distribution is adequate and models and tabulations have been proposed to describe this situation (Hunt, 1984; Kocher & Sjoreen, 1985; ICRU, 1994).

Exposure to beta particles
Beta particles from contaminated sediments present a source of exposure either to skin or to organs of the body near to its surface within the particles' range; the main exposure route in the latter category is of the male gonads. Beta particles of low energy (end-point energy less than about 0.1 MeV) are attenuated in the epidermis and may be neglected. Models and radionuclide tabulations of dose rates from skin contamination can be found in the literature (e.g. Kocher & Eckerman, 1987; Cross et al., 1992). A simple model (Hunt, 1984) is based on energy conservation considerations such that the dose rate D_i (μGy h^{-1}) due to radionuclide i is given by

$$D_i = 0.288 C_i E_i F, \tag{7}$$

where:

C_i is the concentration (Bq g^{-1}) of radionuclide i,
E_i is the mean beta energy for decay (MeV), and
F is a modifying factor to take account of attenuation of the beta particles by the epidermis.

A further allowance can be made for effective dilution of the sediments being entrained on the fishing gear (Hunt, 1984). It should be noted that for skin irradiation the ICRP recommends a deterministic dose limit for the public of 50 mSv y^{-1} (ICRP, 1991); there is also an addition to the effective dose with a weighting factor of 0.01.

Irradiation of the male gonads also requires consideration, as mentioned above, particularly during activities such as angling and wildfowling involving close proximity to sediments. The beta particles are significantly attenuated by clothing and overlying tissues of the gonads. For example, for the relatively hard beta particles from ^{234m}Pa (end-point energy 2.28 MeV) dose rates are reduced to about 2% of their level in air (Hunt, 1992); nevertheless there may still be a significant contribution to overall dose.

Exposure to swimmers and on boats

In this situation similar models to those described above may be used, on the basis that an immersed body will be subject to 4π (i.e. surrounding) rather than 2π (semi-infinite) geometry, and the appropriate concentration to be used is the bulk radionuclide concentration in the water including any mobilised sediment. Shielding by the hull of a boat could be allowed for. Simple models based on these principles have been described by Hunt (1984).

2.6. *Selection of critical groups and the role of habits surveys*

In assessing radiation exposures to members of the public, the concept of the 'critical group' formulated by the ICRP is of basic importance. It is generally not possible, due to a range of variabilities, to determine doses that might be received by members of the public individually (ICRP, 1966). Thus ICRP consider it is feasible to take account of these variabilities by the selection of appropriate critical groups. Such a group should be representative of those individuals expected to receive the highest dose; it is believed reasonable (ICRP, 1966) to apply the dose limit for members of the public to the mean dose to the critical group. The critical group should be small enough to be homogeneous with respect to age, diet and those aspects of behaviour which affect the doses received. Although these criteria were set down in the 1960s, recent recommendations have left them essentially unchanged (ICRP, 1991).

To assess doses due to a particular source it is generally necessary to derive consumption, inhalation and occupancy rates typical of the critical group. The dose to the critical group is the sum of the contributions of dose from all exposure pathways. In some countries, regulations for operational and especially pre-operational assessments specify rates to be employed, usually based on hypothetical, maximising assumptions. Where site-specific data are not available, IAEA have suggested 'default' values which can be used, and examples are given in Table 2. The usual radiological protection principle applies, that if more specific information is available it should be used.

More realistic assessments are based on people's habits which might obtain over the period of the practice, e.g. the discharge authorisation. Surveys of local habits (including consumption and occupancy rates) are the main method of deriving site-specific data. Such site-specific

Table 2
Default values of ingestion, inhalation and occupancy rates for maximally exposed adults given by IAEA (IAEA, 2001)

Food		Consumption rate (kg y^{-1})
Marine fish	(Europe)	50
	(Far East)	60
Shellfish	(Europe)	15
	(Far East)	20
		Breathing rate (m^3 y^{-1})
Inhalation		8400
Activity		Occupancy rate (h y^{-1})
Working/playing over contaminated sediments		1600

information is often important for the marine environment as consumption and occupancy habits tend to be more variable than in the terrestrial situation. Habits surveys are mainly used in the case of retrospective assessments to assess compliance with dose limits. However, they can provide a good guide to consumption and occupancy rates for prospective assessments over timescales of a few years, e.g. the lifetime of a discharge authorisation. Habits surveys also provide another essential function in radiation protection of the public, guiding the formulation of appropriate environmental monitoring programmes. Because the main objective of habits surveys is to identify the critical group, they are essentially different from random, market-research surveys, and concentrate on those people likely to be members of the critical group. Surveys need to be carried out with appropriately framed questions to cover all seasons of the year when, for example, fishing may or may not be carried out. Corrections need to be made for portion sizes and edible fractions of species consumed. Useful validation can be obtained by asking those surveyed to complete a diary of foods eaten at specific times during the year.

The shape of a distribution which might be obtained during a habits survey for consumption rate of a given food (e.g. fish from the vicinity of a nuclear site) is shown schematically in Fig. 2.

Selection of the observations which form the critical group may be based on the ICRP criteria of homogeneity referred to earlier. ICRP have interpreted (ICRP, 1985) that, to satisfy the homogeneity requirement, the range of maximum to minimum observations in the group should be no more than a factor of 3 if the mean effective dose to the group is more than 0.1 times the dose limit. Otherwise a factor of 10 is acceptable. Thus in Fig. 2, in which $C_{\max}$ represents the maximum consumption rate observed, the value of $C_{\min}$ (which describes the lowest rate for membership of the critical group) would be set to $C_{\max}/3$. The value of $\overline{C}$ would then be the mean of all observations greater than $C_{\min}$. The dose corresponding to $\overline{C}$ could be calculated, and following the above principles, if it proved to be less than 0.1 mSv, the calculation could be repeated using $C_{\min} = C_{\max}/10$.

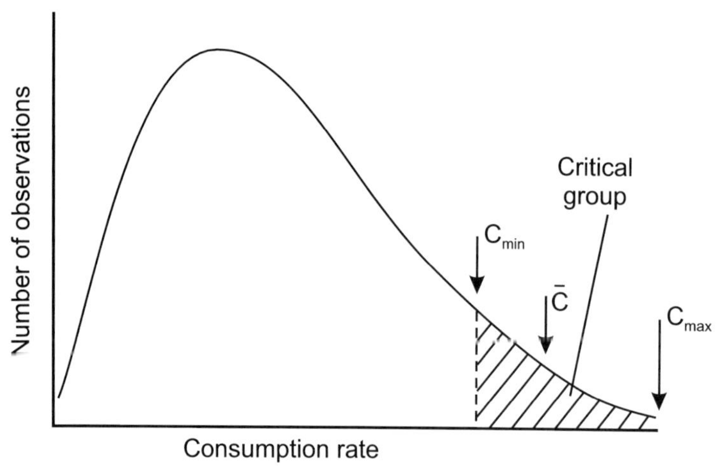

Fig. 2. Schematic distribution of consumption rate data from a habits survey.

The above illustration assumes that doses to the critical group are dominated by one particular pathway. In practice it is usually necessary to consider additive effects due to other contributing pathways, prior to selection of the critical group.

2.7. *Collective dose*

The collective dose (more strictly, post ICRP-60, collective effective dose) S (units: man-Sv) to a population of exposed people can be expressed as:

$$S = \sum_i H_i N_i, \tag{8}$$

where H_i [Sv] is the per caput effective dose to the ith group of people, of size N. The quantity S was recognised to be useful in the 1970s as a consequence of the linear, no-threshold hypothesis; it can express the overall detriment to a group of people, and can be an important input to optimisation as part of the dose limitation system (see Section 1.3). Because the man-Sv corresponds to a detriment which can be calculated from risk factors, it can be ascribed a monetary value. The concept can be used in cost-benefit calculations to assist decision making.

In the marine context, a particular example of use of the concept of collective dose was in identifying levels of interim control of discharges of radioactive waste from Sellafield to the Irish Sea (Handyside et al., 1982). In this example, the detrimental cost associated with collective dose due to discharges of radiocaesium was calculated, along with the cost of installing skips of absorbing zeolite in the fuel storage ponds. By increasing the number of skips, discharges could be reduced, at greater cost. Cost-benefit analysis was used to indicate the optimum number of skips to be installed.

The concept of collective dose remains a useful one but its limitations need to be recognised. First, there is the uncertainty inherent in summing up very small doses, not greatly in excess of natural background, to large numbers of people. Similarly, the summation of doses over timescales in the far future with assumptions on parameters such as the numbers of a population and its habits, will have significant uncertainties. Recent ICRP Publications 77 (ICRP, 1997), 81 (ICRP, 1998) and 82 (ICRP, 1999) have highlighted these uncertainties. In addition, suggested changes to the main recommendations (Clarke, 1999) reflect these developments. The NCRP have published a specific report on principles and application of collective dose (NCRP, 1995).

3. Comparative radiological assessment of sources of radioactivity in the oceans

3.1. *Introduction*

Sources of radioactivity in the oceans may be either natural or anthropogenic; those in the latter category have mostly occurred since the commencement of the nuclear age following the Second World War, and have been both weapons-related and for peaceful purposes. In this section, sources of radioactivity in the oceans are summarised and compared in terms of their radiological effects. Most of the details of these sources and their distribution are described in greater detail elsewhere in this book.

3.2. *Natural radionuclides*

The main natural radionuclides present in the oceans are listed in Table 3 together with their estimated inventories in the world ocean. The nuclides ^{234}U, ^{226}Ra, ^{210}Pb and ^{210}Po are members of decay chains of the long-lived primordial parent ^{238}U. Thorium is essentially insoluble and rapidly removed to the ocean floor, thus concentrations of ^{232}Th and daughters in seawater are very low. By combining representative concentration data in seafoods with maximising consumption rates and relevant dose coefficients, it is possible to estimate the dose to a high-rate consumer. Estimates using recent dose coefficients and values of consumption rates are presented in Table 4. High-rate seafood consumers could receive up to about 2 mSv y^{-1}, with over 95% contributed by ^{210}Po, mainly due to its concentration in molluscan shellfish.

For collective dose purposes, a detailed study of the ^{210}Po component has been undertaken by the international MARDOS group (IAEA, 1995). This group collated data on measured concentrations of ^{210}Po in seawater, fish, crustaceans and molluscs from the major world fishing areas delineated by the Food and Agriculture Organisation (FAO). There were some variabilities, caused by different species caught in different areas, and a shortage of data for some areas. Dose calculations were carried out assuming appropriate concentration factors (IAEA, 1985) combined with seawater concentrations, and, in addition, doses were calculated using the biota data directly. An average delay time between catch and consumption of 1 half-life of ^{210}Po (138 d) was assumed. The Group calculated the world population collective committed effective dose commitment for fish and shellfish caught using FAO landings statistics for 1990. The results lay in the range 27,000–48,000 man-Sv; this gives an average per caput effective dose (i.e. per head of world population of 5.3×10^9 then used by the Group) in the range 5.1–9.1 μSv. It is noted that this dose is to a world average consumer of fish

Table 3
Main naturally-occurring radionuclides in the oceans (Pentreath, 1988)

Radionuclide	Concentration (Bq l^{-1})	Approximate inventory (TBq)
Primordial:		
^{40}K	12	1.6×10^{10}
^{87}Rb	0.11	1.5×10^8
^{234}U	0.046	6.3×10^7
^{238}U	0.040	5.5×10^7
^{226}Ra	0.0035	4.8×10^6
^{210}Pb	0.0030	4.1×10^6
^{235}U	0.0022	3.0×10^6
^{210}Po	0.0020	2.7×10^6
Cosmogenic:		
^{14}C	0.006*	8.0×10^6
^{3}H	0.1*	8.5×10^5

*Surface water values.

Table 4
Representative concentrations of natural radionuclides in seafoods and effective dose rates to high-rate consumers

Radionuclide	Concentration* ($Bq\ kg^{-1}$) (wet)			Dose coefficient† ($Sv\ Bq^{-1}$)	Effective dose rate to high rate consumers ($\mu Sv\ y^{-1}$)	
	Fish	Crustaceans	Molluscs		(1)	(2)
^{210}Po	1.5	25	50	1.2×10^{-6}	1008	1980
^{210}Pb	0.04	0.2	3	6.91×10^{-7}	24	47
^{226}Ra	0.1	0.02	0.3	2.8×10^{-7}	2.6	4.6
^{234}U	0.012	0.12	0.3	4.9×10^{-8}	0.2	0.4
^{238}U	0.011	0.11	0.27	4.84×10^{-8}	0.2	0.4
Total					1035	2033

*Concentration data from (Pentreath, 1988) and references therein; it is to be noted that there are significant variations between species and sea areas (e.g. IAEA, 1995).
†(ICRP, 1996).
Note:
(1) Assumed high rates 60 kg y^{-1} fish, 10 kg y^{-1} crustaceans, 10 kg y^{-1} molluscs given for consumers in the far east (IAEA, 2001).
(2) Assumed high rates 100 kg y^{-1} fish, 20 kg y^{-1} crustaceans, 20 kg y^{-1} molluscs (NRPB, 1998).

and shellfish, as compared with the estimate given above for a high-rate consumer. The per caput results compare with an average effective dose due to natural background of 2.4 mSv (Section 1.2). It is to be noted that since these data were calculated, ICRP have increased the value for the dose coefficient for ingestion of ^{210}Po by a factor of 2.8; this would give an annual collective effective dose up to 134,000 man-Sv and annual per caput effective dose up to 25 μSv. In addition there is evidence to suggest that the dose coefficient for ^{210}Po could be higher by a further factor of 1.6 (Hunt & Allington, 1993).

3.3. *Anthropogenically-enhanced natural radionuclides from non-nuclear industries*

Naturally occurring radioactive materials (NORM) contribute to the discharges from many industrial processes, for example phosphate processing, oil and gas production, mining, ore processing and burning of fossil fuels. A review of these sources has been carried out by UNSCEAR (UNSCEAR, 2000) and for European waters in the MARINA II study (EC, 2002a). The main sources of enhanced natural radioactivity to the marine environment from non-nuclear industries are from the operations of the phosphate industry and the discharges from oil and gas production. These are described in more detail below.

In order to produce phosphoric acid, phosphate ores are acidified mainly with sulphuric acid; this produces phosphogypsum as a precipitate which contains small quantities of the calcium analogue, ^{226}Ra. In some cases the waste phosphogypsum is discharged into surface waters and enters the sea. In many developed countries the discharges have reduced or ceased in the last several decades due both to disposals on land and also to economic conditions, as it may be advantageous to import raw phosphoric acid. Exposures due to these operations have been mainly due to the ^{210}Po product of ^{226}Ra decay, and mainly through consumption of molluscan shellfish. Exposures are generally minor and only of local significance, being dominated at distance by the unenhanced ^{210}Po as described in the previous section. A study of exposures near the phosphate plant at Whitehaven, UK (which ceased operations in 2001 and had used imported phosphoric acid since 1992 following which the discharges reduced significantly) showed that in the period 1962–1992 effective doses above background levels to the critical group of shellfish consumers rose to between 2 mSv y^{-1} and 6 mSv y^{-1}, but have now significantly reduced (Camplin et al., 1996). A survey of non-nuclear discharges to the OSPAR area (OSPAR, 1997) shows that phosphogypsum is released to the mouth of the Rhine by plant in the Netherlands, and to the confluence of the Rivers Odiel and Tinto in Spain. Effective doses above background levels are estimated at between 100–300 μSv y^{-1} to the critical groups, due mainly to ^{210}Po. Releases to the Rhine are estimated to result in a collective effective dose of 170 man-Sv y^{-1} to the Dutch population.

Pumping oil and gas from offshore platforms gives rise to 'produced water' which is contaminated by enhanced levels of natural radionuclides, predominantly ^{226}Ra, ^{228}Ra and ^{210}Pb. Whilst sources of enhanced natural radioactivity are more diffuse than phosphogypsum releases, in some areas they could become more important; in 1999, estimated discharges from oil and gas production accounted for 90% of the discharge of alpha-emitters into the OSPAR region (EC, 2002a). However, collective doses in this area are at present dominated by the effects of mainly past phosphate processing. The collective dose rate received by the EU population in 2000 was 91% due to NORM, of which 61% was due to phosphate processing, the remainder to oil and gas production (EC, 2002b).

3.4. *Weapons-test fallout*

Testing of nuclear weapons took place in the atmosphere from 1945 to 1980; further tests took place underground until 1998. Atmospheric tests have contributed to worldwide radioactivity input to the oceans, and as a greater amount of testing was carried out in the northern hemisphere, deposition amounts were greater there than in the southern hemisphere. Detailed reviews of the deposition have been undertaken by UNSCEAR (e.g. UNSCEAR, 2000), who have also kept worldwide exposures under review. Most of the exposures have been and are due to terrestrial pathways following deposition of radionuclides on land. Exposures are due to external radiation following this deposition and to ingestion of radionuclides taken up into foodstuffs, with a small contribution due to inhalation. Dose estimates by year presented by UNSCEAR (UNSCEAR, 2000) show that world average per caput doses due to weapons-test fallout peaked at about 113 μSv in 1963, reducing to about 6 μSv y^{-1} during the 1990s. Most of this exposure is now contributed by the longer-lived nuclides ^{14}C and ^{90}Sr (via ingestion) and ^{137}Cs (ingestion and external exposure). The contribution from fallout radionuclides in the oceans is relatively small and mainly due to seafood ingestion in the world population diet. It is fallout which has provided the main global source of ^{137}Cs to the world oceans, and an estimate of exposures due to oceanic ^{137}Cs has been given by the MARDOS group (IAEA, 1995). This group, following a similar procedure as for ^{210}Po (Section 3.2) collated data for measured concentrations of ^{137}Cs in seawater, fish and shellfish. Dose calculations were carried out assuming appropriate concentration factors with seawater concentrations and using the biota data directly. Using FAO landings statistics for 1990, the world collective effective dose commitment from ^{137}Cs in fish and shellfish caught that year was about 160 man-Sv using both methods, corresponding to an average per caput dose of 0.032 μSv. This represents only 0.5% of the total per caput dose due to fallout of 6 μSv given above, due inter alia to the proportion of marine fish to world diet. However, the effective dose to a high-rate seafood consumer would be much greater. For example, concentrations of ^{137}Cs in cod from Icelandic waters (remote from effects of nuclear power production) in 2000 was about 0.2 Bq kg^{-1} (FSA & SEPA, 2001). Based on high-rate consumption of fish and shellfish of 80 kg y^{-1} (IAEA, 2001), the annual dose from fallout ^{137}Cs could be up to 0.21 μSv y^{-1}.

It is important to note that as nuclear tests were carried out in relatively remote areas, exposures of local populations are not likely to contribute significantly to world per caput doses from fallout. People who might now or in the future occupy formerly contaminated areas could receive enhanced exposures, and re-evaluation of these sites is being carried out. Some of these sites are in predominantly marine situations and two examples are summarised here. Surveys have been carried out throughout the Marshall Islands in the Pacific (including Bikini and Eniwetak Atolls) (Simon & Graham, 1996; IAEA, 1996). Effective doses from residual contamination to those who might return to Bikini Atoll were estimated at 4 mSv y^{-1} for a mixed diet and 15 mSv y^{-1} for a diet of solely local origin. An international investigation of the present radiological situation at the former test sites of Mururoa and Fangataufa (IAEA, 1998a) was carried out from 1996–1998. It was concluded that, though it is doubtful if these atolls could sustain a permanent population dependent upon local foods, a hypothetical population with a diet of local produce and seafood would not generally receive a dose attributable to the residual contamination exceeding 10 μSv y^{-1}.

3.5. *Operations of the nuclear power industry*

Nuclear electricity generation began in 1956, followed by rapid expansion until the 1990s, since when the rate of expansion has slowed considerably. Operations of the nuclear industry may be divided into mining and milling of the uranium ore; uranium processing, enrichment and fuel fabrication; reactor operation; fuel reprocessing; and solid waste disposal (liquid and airborne waste discharges being considered as part of the other operations). The mining and milling operations produce very little input of radioactive waste directly into the oceans. Solid radioactive waste dumping into the oceans has been prohibited by an amendment in 1993 to the London Convention and this has now been ratified by almost all countries; the effects of past dumping are discussed in Section 3.6. The effects of uranium processing, enrichment and fuel fabrication, reactor operation and fuel reprocessing are discussed in this section. The effects of reprocessing dominate because the discharges are greater in activity terms, as a result of the process which involves dissolution of irradiated fuel followed by separation of plutonium and uranium from the fission products.

Uranium ore is usually supplied as a concentrate which requires further processing to produce uranium metal, oxide or hexafluoride for enrichment. Separation of the uranium produces waste thorium and daughter products, particularly the ^{234}Th daughter of ^{238}U and its own daughter, ^{234m}Pa. The half-lives of these waste products are short (^{234}Th: 24.1 d; ^{234m}Pa: 1.17 min) thus in most cases storage in lagoons allows radioactive decay, and discharges to the wider environment are low. One exception is the Springfields plant, in the United Kingdom, which discharges to the River Ribble, but even here doses to the public due to discharges from the plant are small. In 2000, the critical group of anglers near the site received only 0.02 mSv y^{-1} (Environment Agency, 2001). The process of enrichment involves only very small releases of waste products, especially to the marine environment.

There are a number of different types of nuclear reactor, and the waste discharges from reactor operations have different radionuclide compositions depending on construction materials, type of fuel, cladding, moderator, etc. Discharges tend to be small by comparison with fuel reprocessing, which involves dissolution of the fuel as referred to above. UNSCEAR (2000) have analysed distributions of discharges of radionuclides normalised to energy output for different reactor types and for reprocessing plants, and calculated collective effective doses based on generic assessments. These results include exposures through terrestrial pathways. A large proportion of the world collective dose due to discharges to marine waters is from operations of the Nuclear Industry in Europe. The seas of western Europe receive inputs from some 70 nuclear power stations (including those which discharge to rivers) and three reprocessing plants, namely Sellafield, which discharges to the Irish Sea; Cap de la Hague, which discharges to the English Channel; and Dounreay (where reprocessing has now ceased) which discharges to waters which enter the North Sea. The effects of these plants have been subject to extensive study since the 1960s as part of national and international programmes. A major international study of North European waters was the project MARINA (CEC, 1990). This has recently been reviewed as MARINA II (EC, 2002a), and both studies summarise critical group doses derived mainly from data from national monitoring programmes, using the concentrations of radionuclides in environmental materials combined with data on the habits of the critical groups. A summary of the results from the MARINA studies is shown in Table 5. For comparison are shown data from the 1970–1980s and more recent data, published by the Food Standards Agency and Scottish Environment Protection Agency (2001) and from the extensive study mainly related to the la Hague discharges carried out by the Nord-Cotentin Radioecology Group (2000).

It can be seen that the doses to critical groups have reduced in recent years, and this has been due to reductions in discharges; at Sellafield and la Hague, particularly, these reductions have been due to improved waste management procedures and treatment facilities. For both these sites, reconstruction of the history of doses to the critical groups has been carried out. The la Hague dose reconstruction study was part of the work of the Nord-Cotentin Radioecology Group (2000). Doses due to the Sellafield discharges from the 1950s to the 1990s are shown in Fig. 3 (Hunt, 1997). This reconstruction indicated that average doses to the critical group of fish and shellfish consumers were at a maximum of about 2 mSv y^{-1} in the mid-1970s. The doses were due to the level of releases of radiocaesium and actinides shortly before that time.

Both the original and revised MARINA studies also considered collective doses to the European Union population due to nuclear industry discharges to European waters. The maximum annual collective dose rate was about 280 man-Sv y^{-1} around 1978 due mainly to discharges from Sellafield, which had peaked several years earlier (EC, 2002a). This rate has now reduced to around 14 man-Sv y^{-1}. These data compare with an estimated collective dose rate to the EU population from natural background radiation of about 840,000 man-Sv y^{-1}. The truncated collective dose commitment from the start of nuclear operations up to the year 2500 from discharges to 2000 is calculated to be about 5100 man-Sv (EC, 2002b). The collective dose to the rest of the world due to discharges to EU waters is around an order of magnitude lower than for the EU population (CEC, 1990).

Table 5
Annual individual (critical group) doses from direct inputs to northern European waters

Source	Annual effective dose to critical group (mSv)	
	MARINA studies (data maxima for 1970–1980s)	Recent data
Sellafield	~2	0.15[b]
Dounreay	<0.05	<0.005[b]
La Hague	~0.2	0.026[c]
Other nuclear sites	0.0001–0.3[a]	<0.07[b]

[a]Doses towards the higher end of this range are largely due to the influence of Sellafield.
[b]Data for 2000. Reference: Food Standards Agency and Scottish Environment Protection Agency (2001).
[c]Data for 1996 and apply to the group of Huquets fishermen. Reference: Nord-Cotentin Radioecology Group (2000).

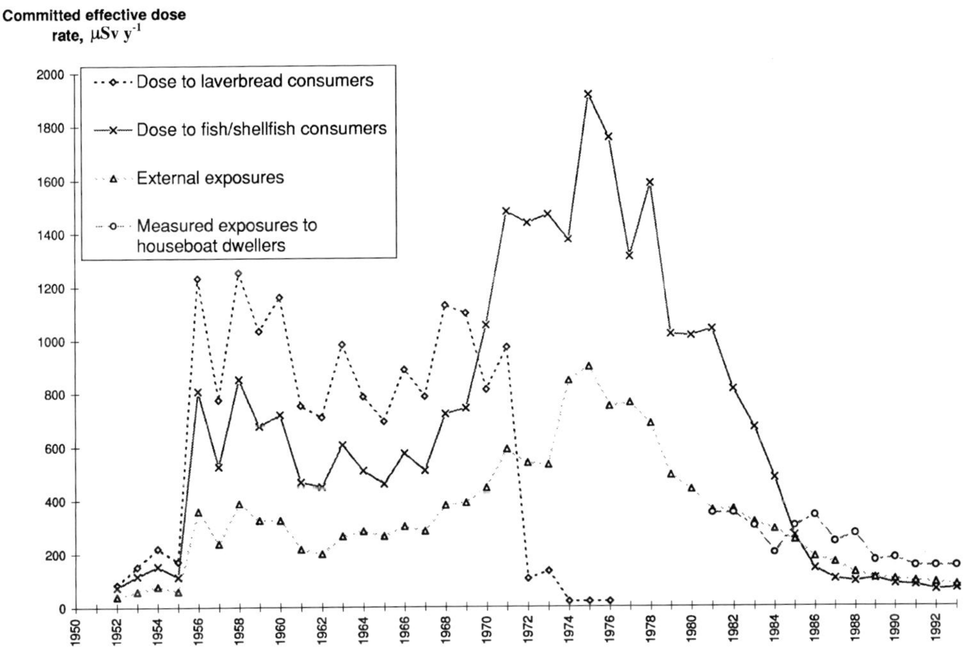

Fig. 3. Doses to critical groups near Sellafield, in 1950–1990 (Hunt, 1997).

3.6. *Ocean dumping of solid radioactive waste*

This practice first took place in 1946 at a site some 80 km off the coast of California. Since then, similar dumping has taken place at a range of sites in the world ocean. Almost all countries have now ratified the 1993 prohibition of this practice by the London Convention. The history of radioactive waste dumping and the inventory of radioactivity dumped are dealt with in greater detail in Chapter 4, and the data have been published by the IAEA (1999c). Totals of 0.55 PBq were dumped in the NE Pacific, 2.94 PBq in the NW Atlantic, 42.32 PBq in the NE Atlantic, 38.37 PBq in the Arctic, and 0.89 PBq in the W Pacific, in each case at a range of different sites. The greatest amounts were dumped in the NE Atlantic and in the Arctic; the effects of the latter disposals are considered in Section 3.7.

Disposals in the NE Atlantic took place from 1949 to 1982 and consisted of waste from a number of European countries; the majority of the operations were in the later years and were coordinated internationally. From 1977 to 1982 operations were carried out under the Multilateral Consultation and Surveillance Mechanism coordinated by the Nuclear Energy Agency (NEA) of the Organisation for Economic Cooperation and Development (OECD). In 1980 a Coordinated Research and Environmental Surveillance Programme (CRESP) was set up by NEA to provide scientific assessment of continuing dumping operations in the NE Atlantic. The radiological impact of the disposals has not been measurable, such that environmental monitoring does not provide a suitable method of radiological assessment, and the emphasis has been on the use of suitable mathematical models to assess the radiological impact. Such an assessment was carried out as part of a review of the continued suitability of the 1974–1982 dumpsite (NEA, 1985). The radiological assessment was carried out in four parts. First, doses to critical groups were calculated for three dumping scenarios:

(a) past dumping (i.e. to 1982);
(b) past dumping plus a further five years of dumping at rates typical of those of 1978–1982;
(c) past dumping plus five year's dumping at ten times the rate of dumping in 1978–1982.

The last two scenarios were chosen to indicate the possible radiological impact rather than being a prediction of what might be dumped. In the second part of the assessment, doses to marine organisms for these scenarios were calculated and compared with natural background levels and levels at which detrimental effects on marine organisms have been observed. In the third part, the sensitivity of calculated doses to critical groups to variations in key parameters was investigated to study uncertainties and identify priorities for future work. The fourth part produced preliminary estimates of collective doses from dumping at this site.

The assessment was based on a system of linked mathematical models. A release-rate model calculated the releases of radionuclides from different types of waste package. Outputs fed a world ocean dispersion model which was of the box type with 92 water compartments, overlaid with a model to represent sediment interactions. The output from the dispersion model fed a radiological model which calculated doses to man and to organisms.

The results of the first part of the assessment showed that even for scenario (c), peak doses would at most be 0.1 $\mu Sv\ y^{-1}$ (to notional high-rate mollusc consumers in the Antarctic). For scenario (a), actual past dumping, the peak individual dose was estimated to be 0.02 $\mu Sv\ y^{-1}$. Doses to marine organisms were, as expected, highest within the dump site area and whilst for scenario (c) they were greater than the natural background, dose rates were at least two orders of magnitude below the level at which deleterious effects have been observed. The sensitivity analyses showed that exposures (whilst still small) were most sensitive to values chosen for sediment interactions and radionuclide concentration factors. Further, if a marine food pathway existed which provided a 'short circuit' from the deep ocean to man, then individual doses might be higher than the more realistic pathways, but would still be low. The collective dose assessment showed that for scenario (c) the highest collective dose commitment would be 3×10^5 man-Sv over the world population, and delivered fairly uniformly over 10^4 years. For scenario (a), the commitment would be 4×10^4 man-Sv over the same time, the collective dose rate being $\sim$4 man-Sv y^{-1} achieved some decades after the start of dumping. As for other collective dose calculations, the results are composed of very low doses spread over large populations and timescales, and are subject to considerable uncertainties.

Overall, the conclusion of the study was that the dump site could continue to be used for dumping of packaged radioactive wastes for a further five years. Dumping rates during this period of ten times the average for 1978–1982 were shown to give insignificant doses to the public and benthic species, but the study recommended that if these rates were exceeded, the suitability of the site should be reconsidered. Despite the results of this review, however, for political reasons dumping in the NE Atlantic was not resumed.

3.7. *Radioactive waste dumping in the Arctic seas*

Following the demise of the Former Soviet Union, it became clear in the early 1990s that for over three decades significant amounts of radioactive wastes had been dumped in shallow waters of the Arctic seas. Several international initiatives for studying the effects of radioactivity in the Arctic region including terrestrial effects have been pursued, but to address this dumping specifically, an international study was proposed by the IAEA, and in 1993 the International Arctic Seas Assessment Project (IASAP) was set up. The objectives were: to assess the risks to human health and the environment associated with the wastes dumped in the Kara and Barents Seas, and (principally at the request of contracting parties to the London Convention) to examine possible remedial actions and to advise on whether they were necessary and justified.

Russian estimates of the amount of radioactive waste dumped indicated some 90 PBq at the time of dumping, but an assessment based on reactor operating histories reduced the estimate of the total activity dumped to 37 PBq. Most of the activity was in high level wastes in reactors with and without spent fuel, mainly from submarines but also from the nuclear-powered icebreaker '*Lenin*' and dumped mainly in the Kara Sea, particularly in the shallower fjords of Novaya Zemlya. The depths of the dumping sites were 12–135 m and in the Novaya Zemlya Trough at depths up to 380 m. Details of the dumping are given in the IAEA Inventory of Radioactive Waste Disposals at Sea (IAEA, 1999c). The inventory in 1994 after radioactive decay since dumping was estimated by IASAP to be 4.7 PBq of which 86% were fission products, 12% activation products, and 2% actinides (IAEA, 1998b).

The IASAP examined the current radiological situation in the Arctic based on data from joint Norwegian–Russian cruises and other international monitoring surveys. Measurements of environmental materials suggested that annual individual doses from artificial radionuclides are in the range of only 1–20 μSv. Whilst there were elevated levels of radionuclides detected in some sediments within a few metres of some low level waste containers, indicating leakage, the outer parts of the fjords did not show measurable increases. Thus at present the wastes have a negligible radiological impact.

Potential risks posed by possible future releases focused on the high level sources dumped. Release rates were estimated and radiation doses to humans and biota were assessed using appropriate models. For each of the dumped high-level waste objects the barriers to release were examined, weak points identified, and corrosion rates taken into account. External events such as collisions and glacial action were also considered. Three release scenarios were considered: a best estimate on the basis of gradual corrosion; a plausible worst case involving catastrophic release of two sources at one dumpsite in the year 2050; and a climate change scenario involving release due to glacial action in the year 3000. A number of different models was used to calculate dispersion of the radionuclides within and from the Arctic Ocean, both of the box model and hydrodynamic type. For estimation of doses to individuals, three population groups were considered:

1. A group whose subsistence is heavily dependent on local Kara Sea fish and other marine produce and activities.
2. A hypothetical group of military personnel patrolling the foreshores of the fjords containing dumped radioactive materials.
3. A group of seafood consumers representative of the population of N Russia on the Kola peninsula eating seafood from the Barents Sea.

The resulting peak doses to these groups under the three postulated scenarios are shown in Table 6 (IAEA, 1998b). The total annual individual doses to seafood consumers (groups 1 and 3) are very small and much less than variations in natural background. Doses to the hypothetical group of military personnel are higher due to proximity to the releases but still comparable to natural background doses.

Collective doses were also estimated for the best estimate scenario. The collective effective dose up to the year 3000 is of the order of 10 man-Sv. Doses to marine organisms were also estimated, the maximum being about 0.1 μGy h^{-1}, well below levels likely to cause detrimental effects to populations.

Potential remediation was considered as part of the Project. A particular example was chosen, that of dealing with the container of spent fuel from the icebreaker '*Lenin*'. Feasible options after screening out of the less practicable ones involved in situ capping with concrete or other suitable material, or recovery for land treatment and disposal. The costs of the marine operations (i.e. not including any subsequent transport and land disposal) were estimated at US $6–10 million. The radiological protection aspects of the justification for remediation were considered. Individual doses resulting from leaving the wastes in place would be small as described above. Nevertheless, when summed over time and over a population some health effects might be predicted, thus the collective dose was considered further. A simple cost-benefit argument was applied which indicated that remedial measures costing in excess of US $200,000 would not appear to offer sufficient benefit to be warranted. Therefore, it was

Table 6
Maximum total annual doses for postulated Arctic groups and scenarios

Scenario	Annual doses (μSv)	
	Seafood consumers (groups 1 and 3)	Military personnel (group 2)
1. Best estimate	<0.1	700
2. Plausible worst case	<1	4000
3. Climate change	0.3	3000

concluded that, on radiological grounds, remedial steps are not justified. However, it was concluded that institutional controls on occupation of beaches and use of coastal marine resources and amenities in the fjords of Novaya Zemlya must be maintained.

3.8. *Effects of the Chernobyl accident*

Radioactive debris from the unit 4 accident at Chernobyl, Ukraine, on 26 April 1986 caused widespread contamination over much of Europe, and was detectable in more distant parts of the northern hemisphere. A number of European sea areas were affected as a result of changes in wind direction following the accident, and subsequent rainfall during passage of the plumes. The main sea areas affected were the Baltic Sea, the Black Sea and the Irish Sea. Effects were also observed in the Mediterranean (e.g. Papucci et al., 1996), the Aegean (e.g. Polikarpov et al., 1991) and the North Sea (e.g. Mitchell & Steele, 1988) but time-averaged concentrations were generally lower and radiological effects were less than in the aforementioned seas. The major cause of long-term radiation dose through marine pathways, mainly via fish and shellfish consumption and external exposure, was airborne deposition and rainout, of mainly ^{134}Cs and ^{137}Cs on sea and coastal areas. This was even so for the Black Sea, where input of Chernobyl-derived radioactivity from the River Dnieper was relatively small (Kanivets et al., 1999). The fallout from Chernobyl was identifiable in many areas by its characteristic $^{137}Cs/^{134}Cs$ ratio of about 2 : 1 in the early stages. A wide range of data following the Chernobyl accident is published in the literature. Much of the data are on concentrations and inventories rather than radiation exposure. What follows is a brief résumé of reported radiation dose effects, or where these can be derived, from some of the larger studies for these sea areas. It should be pointed out that higher individual doses were potentially received due to consumption of freshwater fish from upland lakes where deposition occurred, and indeed other terrestrial food sources.

The Baltic Sea

Following the accident a wide range of nuclides was detected but these were mainly short-lived and the significant components giving rise to radiation dose were ^{134}Cs and ^{137}Cs. The inventory of ^{137}Cs in the Baltic was estimated at 4300–5000 TBq in 1986, reducing to 2100 TBq in 1990 (HELCOM, 1996). The 'Marina-Balt' study (EC, 2000) assessed critical group and collective doses using model calculations taking account of all the major sources to the Baltic Sea. The maximum critical group doses due to the Chernobyl accident were to consumers of seafood from the Bothnian Sea and Gulf of Finland, at 0.2 mSv y^{-1} in 1986, reducing to 0.02 mSv y^{-1} in 2000. The collective dose commitment to the year 2400 to all countries due to Chernobyl radioactivity deposited in the Baltic was estimated to be 1700 man-Sv.

The Black Sea

A similar overall picture to that of the Baltic was experienced; initially, deposition was uneven but becoming more widely distributed, the significant longer term components giving rise to radiation doses being ^{134}Cs and ^{137}Cs. Most of the literature describes measurements of concentrations/inventories in seawater; the mean concentration of ^{137}Cs in the 0–50 m surface layer is reported as 134 Bq m^{-3} for 1986, the peak year (Egorov et al., 1999). On the basis of a concentration factor of 100 for radiocaesium in fish (IAEA, 2001), a critical group consumption rate of 50 kg y^{-1} (IAEA, 2001) and ICRP dose coefficients (ICRP, 1996) the dose to the critical group would have been less than 0.02 mSv in 1986.

The Irish Sea

Deposition of fallout from the Chernobyl accident in some coastal areas of the Irish Sea was elevated because of rainfall during passage of the plume. Short-lived radionuclides were observed particularly in shellfish, but these declined such that, as elsewhere, the more lasting effects were due to ^{134}Cs and ^{137}Cs. Doses due to the radioactivity from Chernobyl were assessed after subtraction of other anthropogenic effects (mainly due to Sellafield) and related to steady consumption over one year after the accident (Camplin et al., 1986). High-rate consumers were estimated to have received an effective dose of 0.054 mSv from marine fish and 0.093 mSv from shellfish, mainly molluscs, a total of 0.15 mSv. The collective dose from seafood consumption in 1986 due to input of Chernobyl-derived activity to UK waters was tentatively estimated to be 30 man-Sv.

The North Sea

Radioactivity from Chernobyl was widely detected in the North Sea during summer 1986, largely by means of its distinctive $^{137}Cs/^{134}Cs$ ratio of about 2 : 1. However, concentrations were much lower than in the Baltic Sea, the Black Sea and the Irish Sea. In many areas, including the Northern North Sea, the deposition from Chernobyl overshadowed the effect of BNFL Sellafield, otherwise the main source of radiocaesium (Mitchell & Steele, 1988). However, compared with some other areas, concentrations were low, such that high-rate North Sea fish consumers were estimated to have received only up to 3 μSv y^{-1} in 1986.

Summary of Chernobyl effects on marine waters

The European MARINA studies (CEC, 1990; EC, 2002a) also considered the radiological effects of the Chernobyl accident on marine waters, including collective doses to the EU population. The collective dose to the EU population committed to the year 2500 from marine pathways due to the effects of the Chernobyl accident was estimated to be about 60 man-Sv. This is a small fraction of the total dose commitment due to the Chernobyl accident, estimated to be some 600,000 man-Sv (UNSCEAR, 1993), which is due predominantly to terrestrial pathways. The collective dose rate to the EU population in 2000 from marine pathways is estimated to be about 0.5 man-Sv y^{-1} (EC, 2002a). This would need to be increased by a factor of about 3 to include non-EU populations (CEC, 1990). Thus the world collective dose rate received in 2000 due to marine pathways from the Chernobyl accident would be about 1.5 man-Sv y^{-1}.

3.9. *Conclusions*

A summary of the current radiological effects of the sources of radioactivity in the oceans is presented for comparative purposes in Table 7 and graphically in Fig. 4 (critical group dose) and Fig. 5 (collective dose). All the data have been mentioned earlier in this text and details of their derivation and references may be found there.

Table 7
Summary of current (~Y 2000) annual doses due to major sources of exposure through marine pathways

Source	Dose to critical group (mSv y^{-1})	Collective dose rate to world population (man-Sv y^{-1})
Natural radionuclides	~2	134,000
Enhanced natural radionuclides	0.1–0.3 (to Dutch and Spanish groups)	200
Weapons-test fallout	0.0002	160
Nuclear industry	0.15 (Sellafield) (peak: ~2 in 1975)	20
Ocean dumping of solid radioactive waste	0.00002	4
Radioactive waste dumping in Arctic seas	<0.0001	<0.01
Chernobyl accident	0.02 (Baltic) (peak: ~0.2 in 1986)	1.5

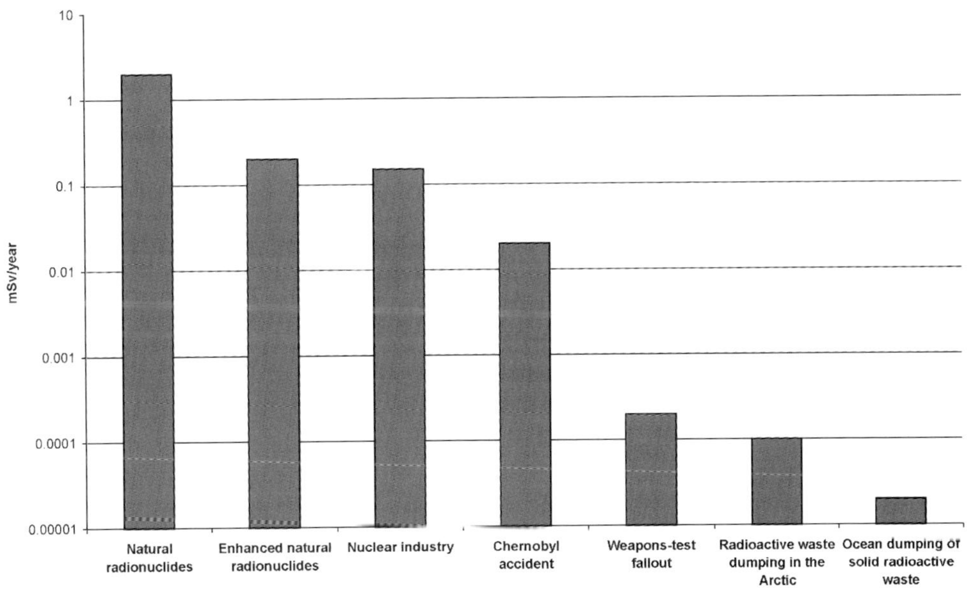

Fig. 4. Comparison of current (~Y 2000) critical group effective dose rates from major sources due to marine pathways (log scale).

The second column of Table 7 and Fig. 4 give the annual effective doses to critical groups currently (i.e. around the year 2000) being received; it is to be noted that the critical groups are not necessarily the same, thus the doses are not necessarily additive. In some cases, for information, the locations of the groups and peak dose rates received in the past are also noted. The third column of Table 7 and Fig. 5 give the current annual collective effective doses to the world population. This is a more effective basis for comparison between sources than integrated commitments into the future which may have also been given in the studies

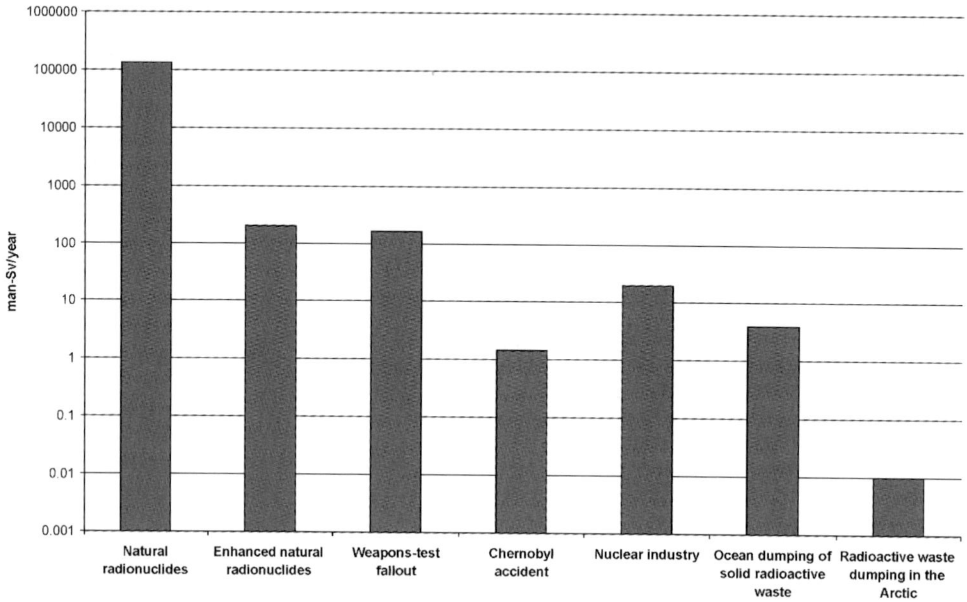

Fig. 5. Comparison of current (~Y 2000) collective effective dose rates to the world population from major sources of exposure through marine pathways (log scale).

referenced, because the timescales used in these studies have often been different, and the results are difficult to extrapolate to a common basis. The reader is referred to the individual studies for more detail, including descriptions of the actual populations exposed.

It is to be noted first that all the exposures are, with one exception, additive to those due to natural background and medical exposures given in Section 1.2, the world average being about 2.8 mSv y^{-1}. The exception is for the collective dose due to average seafood consumption which is considered to be included in the 2.8 mSv y^{-1}. However, the inclusion due to average seafood consumption is small, and indeed derivable by dividing the collective dose of 134,000 man-Sv y^{-1} by the world population of 6×10^9 (UNSCEAR, 2000), i.e. 0.02 mSv y^{-1}.

For individuals the most significant potential addition from marine pathways to the average dose of 2.8 mSv y^{-1} is due to natural radionuclides from high consumption rates of seafood, especially molluscan shellfish; a high-rate consumer could receive an additional ~2 mSv from this source. This dose is overwhelmingly due to ^{210}Po.

Enhanced natural radionuclides due to operations of the phosphate industry are probably the next most significant exposures, with doses to high-rate seafood consumers of 0.1–0.3 mSv y^{-1}. This is small compared with doses due to unenhanced natural radioactivity, potentially to the same individuals. The collective dose is very small compared with that from unenhanced natural radioactivity.

In collective dose rate terms, nuclear weapons-test fallout is the next most significant contribution, due to its wide distribution. However, doses to actual high-rate seafood consumers due to this source are very small.

The highest critical group dose from marine pathways due to operations of the nuclear industry is currently about 0.15 mSv y^{-1}, near Sellafield. This is an order of magnitude lower than from natural radionuclides (potentially to the same critical group). The collective dose consequences of nuclear industry discharges will mostly be in European waters due to the large number of nuclear industry sites and their discharges. Collective doses there from this source, however, are now only of the order of 14 man-Sv y^{-1}, some 4 orders of magnitude lower than from natural radionuclides. An estimate of 20 man-Sv y^{-1} is entered in Table 7 to allow for other world sources.

Ocean dumping has caused concerns about its effects but the potential doses, calculated from modelling studies, are likely to be very small, 4–5 orders of magnitude lower than for natural radionuclides in critical group and collective dose terms.

The radiological implications of radioactive waste dumping in Arctic seas have also been studied by the use of suitable models. Potential critical group doses are of the same order as for the North Atlantic dumping, with collective dose being much less, inter alia because of lower populations and seafood production.

Lastly, the remains of the Chernobyl accident give rise to a critical group dose rate of about 0.02 mSv y^{-1} in the northern Baltic area. Collective dose rates due to marine pathways from Chernobyl deposition are currently of the order of 1.5 man-Sv y^{-1}.

References

Aldridge, J. N. (1998). CSERAM: A model for prediction of marine radionuclide transport in both particulate and dissolved phases. *Radiation Protection Dosimetry*, *75* (1–4), 99–103.

Beck, H. & de Planque, G. (1968). The radiation field in air due to distributed gamma ray sources in the ground. Tenn. HASL-195. Oak Ridge: US Health and Safety Laboratory.

Camplin, W. C. et al. (1986). Radioactivity in surface and coastal waters of the British Isles. Monitoring of fallout from the Chernobyl reactor accident (49 pp.). Aquatic Env. Mon Report No 15. Ministry of Agriculture, Fisheries and Food, Lowestoft.

Camplin, W. C., Baxter, A. J. & Round, G. D. (1996). The radiological impact of discharges of natural radionuclides from a phosphate plant in the United Kingdom. *Environment International*, *22*, *Suppl.* 1, 5259–5270.

CEC (1990). The radiological exposure of the population of the European Community from radioactivity in North European marine waters: Project MARINA (571 pp.). EUR 12483 EN. Luxembourg: European Commission.

Clarke, R. (1999). Control of low-level radiation exposure: Time for a change? *Journal of Radiological Protection*, *19* (2), 107–115.

Cross, W. G., Freedman, N. O. & Wong, P. Y. (1992). Beta ray dose distributions from skin contamination. *Radiological Protection Dosimetry*, *40* (3), 149–168.

EC (1994). Methodology for assessing the radiological consequencies of routine releases of radionuclides to the environment. Radiation protection 72 (362 pp.). XI-5026/94-EN. Brussels: European Commission.

EC (2000). The radiological exposure of the population of the European Community to the radioactiviry in the Baltic Sea. MARINA-Balt Project: Radiation Protection 110. UER 19200 EN. Luxembourg: European Commision.

EC (2002a). Update of the MARINA Project on the radiological exposure of the European Community from radioactivity in north european marine waters. Brussels: European Commission.

EC (2002b). MARINA II. Update of the MARINA Project on the radiological exposive of the European Community from radioactivity in north european marine waters. Radiation Protection 132. Luxembourg: European Commision.

Egorov, V. N. et al. (1999). Strontium-90 and Caesium-137 in the Black Sea after the Chernobyl accident: Inventories, balance and tracer application. *Journal of Environmental Radioactivity*, *43*, 137–155.

Environment Agency (2001). *Radioactivity in the Environment* (156 pp.). Lancaster, UK: Environment Agency.

FASSET (2003). Framework for assessment of environmental impact. EC 5th Framework Programme FIGE-CT-2000-00102. Deliverables 1–6. Available at *www.fasset.org*.

Food Standards Agency and Scottish Environment Protection Agency (2001). Radioactivity in food and the environment, 2000 (184 pp.). RIFE-6. London and Stirling: Food Standards Agency and Scottish Environment Protection Agency.

FSA & SEPA (2001). Radioactivity in food and the environment, 2000 (183 pp.). RIFE-5. London and Stirling: Food Standards Agency and Scottish Environment Protection Agency.

Handyside, I., Hunt, G. J. & Partington, C. (1982). Control of radiocaesium in discharges to the Irish Sea: ICRP-26 in practice. In *Proc. Symp. IAEA on the Dose Limitation System in the Nuclear Fuel Cycle and in Radiation Protection* (pp. 325–345). Madrid, 19–23 October 1981. Vienna: IAEA.

HELCOM (1996). Radioactivity in the Baltic Sea, 1984–1991. Joint Evaluation Report of the Expert Group MORS. Baltic Sea Environment Proceedings. Helsinki.

Hunt, G. J. (1984). Simple models for prediction of external radiation exposure from aquatic pathways. *Radiation Protection Dosimetry*, *8* (4), 215–224.

Hunt, G. J. (1992). External doses to the public from beta emitters in the aquatic environment near Springfields and Sellafield. *Journal of Radiological Protection*, *12* (4), 233–238.

Hunt, G. J. (1997). Radiation doses to critical groups since the early 1950s due to discharges of liquid radioactive waste from Sellafield. *Health Physics*, *72* (4), 558–566.

Hunt, G. J. & Allington, D. J. (1993). Absorption of environmental polonium-210 by the human gut. *Journal of Radiological Protection*, *13* (2), 119–126.

IAEA (1961). *Radioactive Waste Disposal into the Sea*. IAEA Safety Series 5. Vienna: IAEA.

IAEA (1983). *Control of Radioactive Waste Disposal into the Marine Environment*. IAEA Safety Series 61. Vienna: IAEA.

IAEA (1985). Sediment k_ds and concentration factors for radionuclides in the marine environment. Technical Reports Series 247. IAEA, Vienna.

IAEA (1986). *Principles for Limiting Releases of Radioactive Effluents into the Environment*. IAEA Safety Series Report 77. Vienna: IAEA.

IAEA (1992). *Effects of Ionizing Radiation on Plants and Animals at Levels Implied by Current Radiation Protection Standards*. Technical Reports Series 332. Vienna: IAEA.

IAEA (1995). Sources of radioactivity in the marine environment and their relative contributions to overall dose assessment from marine radioactivity (MARDOS) (55 pp.). IAEA-TECDOC-838. Vienna: IAEA.

IAEA (1996a). *International Basic Safety Standards for Protection Against Ionizing Radiation and for the Safety of Radiation Sources* (353 pp.). Safety Series 115. Vienna: IAEA.

IAEA (1996b). Radiological conditions at Bikini Atoll: Prospects for resettlement. Report of an Advisory Group. Vienna: IAEA.

IAEA (1998a). The radiological situation at the atolls of Mururoa and Fangataufa (97 pp.). Report by an International Advisory Committee. STI/PUB/1028. Vienna: IAEA.

IAEA (1998b). Radiological conditions of the Western Kara Sea: Assessment of the radiological impact of the dumping of radioactive waste in the Arctic Seas (124 pp.). Report on the International Arctic Seas Assessment Project (IASAP). Radiological Assessment reports series. Vienna: IAEA.

IAEA (1999a). Application of radiological exclusion and exemption principles to sea disposal. IAEA-TECDOC-1068. Vienna: IAEA.

IAEA (1999b). Protection of the environment from the effects of ionizing radiation. IAEA-TECDOC-1091. Vienna: IAEA.

IAEA (1999c). Inventory of radioactive waste disposals at sea (122 pp.). IAEA-TECDOC-1105. Vienna: IAEA.

IAEA (2000). *Regulatory Control of Radioactive Discharges to the Environment*. IAEA Safety Guide WS-G-2.3 Vienna: IAEA

IAEA (2001). *Generic Models for Use in Assessing the Impact of Discharges of Radioactive Substances to the Environment*. Safety Reports Series No. 19 (216 pp.). Vienna: IAEA.

IAEA (2004). Sediment distribution coefficients and concentration factors for biota in the marine environment. Technical Reports Series No. 422. Vienna: IAEA.

ICRP (1966). Recommendations of the international commission on radiological protection (27 pp.). *ICRP Publication 9*. Oxford: Pergamon.

ICRP (1977). Recommendations of the international commission on radiological protection. *ICRP Publication 26. Annals ICRP, 1* (3), 1–53.

ICRP (1979). Radionuclide release into the environment: Assessment of doses to man. *ICRP Publication 29. Annals ICRP, 2* (2), 1–76.

ICRP (1985). Principles of monitoring for the radiation protection of the population. *ICRP Publication 43. Annals ICRP, 15* (1), 1–19.

ICRP (1986). The metabolism of plutonium and related elements. *ICRP Publication 48. Annals ICRP, 16* (2/3), 1–98.

ICRP (1991). 1990 Recommendations of the International Commission on Radiological Protection. *ICRP Publication 60. Annals ICRP, 21* (1/3), 1–201.

ICRP (1996). Age dependent doses to members of the public from intake of radionuclides: Part 5. Compilation of ingestion and inhalation dose coefficients. *ICRP Publication 72. Annals ICRP, 26*, 1–91.

ICRP (1997). Radiological protection policy for the disposal of radioactive waste. *ICRP Publication 77. Annals ICRP 27 Supplement.* Oxford: Pergamon.

ICRP (1998). Radiation protection recommendations as applied to the disposal of long-lived solid radioactive waste. *ICRP Publication 81. Annals ICRP, 28* (4), 1–25.

ICRP (1999). Protection of the public in situations of prolonged radiation exposure. *ICRP Publication 82. Annals ICRP, 29* (1/2), 1–109.

ICRP (2001). Doses to the embryo and fetus from intakes of radionuclides by the mother. *ICRP Publication 88. Annals ICRP, 31* (1/3), 1–518.

ICRP (2003). A framework for assessing the impact of ionising radiation on non-human species. *ICRP Publication 91. Annals ICRP, 33* (3), 201–266.

ICRU (1994). Gamma-ray spectrometry in the environment (84 pp.). ICRU Report 53. Bethesda, MA: International Commission on Radiation Units and Measurements.

Kanivets, V. V. et al. (1999). The post-Chernobyl budget of ^{137}Cs and ^{90}Sr in the Black Sea. *Journal of Environmental Radioactivity, 43* (2), 121–135.

Kocher, D. C. & Eckerman, K. F. (1987). Electron dose rate conversion factors for external exposure of the skin from uniformly deposited activity on the body surface. *Health Physics, 53* (2), 135–141.

Kocher, D. C. & Sjoreen, A. L. (1985). Dose rate conversion factors for external exposure to photon emitters in soil. *Health Physics, 48*, 193–205.

Mitchell, N. T. & Steele, A. K. (1988). The marine impact of caesium-134 and -137 from the Chernobyl reactor accident. *Journal of Environmental Radioactivity, 6*, 163–175.

NCRP (1995). Principles and application of collective dose in radiation protection (106 pp.). NCRP Report No 121. USA: Bethesda, MD.

NEA (1985). Review of the continued suitability of the dumping site for radioactive waste in the North-east Atlantic (448 pp.). Paris: Nuclear Energy Agency, Organisation for Economic Co-operation and Development.

Nord-Cotentin Radioecology Group (2000). Estimation of exposure levels to ionising radiation and associated risks of leukaemia for populations in the Nord-Cotentin. Synthesis. Institut de Protection et Sureté Nucléaire, Paris (353 pp.).

NRPB (1998). Revised generalised derived limits for radioisotopes of strontium, ruthenium, iodine, caesium, plutonium, americium and curium. Documents of the NRPB 9 (1). Chilton: National Radiological Protection Board.

OSPAR (1997). Report on Discharges of Radioactive Substances by Non-Nuclear Industries (28 pp.). London: OSPAR.

Papucci, C. et al. (1996). Time evolution and levels of man-made radioactivity in the Mediterranean Sea. In P. Gueguenit, P. German & H. Metivier (Eds), *Radionuclides in the Oceans. Inputs and Inventories.* ISBN 2-86883-285-7. France: Les Editions de Physique.

Pentreath, R. J. (1988). Radionuclides in the aquatic environment. In *Radionuclides in the Food Chain* (pp. 99–119). Berlin: Springer-Verlag.

Pentreath, R. J. & Woodhead, D. S. (1988). Towards the development of criteria for the protection of marine fauna in relation to the disposal of radioactive waste into the sea. In *Proc. Conf. IAEA on Radiation Protection in Nuclear Energy, Sydney, 1988* (pp. 213–243). Vienna: IAEA.

Pentreath, R. J. & Woodhead, D. S. (2001). A system for protecting the environment from ionising radiation: selecting reference fauna and flora, and the possible dose models and environmental geometries that could be applied to them. *Science of the Total Environment, 277*, 33–43.

Perianez, R. & Reguera, J. (1999). A numerical model to simulate the tidal dispersion of radionuclides in the English Channel. *Journal of Environmental Radioactivity, 43*, 51–64.

Polikarpov, G. G. et al. (1991). ^{90}Sr and ^{137}Cs in surface waters of the Dnieper River, the Black Sea and the Aegean Sea in 1987 and 1988. *Journal of Environmental Radioactivity, 13* (1), 25–38.

Salomon, J. C., Breton, M. & Gueguenia, P. (1993). Computed residual flow through the Dover Strait. *Oceanologica Acta, 16* (5–6), 449–455.

Simon, S. L. & Graham, J. C. (1996). Dose assessment activities in the Republic of the Marshall Islands, *Health Physics, 71* (4), 438–456.

Smith, D. L., Knowles, J. F. & Winpenny, K. (1998). The accumulation and distribution of ^{95m}Tc in crab (*Cancer pagurus* L.) and lobster (*Homarus gammarus* L.): A comparative study. *Journal of Environmental Radioactivity, 40* (2), 113–135.

UNCED (1992). *United Nations Conference on Environment and Developments, Rio Declaration on Environment and Development.* New York: UN.

UNSCEAR (1993). Sources and effects of ionizing radiation. UN Scientific Committee on the Effects of Atomic Radiation, 1993 report to the General Assembly. New York: UN.

UNSCEAR (1996). Effects of radiation on the environment. Annex to Sources and Effects of Ionizing Radiation, 1996 report to the UN General Assembly. UN Scientific Committee on the Effects of Atomic Radiation. New York: UN.

UNSCEAR (2000). Sources and effects of ionizing radiation. UN Scientific Committee on the Effects of Atomic Radiation, 2000 report to the General Assembly. New York: UN.

Woodhead, D. S. (1979). Methods of dosimetry for aquatic organisms. In *Methodology for Assessing Impacts of Radioactivity in Aquatic Ecosystems* (pp. 43–96). Technical Reports Series 190. Vienna: IAEA.

MARINE RADIOACTIVITY
Hugh D. Livingston (Editor)

Chapter 8

Developments in analytical technologies for marine radionuclide studies

Pavel P. Povinec

International Atomic Energy Agency, Marine Environment Laboratory, MC 98012, Monaco

Recent developments in sampling, sample preparation and measuring techniques using radiometrics and mass spectrometry methods for analysis of natural and anthropogenic radionuclides in the marine environment are presented and discussed. As the levels of anthropogenic radionuclides observed at present in the marine environment are very low, high sensitive analytical systems are required for carrying out oceanographic investigations. The developments in sampling and measuring techniques are illustrated by several examples of marine radioactivity studies, which include both contaminated sites (e.g. the Pacific weapons testing sites and Arctic dumping sites), open ocean sites and also the analysis of IAEA marine reference materials. As the topic is very wide, it has not been possible to cover all aspects of sampling and radionuclide analyses in detail; the emphasis has been on recent developments in the field.

1. Introduction

Oceanic investigations with radionuclide tracers began about 45 years ago with application of ^{226}Ra (Koczy, 1958), and they have always been limited by the techniques available for sampling and analysis of radionuclides in the marine environment (Livingston & Povinec, 2002). The water sampling has developed from the reversing Nansen bottle into the present robotic systems based on ROVs (remotely operating vehicles) and AUVs (autonomous underwater vehicles), and other sophisticated sampling technologies, which use satellite views of sea areas for the optimisation of sampling. In the field of nuclear technology, investigations started with simple radiochemical methods and gas counters, and then developed into the age of robotic radiochemical technologies, sophisticated detectors working on line with powerful computers (sometimes with satellite data transmission from continuous underwater radionuclide monitors), and changing philosophy of radionuclide analysis from the concept of counting decays to the counting of atoms using high sensitive mass spectrometers working with low energy ions (e.g. thermal ionisation mass spectrometry (TIMS), inductively coupled plasma mass spectrometry (ICPMS), resonance ionisation mass spectrometry (RIMS) or accelerated ions to levels of tens or hundreds of MeV in accelerator mass spectrometry (AMS) systems.

This move from simple analytical techniques to the present sophisticated state of the art technologies has been accompanied by a considerable change in the philosophy of oceanic studies as well (Livingston & Povinec, 2002), from institutional investigations from the fifties to the seventies, to global international oceanic investigations carried out in recent years (e.g. WOCE, JGOFS and other programmes).

The levels of anthropogenic radionuclides observed at present in the marine environment are very low, therefore high sensitive analytical systems are required for carrying out oceanographic investigations (Livingston & Povinec, 2000). In the present review we discuss some recent developments in ocean sampling techniques, sample pre-treatment and preparation techniques, and measuring techniques using radiometrics and mass spectrometry methods. These developments are illustrated by several examples of marine radioactivity studies, which include both contaminated sites (e.g. the Pacific weapons testing sites and Arctic dumping sites), open ocean sites and also the analysis of IAEA marine reference materials.

As the topic is very wide, it has not been possible to cover all aspects of sampling and radionuclide analyses in detail; the emphasis has been on recent developments in the field. Similarly, the list of references, although very comprehensive, could not cover all the work done, but again it is listing only the recent publications. Several conferences on low-level counting and spectrometry have been organised in past, where more detailed information can be found (e.g. Povinec & Usacev, 1977; IAEA, 1980, 1999; Povinec, 1982, 1986, 1987, 1991a, 1992; Garcia-Leon & Madurga, 1991; Garcia-Leon & Garcia-Tenorio, 1994; Holm et al., 1996, 2000; Cook et al., 1996; Courser & Taylor, 2000; Kutchera et al., 2000).

2. Marine sampling for radionuclide studies

2.1. *Water sampling*

Water sampling for traditional radionuclide analysis by radiometrics techniques (RMT) requires the collection of large volume samples (100–500 l). This is very laborious and time consuming, especially if water has to be raised from several kilometres depth and taken on deck. The operation with large volume (around 100 l, 200 l, 500 l) samplers (e.g. Gerard–Ewing type) usually made from glass fibre reinforced plastic is very difficult and time consuming. Water after collection is filtered (0.45 μm) and transferred by a pump to precipitation tanks for a pre-treatment.

Therefore, a decrease in water volume would be a great advantage for facility of operation and time saving. To do this, far more sensitive methods of radionuclide analysis are needed. As will be discussed later, if AMS, or TIMS are used, only small volume samples are needed for the analysis of several radionuclides in the water column. For example, ^{14}C or ^{129}I analysis by AMS requires only 0.5 l of water; ^{239}Pu or ^{240}Pu analysis by AMS or TIMS requires only a few litres of water. For other radionuclides, analysed by γ-ray spectrometry (e.g. ^{137}Cs) or β-ray spectrometry (e.g. ^{90}Sr), measurements in underground laboratories would seem to be the solution for decreasing sample volumes to about 10 l (Povinec et. al., 2001a–c).

Fig. 1. CTD/RMS sampling device with 24 Niskin bottles, 2.5 l each.

Consequently, the most important development in seawater sampling has been the introduction of the CTD/Rosette Multi-bottle Samplers (CTD/RMS) (Fig. 1). They have been employed for the collection of both small and medium volume samples by using one bottle per depth layer or several bottles at the same depth. The volume and number of Niskin bottles used in a CTD/RMS sampler vary from type to type. Further advantages of the CTD/RMS sampler are real time monitoring of oceanographic parameters, e.g. pressure, temperature, conductivity/salinity, dissolved oxygen, sound velocity, etc. Twenty-four sets of CTD data can be collected every second. The CTD/RMS system is submerged during operation at a vertical speed of 1–1.5 m s^{-1} and the CTD data are monitored in real time in a dry laboratory on shipboard. The data obtained during downcast are usually used as the CTD profile data, seawater samples are collected during upcast. In general, two casts, one shallow (up to about

1000 m) and one deep, are made for deep water sampling. This way the whole water column can be sampled more accurately and with only two casts (IAEA, 2002).

2.2. *Sediment sampling*

Sediment is usually sampled by box corer or multiple corer down to a depth of around 60 cm. Both types of corers have their advantages and disadvantages. Box corers allow sub-sampling of cores, are easier to control and show a relatively large undisturbed sediment surface. A pinger fixed around 50 m above the corer is used for reading the relative distance of the box core from the seafloor, as registered by an echo-sounder on board. The last 50–60 m from the bottom are recorded by a Deep Sea Echo Sounder and the winch cable is slowly lowered until the corer hits the bottom. The changes in cable tension are continuously recorded on an X–Y chart recorder. A Kullenberg corer is used for sampling sediment layers a few meters deep. As surface layers of sediment are usually compressed during sampling, this type of corer, which acts as a gravity corer, is not suitable for sampling surface sediment layers.

After successful retrieval of a box core, Plexiglas tubes are used to subcore sediments by slow and careful penetration into the collected sediment. Special care must be taken not to compress the sediment column by modifying the force while maintaining equal sediment surface levels inside and outside the wall of the Plexiglas tube. When sediments are too cohesive, a vacuum pump should be used to ease the core liner into the sediment. The tubes containing sediment are capped on top and plugged at the bottom with a PWC stopper with silicone O-rings to make the plugging water tight. After X-ray investigations for studying micro and macro sediment structures (which can also be done later in the home laboratory on an extra collected tube), the tubes are sliced into pre-determined sections in the wet laboratory on board. Each sediment section is placed in a pre-weighed plastic container and kept at 4°C until the appropriate conditions for wet weighing are encountered. The samples are freeze dried in the home laboratory and weighed again, thus determining the total sediment weight. This information is used to determine water content, bulk density, porosity and the mass depth of the sediment column sampled, which are the sediment parameters needed for the correct interpretation of radionuclide results. The sediment remaining in the box after sub-sampling, usually from 0–5 cm, is squeezed to obtain pore water. The pore water is filtered through a 0.45 μm filter, and stored at 4°C in small bottles for later analyses in the home laboratory.

2.3. *Biota sampling*

In order to study the partitioning of radionuclide input in the food chain, plankton, seaweed, fish and shellfish are often sampled during oceanographic expeditions. As very traditional sampling methods are used for seaweed, fish and shellfish sampling, we shall concentrate on plankton sampling in detail. Planktic invertebrates are collected using plankton nets generally of four different mesh sizes (1000 μm, 500 μm, 250 μm and 100 μm). The ring aperture is around 1 m. The nets are towed both horizontally or vertically (rarely). When towing horizontally at a depth of 10–20 m, it is possible to use nets with different meshes simultaneously (except for 100 μm nets which are very fragile and with which special precautions must be taken). The ship's speed is maintained at about 1 knot to avoid damaging both the nets and the animals collected. The sampling period is usually a few hours, the total volume of water passed through the nets is estimated by a fixed current meter centred at the opening of one of the plankton nets. Upon retrieval of the net from the water, the collected animals are placed in labelled plastic containers and studied. Finally, they are frozen in plastic bags and during transportation to the home laboratory they are cooled in a box cooler containing dry ice.

2.4. *Sampling of particulate matter*

Large, rapidly sinking particles have been collected in sediment traps, by free floating or moored systems remaining in the water column for periods of months and years, thus enabling seasonal and annual changes in particle fluxes to be recorded. These studies have broadened our knowledge of ocean particulate dynamics and their links with the ocean carbon cycle (Fowler et al., 1983, 2000; Deuser, 1986; Cochran et al., 1993). The most important development in the sediment trap sampling technique is microprocessor controlled sampling using sediment traps with sampling bottles changed according to pre-programmed time intervals thus enabling the development of a continuous time series over a number of years. A formaldehyde solution in the bottles prevents the deterioration of sampled particulate matter throughout the sampling period.

3. Preparation of samples

Generally, four types of samples, water, sediment, biota and particulate matter, are used for the investigation of radionuclides in the marine environment. As the preparation chemistry is very complex, we shall deal only with water samples in this review. For a recent review on advanced speciation techniques for radionuclides associated with colloids and particles see Salbu et al. (2001).

Two kinds of water samples are collected during cruises. Small volume samples (e.g. 1 l for ^{3}H analysis, 0.5 l for ^{129}I and 0.5 l for ^{14}C analysis (poisoned with mercury chloride to prevent biological activity)) are normally filtered only (0.45 μm membrane filter) and then bottled so as to prevent any exchange of air with water. All the preparation chemistry is carried out in the home laboratory.

As the sensitivity often demands the analysis of large samples (e.g. hundreds of litres of seawater) in order to obtain sufficient amounts of the radionuclides of interest to permit analysis by γ-ray or β-ray spectrometry, the need for accurate and timely radiochemical measurements is the driving force behind the search for rapid methods with high selectivity, excellent purification factors and good chemical recoveries. Large volume samples are too large to be transported to the laboratory, and pre-concentration is carried out on shipboard. We shall illustrate the preparation chemistry for three of the most frequently analysed radionuclides (^{90}Sr, ^{137}Cs and Pu isotopes), using the methods adopted at IAEA-MEL (La Rosa et al., 2001; Lee et al., 2001; IAEA, 2002) in order to demonstrate the present state of the art techniques.

3.1. *Shipboard chemistry*

Pre-concentration chemistry is carried out usually in three steps for transuranics, caesium and strontium (Fig. 2).

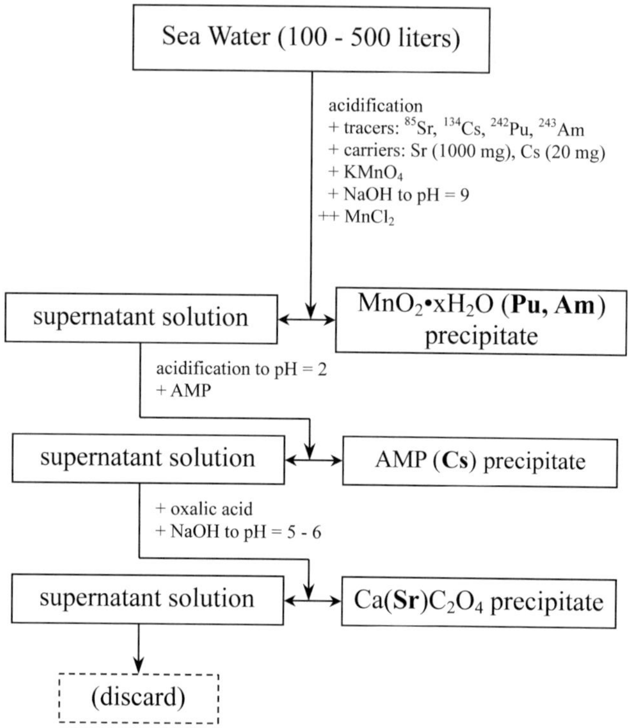

Fig. 2. Pre-concentration of radionuclides in seawater on shipboard (IAEA, 2002).

Pre-concentration of transuranics

Samples of about 400 l of filtered (0.45 μm cut-off) seawater transferred into precipitation tanks are acidified by 250 ml of concentrated HCl followed by the addition of tracers (e.g. ^{85}Sr, ^{134}Cs, ^{242}Pu, ^{243}Am) and carriers. The samples are then stirred by bubbling for 15 minutes to homogenise tracers and carriers with the sample. Then 125 ml of saturated $KMnO_4$ is added and the bubbling is continuing for 15 minutes. 250 ml of 0.5 M $MnCl_2$ is then added followed by 150 g of NaOH to adjust the pH to 6–7 and finally 2–3 ml of H_2O_2 is added to transform the $Mn(OH)_2$ to MnO_2 ($Mn(OH)_2 + H_2O_2 \geqslant MnO_2 + 2H_2O$) which is then stirred for 30 minutes. The sample is then left to stand for 4–5 hours to enable the separation of the precipitate from the sample solution. The supernate is then transferred to a different plastic tank for pre-concentration of caesium with ammonium molybdophosphate (AMP). The precipitate is transferred to a plastic container for further separation and determination of transuranics in the home laboratory.

Pre-concentration of caesium

The supernate from the previous step is acidified (pH 1.5–2.0) with 350 ml of concentrated HCl and mixed by bubbling air for 15 minutes. 2 ml of H_2O_2 is then added to dissolve traces of MnO_2 solids. 30 g of AMP, slurred in 100 ml of demineralised water is added and the sample bubbled for 30 minutes, then left to stand for 4–5 hours to allow the precipitate to settle. It is then transferred to another plastic tank to separate strontium by oxalate co-precipitation. The AMP precipitate is transferred to a plastic container for further separation and determination of caesium in the home laboratory.

Pre-concentration of strontium

500 g of oxalic acid, dissolved in 1 l of boiling demineralised water is added to the supernate, which is stirred by bubbling air for 15 minutes. 400 g of NaOH is then added to adjust the pH to 6–7 resulting in white oxalate precipitate. The sample is stirred for 30 minutes and left to stand for 4–5 hours to allow the precipitate to settle. The supernate sample is then discarded and the oxalate precipitate collected and stored for further separation and determination of strontium in the home laboratory.

Shipboard analysis of short-lived radionuclides

^{234}Th

As the half-life of ^{234}Th (24 days) is very short, sampling and analytical work are usually carried out on shipboard. ^{234}Th has been used as a particle scavenging tracer to determine particle residence time (export) and transformation rates in seawater. The export rate of particulate organic carbon (POC) has been determined from the export flux of Th using ^{234}Th/^{238}U disequilibrium (Buesseler et al., 1992).

Measurements of ^{234}Th are carried out on shipboard using β-ray or γ-ray spectrometers on samples scavenged with $Fe(OH)_3$ or MnO_2. 20–30 l of water is filtered using an 0.6 μm polycarbonate filter (^{234}Th-particulate) and thorium is scavenged in its dissolved form using $Fe(OH)_3$ precipitation (Van der Loef et al., 1997). Thorium-229 is added as yield tracer. An aliquot of the precipitate is recovered by filtration. The precipitate is then mounted on a plastic disk and counted in a multiple gas β-ray detector.

Another method is using MnO_2 impregnated cartridges with in situ pumps which are also fitted with filters to collect particles bearing ^{234}Th. After sampling (1–2 h of pumping, flow rate 4–25 l min^{-1}), the cartridges are ashed (at 550°C for 8 h), placed onto petri dishes of known geometry, and γ-counted on an HPGe detector.

The samples are re-counted later in the home laboratory. Thorium-234 activities are corrected to the dates and times of collection. The scavenging efficiency of the MnO_2 cartridges is 70–85%. There has been a reasonable agreement between ^{234}Th concentrations measured by β and γ-ray spectrometry (Povinec et al., 2001b).

Ra *isotopes*

Another elegant technique has recently been developed for shipboard measurements of short-lived radium isotopes ^{223}Ra ($T_{1/2} = 11.7$ days) and ^{224}Ra ($T_{1/2} = 3.6$ days), which can be extended for ^{228}Ra ($T_{1/2} = 5.75$ y) and ^{226}Ra ($T_{1/2} = 1602$ y) measurements in laboratory conditions (Moore & Arnold, 1996). Radium isotopes have been widely used in coastal water dynamics studies including submarine groundwater discharge (Moore, 1996, 1999). Radium is sampled from water by its adsorption onto a column of MnO_2 coated fibre. The short-lived radon daughters of ^{223}Ra (^{219}Rn and ^{215}Po) and ^{224}Ra (^{220}Rn and ^{216}Po) which recoil from the Mn fibre are swept into a scintillation detector where α-decays of Rn and Po occur. Signals from the detector are sent to a delayed coincidence unit, which discriminates decays of the ^{224}Ra daughters from decays of the ^{223}Ra daughters. The method is extendible to ^{228}Ra (through its decay product ^{228}Th) as well as to ^{226}Ra by its analysis on the Mn fibre. This technique provides a highly effective and simple way to analyse the Ra isotopic quartet in water without chemical treatment of the Mn fibre.

3.2. *Laboratory chemistry*

Transuranics

Manganese dioxide containing Pu and Am isotopes obtained from shipboard pre-concentration is prepared from the suspension in the home laboratory (after checking the pH of the liquid) using a combination of settling, decantation/siphoning of the supernatant liquid and centrifugation. The procedure is described in Fig. 3 (Lee et al., 2001).

Further separation of Pu and Am from the iron matrix is carried out by separation of Pu(IV) using an anion exchange resin column from 8M HNO_3 followed by the recovery of Am from the column eluate. An alternative method has recently been developed by Burnett et al. (1997) using the EiChrom UTEVA + TRU double column. The flow chart for this method as adapted for mass spectrometry measurements of small volume water samples (less than 20 l) is shown in Fig. 4 (Lee et al., 2001).

With the traditional method, the typical chemical recoveries of Pu and Am from several hundred litres of seawater are in the range of 40–60% and 60–80%, respectively. The resulting α-ray spectra of Pu and Am electrodeposited sources show good radiochemical purity and well-resolved peaks.

With the extraction chromatography double column method, the Am and Pu sources have high chemical recoveries (ca. 90% in each case) and the excellent quality α-ray spectra show no visible foreign peaks. In large volume sea water samples, it is expected that additional purification of the Pu and Am fractions from the TRU column would be necessary; however, the small working volumes permit easy and fast chemical operations with less time spent in evaporation steps. In particular, Pu and Am may be retained together on the same TRU column (unlike in the traditional method). Small volume water samples suitable for ICPMS or AMS measurements of Pu isotopes show very good results. In Am separation and purification, sources produced by both methods show very good chemical recoveries, purity and spectral resolution. In the reprocessing of Am electrodeposition onto disks, one cycle of TEVA + TRU columns gives excellent decontamination from Po with very little loss of Am.

Caesium

The AMP suspension (10–20 l) is allowed to settle and the supernatant solution is decanted or siphoned away. The remaining suspension is centrifuged and the AMP solid left is washed with water acidified to pH 2. The flow chart for caesium procedures is given in Fig. 5.

Recently, a comparison of two sorbents for the collection of caesium (ANFEZH and KCFC) with the AMP was made (La Rosa et al., 2001). It shows that AMP and KCFC quantitatively collect Cs tracer in 20 minutes, whereas ANFEZH does so more slowly and incompletely after 180 minutes. Experimentally-derived Cs capacities of the sorbents show that AMP has the highest capacity for Cs sorption and ANFEZH, the lowest.

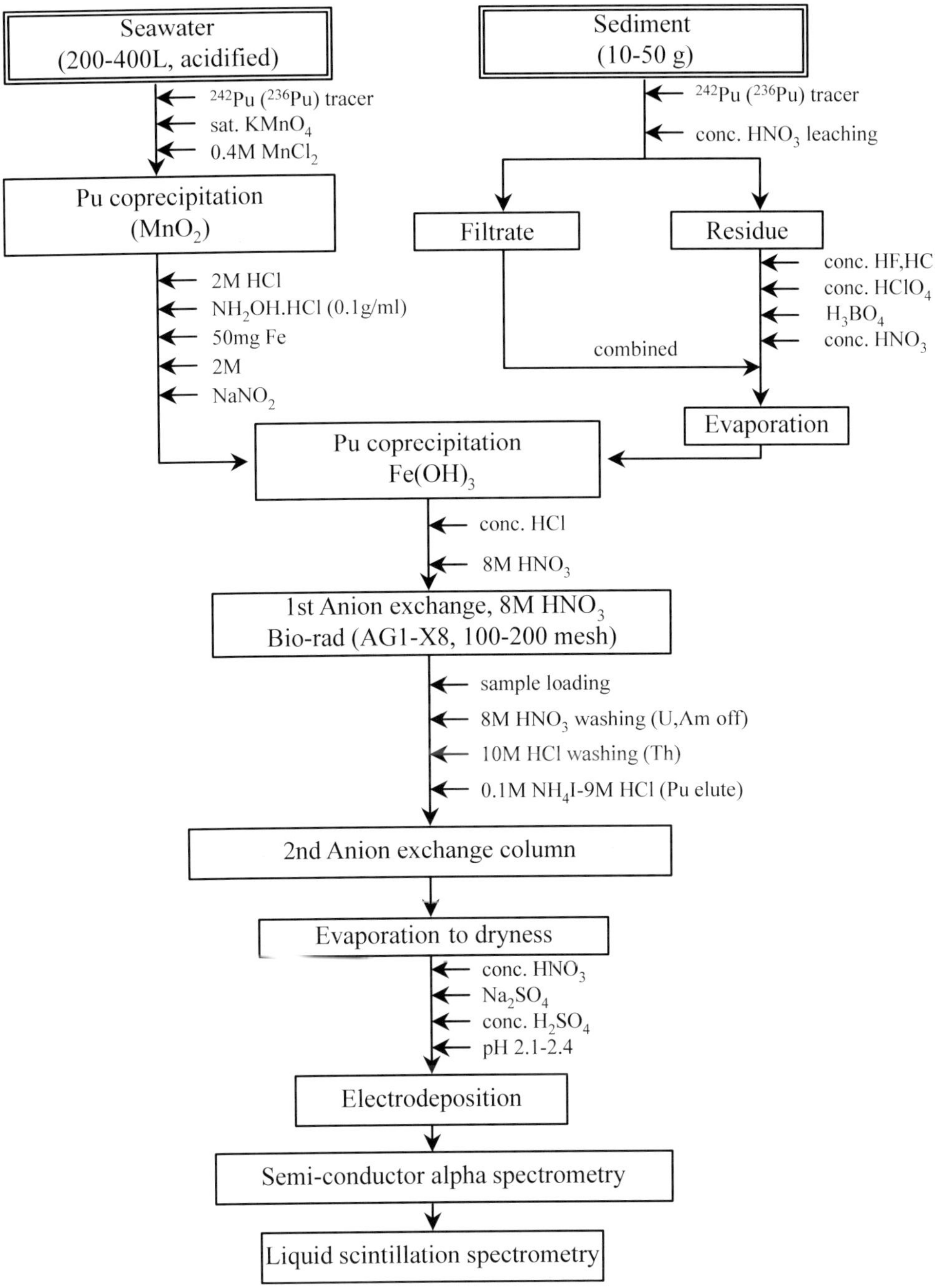

Fig. 3. Separation of Pu by oxidation state adjustment and anion exchange chromatography (IAEA, 2002).

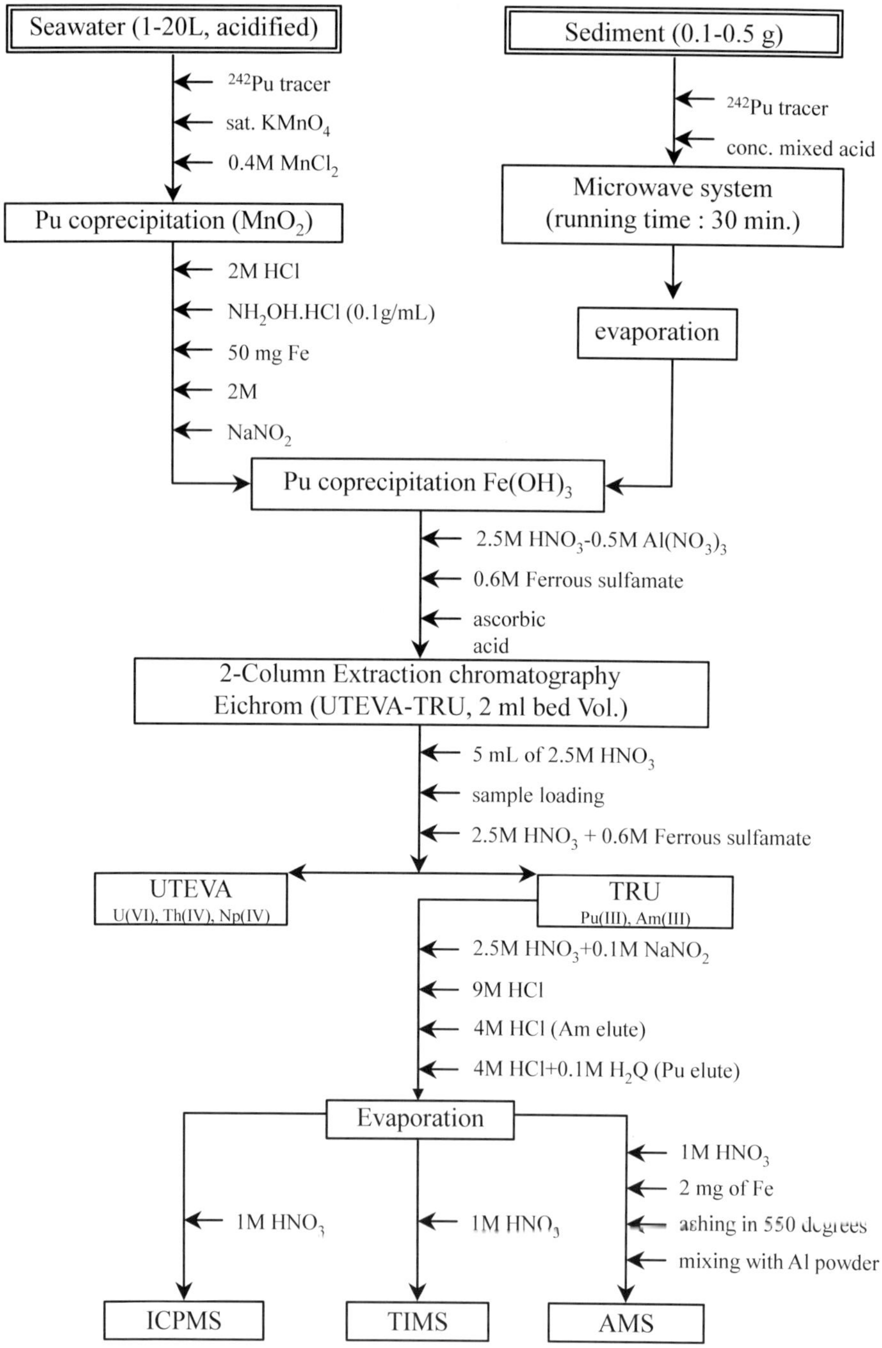

Fig. 4. Separation of Pu for mass spectrometry analysis (IAEA, 2002).

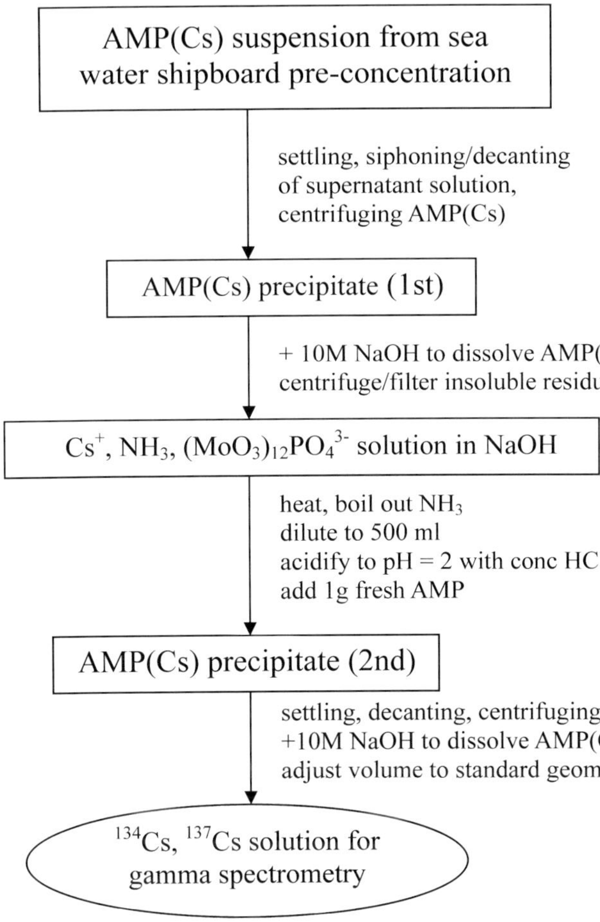

Fig. 5. Separation of Cs for γ-ray spectrometry measurements (IAEA, 2002).

Strontium

The strontium chemical separation and purification procedure is illustrated in Fig. 6. The calcium oxalate suspension is transferred to a large beaker and allowed to settle. The supernatant solution is siphoned off, and the precipitate washed several times with water to remove the seawater. It is then dried and ashed at 600°C to convert the oxalate to carbonate. Further steps are described in detail in Fig. 7. After separation, ^{90}Sr is measured by liquid scintillation spectrometry (LSS) (chemical recovery 67%, ^{90}Sr counting efficiency 92% and background 1.2 min^{-1} (Liong Wee Kwong et al., 2001) or by milking ^{90}Y from the purified strontium fraction and ^{90}Y is then counted in a gas flow β-ray counter or by LSS. Typical parameters are: 50% Sr chemical recovery, 90% Y chemical recovery, 48% ^{90}Y counting efficiency, 0.1 min^{-1} the counter background, and a detection limit of 3 mBq (or 30 μBq l^{-1} for 100 l seawater samples).

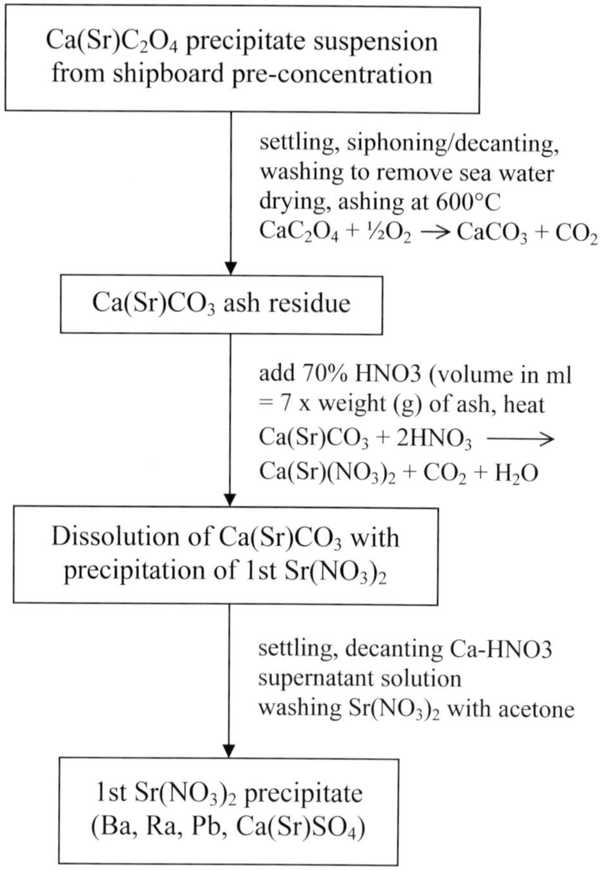

Fig. 6. Separation of Sr, part 1 (IAEA, 2002).

In conclusion on the radiochemistry topic, we can say that the use of extraction chromatography (e.g. EiChrom materials TEVA, TRU, UTEVA and Sr resin) in environmental analyses supplements or replaces existing radiochemical procedures. In particular, the analysis of actinides and ^{90}Sr have greatly benefited. Group actinide separations from common matrix elements are possible with TRU resin for III-, IV- and VI-valent species. Purification of Am can be carried out with TEVA and TRU columns. Uranium separation and purification is easily achieved with UTEVA after preliminary separation with anion exchange resin in strong HCl. Final strontium purification with crown ether extractant (Sr resin) enables excellent separation from Ca and other matrix elements. Intermediate pre-concentration steps are being developed for large samples (required for sensitivity) to allow the ultimate use of small extraction chromatography columns. Several sorbents are available for radiocaesium concentration from seawater; time, cost and efficiency constraints may determine which material is used. A consistent and streamlined set of methods can be further developed to meet the growing analytical demands of the wide variety of samples expected in the near future.

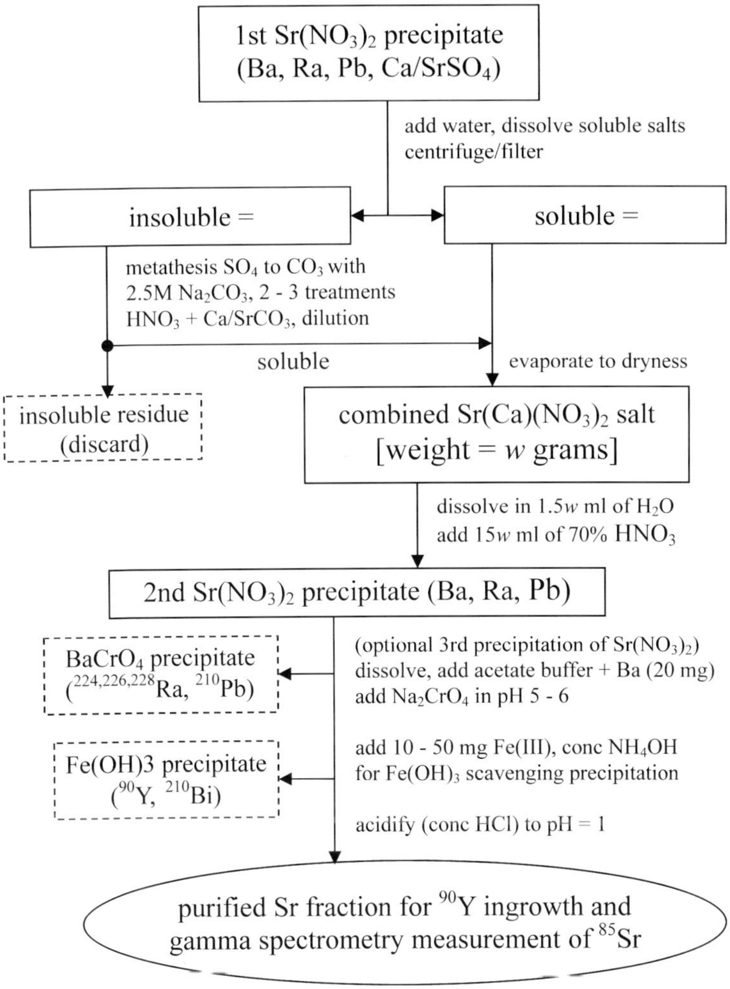

Fig. 7. Separation of Sr, part 2 (IAEA, 2002).

4. Marine radionuclide measurement methods

4.1. *Radiometrics methods for radionuclide measurements*

Gamma-ray spectrometry

Low-level γ-ray spectrometry using HPGe detectors is an effective tool in the study of marine radioactivity. The reasons are an excellent energy resolution that permits the analyses of various radionuclides in composite samples, selectively and very often non-destructively (e.g. in sea sediments), and the high efficiency of recently produced HPGe detectors (up to 200% relative to 75 mm dia × 75 mm long NaI(Tl) crystal). Although the most effective way of increasing the sensitivity of a spectrometer is to increase counting efficiency and the amount

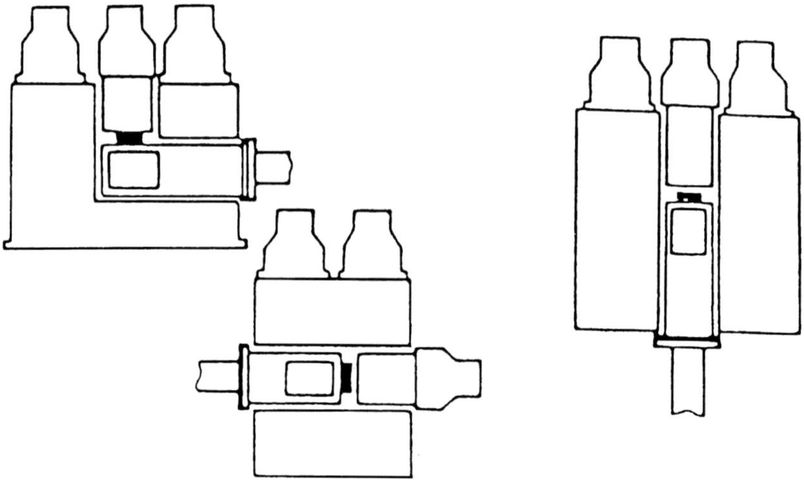

Fig. 8. Configuration of anti-compton spectrometers.

of sample, frequently, the only possible way is to decrease its background. The background components in a typical low-level HPGe detector, not situated deep underground, are cosmic radiation (cosmic muons, neutrons and material activation), environmental radioactivity, radioactivity of shield construction materials, radioactivity of detector construction materials, radon and its progenies. For present-day, carefully designed low-level HPGe γ-ray spectrometers, operating not deep underground, the dominating background component is cosmic rays, mainly cosmic ray muons (Heusser, 1995; Vojtyla & Povinec, 2000).

However, in the majority of marine applications, a simple γ-ray spectrometer with an HPGe detector of about 100% efficiency (especially if the sample size permits the use of a well detector) is the usual choice. In single HPGe spectrometers there is not any protection against cosmic ray muons, therefore a spectrometer with an anticosmic shielding will provide much better sensitivity. The anticompton spectrometer is a powerful technique for reducing the detector's background because it combines both the anticosmic and the anticompton suppression of background (Cooper & Perkins, 1972; Debertin & Helmer, 1998; Reidl et al., 2000). It rejects electronically principal detector pulses that coincide with signals originating in detectors surrounding the principal detector (the active shield). The coincident electronic signals from the active shield and the principal detector are generated by compton scattered γ-rays from the detector, and by cosmic rays, which give rise to a continuous background spectrum and raise detection limits for the γ-rays with energies in this spectral region (e.g. in the case of analysis of ^{137}Cs in the presence of ^{40}K). The scattered photons escaping from the principal detector volume can be detected by a sufficiently large NaI(Tl) or BGO crystal surrounding the detector (Fig. 8). The simplest anticompton spectrometer can be made using a Ge detector and an NaI(Tl) crystal with a well. Using a Ge detector in a large NaI(Tl) crystal (about 30 cm dia $\times$ 30 cm long), the compton suppression factor (Fig. 9) should reach values above 10 (for $E_\gamma \leqslant 1.5$ MeV) and thus decreasing detection limits more than three times (Zvara et al., 1994).

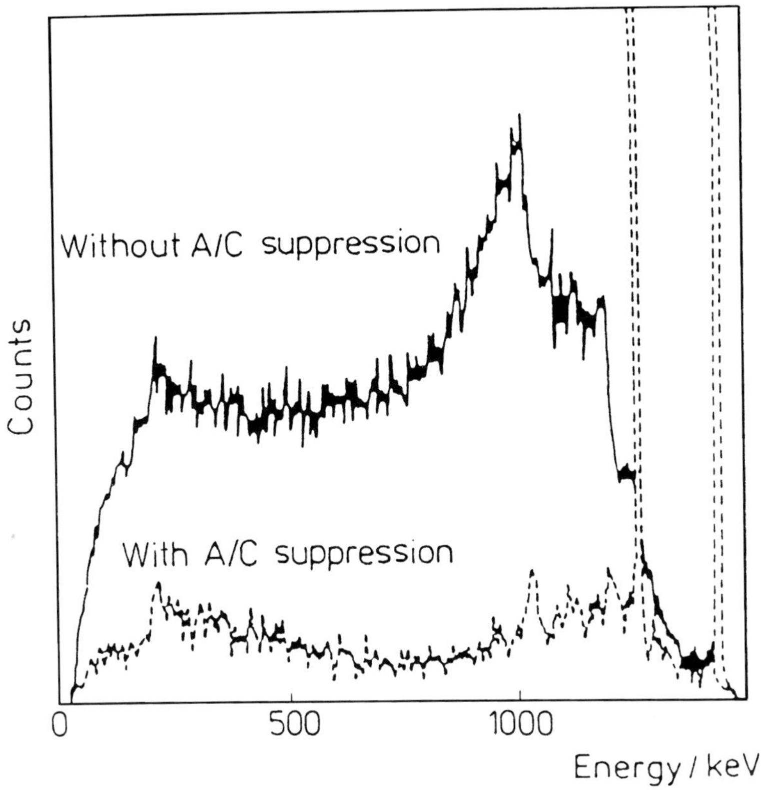

Fig. 9. Anticompton suppression of ^{60}Co spectrum measured with an HPGe-NaI(Tl) spectrometer.

Usually, a well designed anticompton spectrometer can be used simultaneously as a coincidence and anticompton spectrometer, and by adding a β-ray detector, as a triple coincidence spectrometer (e.g. for radionuclides emitting γ-rays in cascades (like ^{60}Co) or for positron emitters (like ^{22}Na, ^{26}Al)). Although detection efficiency is reduced in coincidence spectrometers, a considerable decrease in background (by about two orders of magnitude) makes these spectrometers superior for very low-level γ-ray spectrometry (Zvara et al., 1994).

The most sophisticated spectrometric system is the multidimensional γ-ray spectrometer (Cooper & Perkins, 1971; Povinec, 1981) in which signals from the analysing detectors create three-dimensional spectra (volumetric peaks) which can contain both coincidence and non-coincidence peaks. Analysing electronics, if two HPGe detectors are used, require 8000×8000 channels, which, with present state of the art computer electronics, is not difficult. The background can be reduced by about two orders of magnitude and three-dimensional spectra enable better identification of the peaks registered (Cooper & Perkins, 1971).

Specific applications (a small sample size, better precision) require anticompton or coincidence/anticoincidence spectrometers, often placed underground (see Section 4.1.4) in order to decrease the fluctuating cosmic ray background (Brodzinski, 1991; Heusser, 1992, 1994; Povinec, 1994).

Beta-ray spectrometry

There are several radionuclides (both natural (e.g. cosmogenic ^{32}P, ^{32}Si) as well as anthropogenic (e.g. ^{3}H, ^{85}Kr, ^{90}Sr, ^{241}Pu)) found in the marine environment which are still analysed by β-ray counting or β-ray spectrometry, mainly because their half-lives are too short to be analysed, e.g. by AMS. For some β-ray emitters, like ^{14}C and ^{129}I, which were previously analysed by β-ray spectrometry, AMS is at present the most widely used technique. Other radionuclides, like ^{99}Tc are still analysed by β-ray spectrometry, although ICPMS (and AMS) have been successfully applied as well, and in the future, mass spectrometry techniques will prevail for these radionuclides.

The most important breakthrough in β-ray spectrometry was the introduction of low background Liquid Scintillation Spectrometers (LSS) (Theodorsson, 1996), which is now mainly used for the analysis of short-lived β-ray emitters in the marine environment. The advantage over low-level gas counting (Povinec, 1991b) is their higher efficiency as well as that LSS can in fact operate as a β-ray spectrometer, registering only those β-electrons with energies inside the β-ray spectrum, and not all β-electrons registered by e.g. Geiger–Müller counters.

We shall illustrate the possibilities of LSS techniques using a few examples – ^{3}H (maximum energy of β-electrons 18.6 keV; $T_{1/2} = 12.3$ y) and ^{241}Pu (maximum energy of β-electrons 20.8 keV; $T_{1/2} = 14.4$ y). ^{3}H analysis by LSS without enrichment has a detection limit of about 2 TU (1 tritium unit (TU) is the ratio of 1 atom of ^{3}H/10^{18} atoms of ^{1}H). Electrolytic enrichment can lower the detection limit to 0.01 TU, which is similar to the value obtained by ^{3}He ingrowth mass spectrometry. However, there are only a few laboratories in the world which can reach such low detection limits (e.g. Ostlund & Dorsey, 1975; Taylor, 1977). For ^{3}H analysis of surface waters contaminated by discharges, simple LSS is the best solution. Figure 10 shows a typical ^{3}H spectrum of a water sample (4 TU), a background sample (0.4 min^{-1}) and a ^{3}H standard (about 20% efficiency) as obtained by WALLAC's Quantulus spectrometer (Povinec et al., 1996a). The Quantulus spectrometer benefits from the low background resulting from the anticoincidence shielding of the counting vial. Similar counting parameters are obtained by the Packard LSS, which uses the pulse shape discrimination method for decreasing the spectrometer's background.

Recently an electrolysis apparatus utilising Solid Polymer Electrolyte (SPE) has been developed (which greatly simplifies the enrichment procedure) and with LSS ^{3}H concentrations in water down to 0.01 TU can be measured (Saito et al., 2000).

Plutonium-241 is another good example of the successful use of LSS. As it is a pure β-ray emitter, it cannot be analysed simultaneously with other plutonium isotopes using α-ray spectrometry. However, the disk with electrodeposited Pu after α-ray spectrometry measurement can be placed in a polyethylene vial containing a liquid scintillator (e.g. the Insta-Gel cocktail, Cook & Anderson, 1991; Ryan et al., 1993). Using Pulse Shape Analysis (PSA) method α-rays can be separated from β-rays registered by LSS (Fig. 11). Typical counting parameters are: 5 ml of cocktail, 12% efficiency, 0.3 min^{-1} background and detection limit 0.03 Bq (Liong Wee Kwong et al., 2001). Although compared to other Pu isotopes, ^{241}Pu has highest activity concentrations in marine samples, direct counting is not sufficiently sensitive for many applications. Therefore, the more preferable method is leaching plutonium from the stainless steel disk after α-ray spectrometry (or to separate a portion of Pu sample before its electrode-

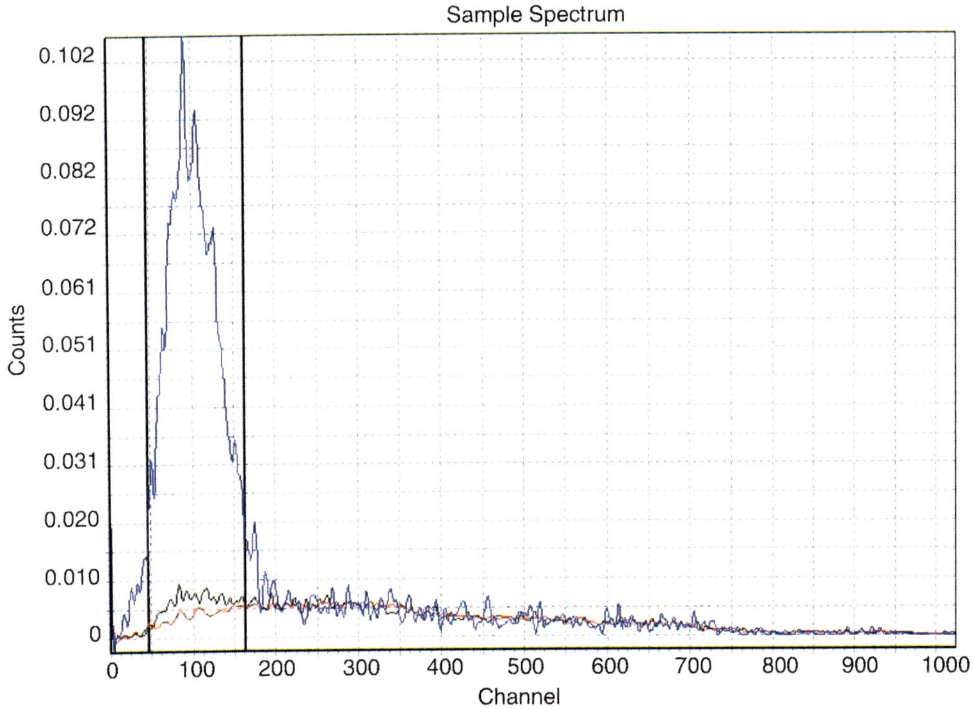

Fig. 10. ^{3}H spectra (top: standard; middle: a sample; bottom: background) measured by LSS using Insta-Gel Plus cocktail (11 ml of water and 9 ml of scintillator).

position on the disk) and dissolving the sample in the counting vial, enabling 4π geometry measurements, thus increasing detection efficiency by a factor of two (Liong Wee Kwong et al., 2001).

Alpha-ray spectrometry

Semiconductor alpha-ray spectrometry (SAS) has become a well matured technique for the analysis of α-ray emitters (mainly actinides, but also ^{210}Po, ^{226}Ra and others) subsequent to the introduction of semiconductor silicon detectors. As the typical energy range of α-rays from investigated α-ray emitters is between 3 and 6 MeV, SAS has been widely used in marine studies. Other techniques, like ionisation chambers and track detectors have not been used in the analysis of marine samples, mainly because of poorer resolution.

Therefore, the most important developments in SAS have been in the radiochemistry sector, as previously discussed. However, one development we would like to mention here, as the major disadvantage of high resolution SAS compared to mass spectrometry techniques is the difficulty in resolving ^{239}Pu and ^{240}Pu peaks. Large surface (about 10 mm in diameter) silicon detectors frequently used in SAS cannot resolve this duplet. However, by using high resolution silicon detectors and special deconvolution software, this problem can be solved (Leon Vintro et al., 1996). The disadvantage here is that because of the smaller detector surface, sensitiv-

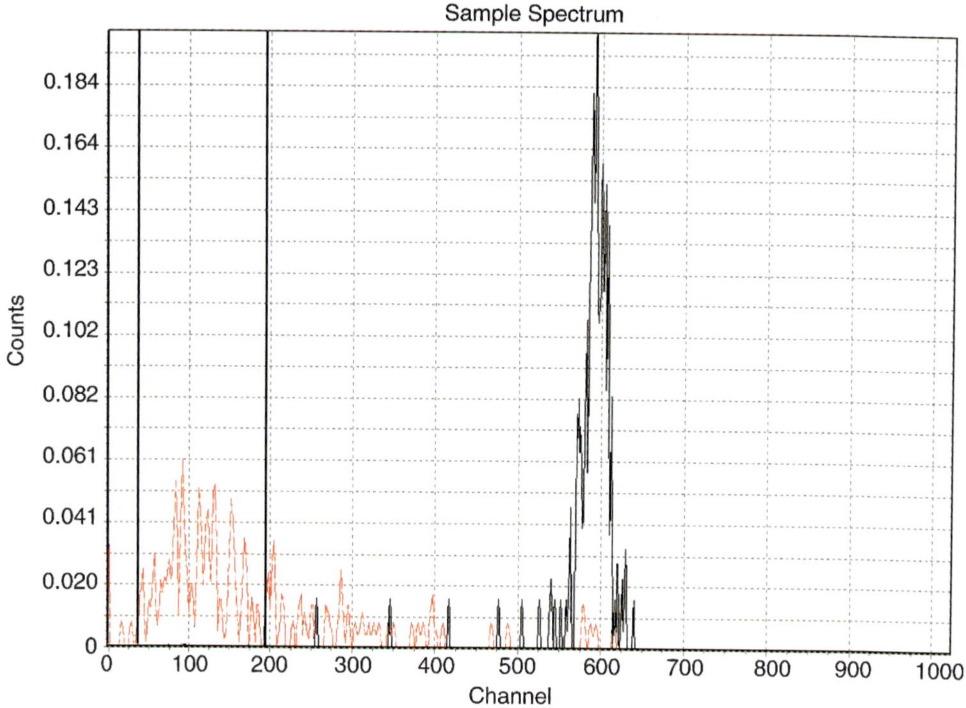

Fig. 11. ^{241}Pu β-ray spectrum (left) and ^{238,239,240}Pu α-ray spectrum (right) of Irish Sea sediment.

ity is lower and only samples with relatively higher concentrations of α-ray emitters can be analysed. This is one of the reasons why mass spectrometry techniques are more and more widely used for analysis of Pu isotopes in the marine environment.

Underground counting laboratories

Muon-induced background becomes important for large volume HPGe detectors as the most prominent peaks observed (e.g. annihilation peak, neutron activation peaks) have been due to cosmic ray interactions (Brodzinski, 1991; Heusser, 1992, 1994). Alpha-ray spectrometry is free of such interference as the detector medium is very thin and light (silicon) and the registered energy is relatively high (typically between 3 MeV and 6 MeV). Beta-ray spectrometry is somewhere between – the detection medium (gas, liquid scintillator) is usually of low density, although typical energies are below 1 MeV. The cosmic ray background will be especially important for low energy β-ray emitters (like ^{3}H), therefore, tritium detectors should also benefit from underground operation. Due to limited space we shall only discuss underground low-level γ-ray spectrometry in this review.

Low levels of radionuclides presently observed in the marine environment require the processing of large volume samples (e.g. in the case of analysis of ^{137}Cs in seawater) and lengthy counting times. Thus, the background of a low-level counting system is a limiting factor in many new applications. A detailed analysis of background (Heusser, 1994;

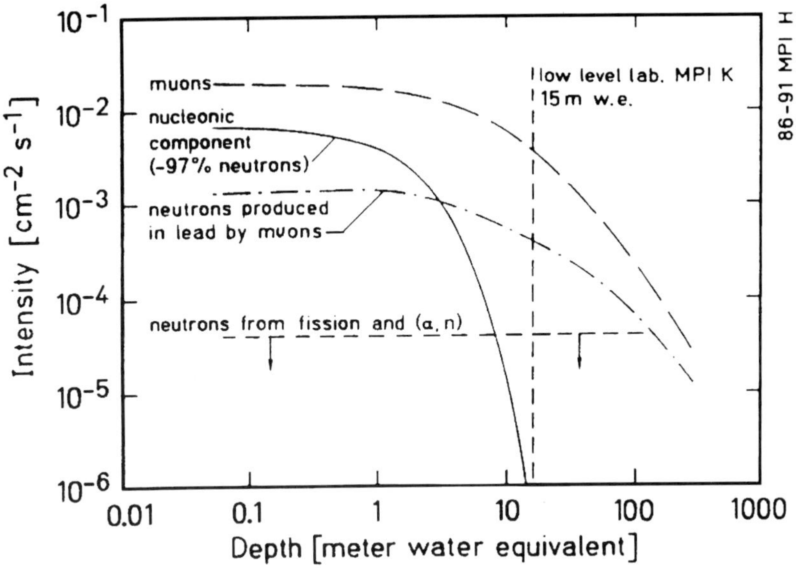

Fig. 12. The flux of secondary cosmic rays and neutrons from fission and (α, n) reactions at various depths (Heusser, 1992).

Povinec, 1994) has shown that many counting systems have not been optimised in relation to background interferences. Even with sophisticated anticompton or anticosmic spectrometers placed in surface laboratories, the background is still too high to substantially decrease the size of samples, to be comparable to the size of samples used for mass spectrometry.

The reason is that in a well-designed passive shield, a cosmic ray-produced background plays a dominant role. Figure 12 compares fluxes of secondary cosmic ray particles under different shielding layers (Heusser, 1992). It can be seen that while the nucleonic component (mainly neutrons) is attenuated at 10 meters water equivalent (w.e.) by about 4 orders of magnitude, the muonic component remains at this shielding layer without notable reduction (by a factor of 2 only). To decrease the muonic component by a factor of 10, shielding of over 50 m w.e. is needed. At 1000 hg cm^{-2} depth, the reduction in muon flux can be as high as 10^6 (Povinec, 1994).

The behaviour of muons makes the design of an optimum shield for low-level counting very difficult. In a surface laboratory, if the lead is too thick muons will produce neutrons and thus increase the quantity of neutrons in the total background. The best solution to this complex problem is to install the detection system underground at sufficient depth for the muon component to be relatively low. Limestone or sandstone are preferable to granite for radionuclide contamination (^{40}K, Th, U series) and will have lower (by a factor of 10) neutron production due to (α, n) reactions.

Monte Carlo simulations of background

The GEANT code system developed at CERN for high energy physics was adopted for the simulation of a cosmic ray muon-induced background (Vojtyla & Povinec, 2000). Three sizes

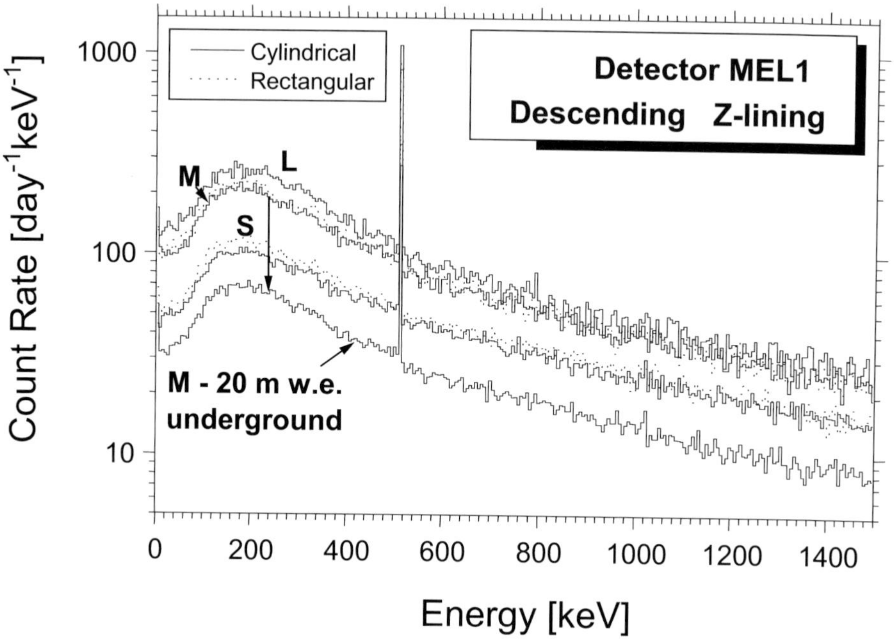

Fig. 13. Computed background spectra for an HPGe coaxial detector (170% relative efficiency) in a shield with descending Z lining (Vojtyla & Povinec, 2000).

of shields were used for the simulations. The small shield was placed close to the detector and only a minimum of space was available inside; the medium size shield was intended for 1 l samples and the large shield was for even larger samples. The optimum shielding thickness for large volume HPGe detectors situated at sea level or at shallow depths underground was found to be 15 cm. If thicker shielding was used in simulations, the background was higher due to interactions of muons with the shield.

The resulting background spectra for the HPGe coaxial 170% relative efficiency detector, in the energy region of 0–1500 keV, are shown in Fig. 13 for the descending-Z lining (150 mm Pb, 1 mm Cd, 2 mm Cu). Simulations with various shields showed that the background depends mainly on the thickness, dimensions and lining of the shield. The shape of the shield (cylindrical or rectangular) has almost no effect on the background provided that the inner dimensions are similar. The best results were obtained for the smallest shield with the least low-Z materials in the lining. The background characteristics most affected are the maximum of the background continuum with a count rate of 450 d^{-1} keV^{-1} in a shield lined with 1 cm of copper compared to the count rate of 190 d^{-1} keV^{-1} in a lead only shield. It is recommended to use shields with removable linings for applications where lead X-rays do not have a disturbing effect. It is inadvisable to build an unnecessarily large shield. In a medium size shield, the integral background count rate (0–1500 keV) is expected to reach count rates of about 85 min^{-1}, the count rate in the annihilation peak is estimated to be about 4400 d^{-1} keV^{-1}, and the maximum count rate in the background continuum should be approximately 210 d^{-1} keV^{-1} (at sea level). A well-type detector intended for the counting of

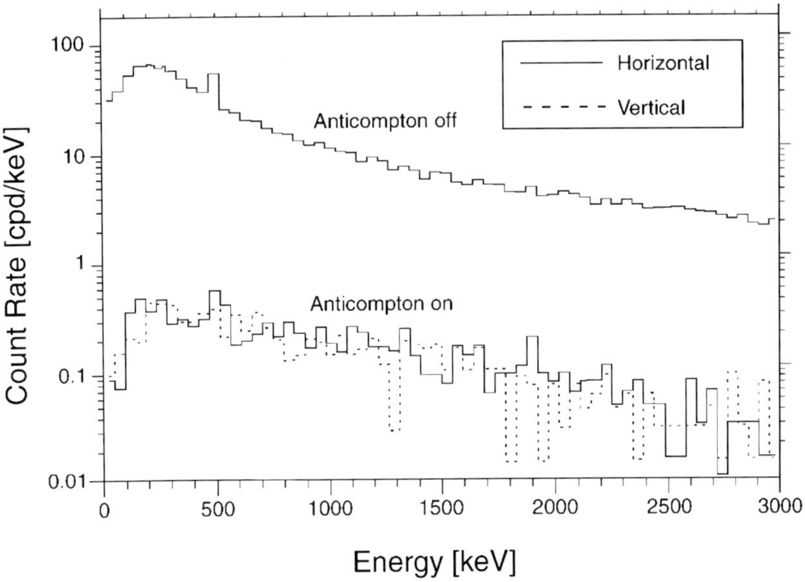

Fig. 14. Computed background spectra of the anticompton spectrometer in horizontal and vertical positions 20 m w.e. underground.

small samples inside the well should be placed into a tight shield without lining. In this case, the integral background count rate can be as low as 50 min^{-1}, the count rate in the annihilation peak can be reduced to 900 d^{-1} keV^{-1}, and the maximum counting rate in the background continuum minimised to only about 100 d^{-1} keV^{-1}. Simulations 20 m w.e. underground showed that the features of simulated background spectra remain unchanged, except that all the count rates scaled down by a factor of about 3.

Background characteristics of the anticompton spectrometer consisting of an annular NaI(Tl) crystal (inner dia 11 cm, outer dia 30 cm and 40 cm high), closed from one side by another cylindrical NaI(Tl) crystal 75 mm in diameter × 75 mm high, were also simulated. The entire detection system is placed inside a cylindrical lead shield 15 cm thick, 70 cm dia × 90 cm high. The HPGe detector (coaxial, 170% relative efficiency) is placed deep inside the anticompton shield to allow a free space (11 cm dia × 5 cm high) for samples. There are two possible ways to position the shield, either vertically (the NaI(Tl) stopcock on the top) or horizontally. Simulations of background spectra of the anticompton spectrometer (Fig. 14) in horizontal and vertical positions showed negligible differences in the lower energy region (differences can only be found in the energy region above 10 MeV which is not of interest for low-level counting of environmental samples). Anticoincidence pulse rejection is quite strong, at least 40 at 1500 keV and on average about 90. The maximum in the background continuum of about 200 $d^{-1}\text{keV}^{-1}$ is reduced to only about 1.5 d^{-1} keV^{-1}. At this level of anticoincidence reduction, other background components like the contamination of the lead shield and the NaI(Tl) detectors and their photomultipliers with primordial radionuclides could dominate the background.

In some applications a simple anticosmic shielding (made of flat gas flow counters or a plastic scintillator) suppressing muons and secondary particles passing through the lead shield could be enough to reach the required sensitivity (Heusser, 1992; Vojtyla et al., 1994; Povinec et al., 2004). Underground γ-ray spectrometry with anticompton or anticosmic shielding would enable to decrease, for example, a water sample necessary for ^{137}Cs analysis by about a factor of 10 (to about 20 l), making it comparable in size to samples used, e.g. for the analysis of actinides using mass spectrometry techniques.

In situ underwater γ-ray spectrometry
As we have seen, measurements of the concentrations of anthropogenic and/or natural radionuclides in the marine environment require the use of very sensitive radioanalytical methods, usually consisting of complicated and laborious sampling, pre-concentration and separation techniques and long-term measurements using low-level counting techniques. In some applications it can be better to use in situ γ-ray spectrometry, which has many advantages compared to traditional sampling and laboratory analysis for several aspects of marine radioactivity monitoring, e.g. in

(i) investigation of radionuclide levels around dumped or sunken nuclear objects/wastes;
(ii) long-term monitoring of liquid discharges of radionuclides from nuclear plants;
(iii) mapping of large areas of sediments to assess the distribution of investigated radionuclides;
(iv) traditional marine radioactivity research when stationary monitoring can replace sporadic sampling;
(v) optimisation and focusing of conventional sample collection in areas expected to be either representative or the most contaminated.

Traditionally, NaI(Tl) detectors (Miller et al., 1982; Thomas et al., 1983; Jones et al., 1988, 1999; Tyler et al., 1996; Johnson et al., 1997; Noakes et al., 1999; Osvath et al., 1999; Jones, 2001; Osvath & Povinec, 2001) have been used in seabed mapping of radionuclides emitting γ-radiation (both natural (^{40}K, ^{232}Th and ^{238}U series) as well as anthropogenic (^{60}Co, ^{137}Cs) radionuclides). However, recent developments in cryogenic physics have enabled the operation of HPGe detectors without liquid nitrogen cooling in deep underwater conditions as well (Povinec et al., 1995; Kobayashi et al., 1999; Sokolov et al., 1999). This has been acknowledged as a considerable improvement over in situ NaI(Tl) γ-ray spectrometry, especially in the investigation of multi-radionuclide sources (Povinec et al., 1997).

Seabed γ-ray spectrometry. Both NaI(Tl) and HPGe detectors have been used in γ-mapping exercises. The advantage of HPGe detectors is clearly their very good resolution, although their operation is much more difficult, especially at sea. The spectrometers usually operate in a towing mode where a highly ruggedised system is towed over the seabed by a ship. However, in areas where the seabed is very rocky or covered by corals, point to point deployment has been used (e.g. Mururoa and Fangataufa lagoons, Osvath et al., 1999). Gamma-mapping has been successfully applied in mineral exploitation (where natural γ-ray emitters of U and Th series were registered (Thomas et al., 1983) and radioactive contamination studies, where anthropogenic radionuclides of interest were registered (Jones et al., 1984, 1988, 1999; Tyler et al., 1996; Noakes et al., 1999). A survey of Mururoa and Fangataufa lagoons was carried out in 1996 (Osvath et al., 1999) using a ruggedised NaI(Tl) detector (55 mm dia, 152 mm long). ^{60}Co and ^{137}Cs were the main γ-ray emitters present in lagoon sediment. The in situ sediment measurements were used to guide the sediment sampling campaign in the most contaminated areas of the lagoons where surface tests of nuclear weapons had taken place. A good correlation ($R^2 = 0.8$) found between ^{60}Co and ^{239,240}Pu concentrations (^{239,240}Pu is the radionuclide of main interest in lagoon sediments) enabled the estimation of ^{239,240}Pu inventories on the basis of ^{60}Co mapping and analysis of ^{239,240}Pu in sediments (Osvath et al., 1999).

An underwater γ-spectrometer consisting of separately housed HPGe and NaI(Tl) detectors with electronics, data acquisition and processing electronics located with the detectors, communicating with a shipboard PC through a modem link has been used for the investigation of dumping sites in the bays of Novaya Zemlya. The HPGe detector has an efficiency of 20% relative to a 75 × 75 mm NaI(Tl) crystal. The deployment time is about 20 h. Further deployments of the detector requires repeated external cooling of the propane in the cryogenic unit of the detector by the helium cooler (for about 6 h). The second γ-detector was made of a 'ruggedised' NaI(Tl) scintillator, 100 mm dia and 150 mm long. Figure 15 shows HPGe and NaI(Tl) spectra recorded at one of the stations in Stepovoy Bay (Kara Sea – Povinec et al., 1996a, b). The far superior resolution of the HPGe detector has enabled observation of a low ^{137}Cs concentration which was not visible in the NaI(Tl) spectrum. The spectra obtained with the HPGe detector represent the first set of high resolution seabed γ-ray spectra ever recorded in situ.

The same NaI(Tl) spectrometer was used for seabed γ-mapping of radionuclides (mainly ^{137}Cs) in the Irish Sea close to the Sellafield reprocessing facility (Osvath & Povinec, 2001). Figure 16 shows ^{137}Cs concentrations in surface sediment as observed in 1995, which are lower by 40–70% than the results obtained by Jones et al. (1988) in the survey carried out in 1985. The observation is consistent with the reported change in ^{137}Cs levels in sediment on a wider scale in the Irish Sea related to its remobilisation from sediment.

Stationary underwater γ-spectrometry. Stationary monitoring systems have several advantages over traditional sampling systems, e.g.:

(i) real time reporting of data,
(ii) searching for temporal changes, and
(iii) development of time series.

As released radionuclides are usually accompanied by γ-ray emitters (e.g. in discharges from reprocessing plants and nuclear power stations, from dumped radioactive wastes) a γ-ray monitoring system would provide an excellent solution for monitoring of radionuclides in the marine environment (Povinec et al., 1996b; King et al., 1997; Soukissian et al., 1999), although β-rays emitters like ^{90}Sr could be monitored in water through Cherenkov radiation as well (Chernyaev et al., 1999).

A large volume scintillation detector of high efficiency and reasonable resolution would be a good choice for long-term γ-monitoring of the marine environment. The advantage of NaI(Tl)-based systems is primarily the high detection efficiency of NaI(Tl) crystals and their far lower cost than the equivalent Ge crystals. Moreover, NaI(Tl)-based systems can be built to be sufficiently robust for long-term underwater deployment (Povinec et al., 1995, 1996b; King et al., 1997; Harms & Povinec, 1999). HPGe-based systems have the advantage of

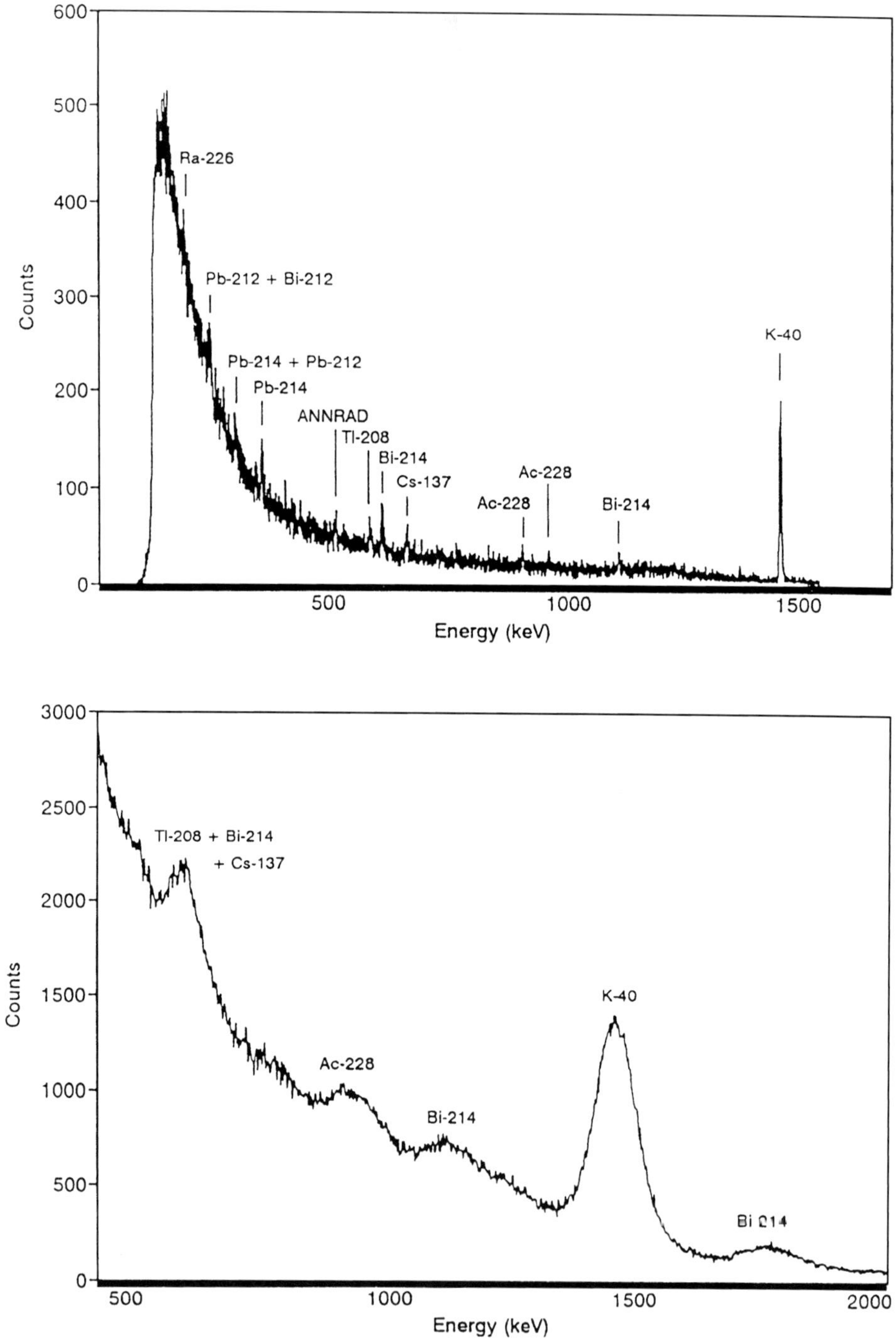

Fig. 15. Gamma-ray spectra recorded by HPGe and NaI(Tl) detectors in Stepovoy Bay, Novaya Zemlya (Povinec et al., 1996b).

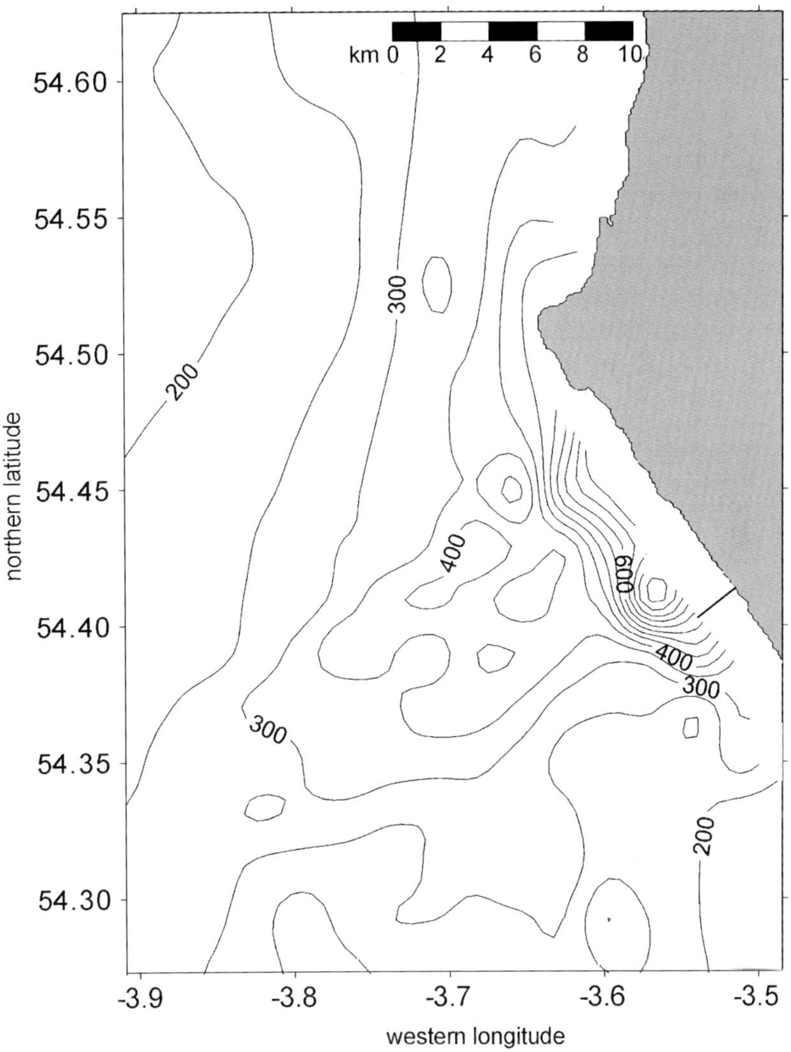

Fig. 16. ^{137}Cs concentrations in surface sediments (Bq kg^{-1}) in the Irish Sea close to Sellafield discharge pipe line (Osvath & Povinec, 2001).

good energy resolution and hence excellent radionuclide identification capability. Recently, electrically-cooled HPGe γ-ray detectors have been developed (Katagiri et al., 1995). The size of a detector assembly like this including the crystal and the refrigerator is very small (38 cm $\times$ 12 cm) and its power consumption extremely low (of the order of 1 W). A submersible electrically-cooled HPGe γ-detector system of this type, operational to 500 m depth, was successfully tested by Kobayashi et al. (1999).

Figure 17 shows γ-ray spectra obtained from the NEMO (Nautique Environnement Marin Observatoire) monitoring system, equipped with an NaI(Tl) detector (75 mm dia $\times$ 75 mm

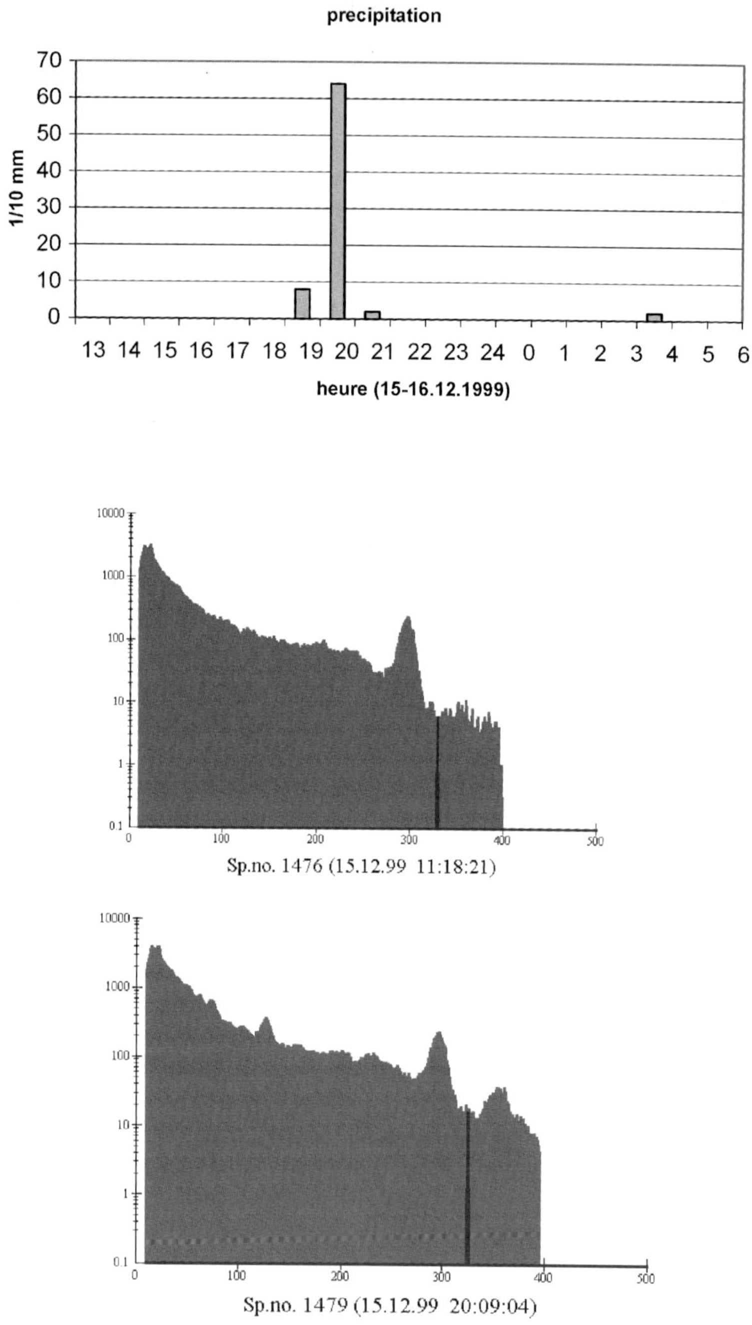

Fig. 17. Gamma-ray spectrum of water (4 m depth) taken with the NEMO observatory in Monaco Bay (top: the precipitation record; middle: after the rain; bottom: before the rain; Povinec et al., 2001b).

long), current, temperature and conductivity/salinity meters with satellite data transmission, deployed in Monaco Bay, before and after a heavy rain (Povinec et al., 2001b). The figure clearly shows the effect of precipitation on the transport of ^{222}Rn daughter products (^{214}Bi) to the sea. The dominant peak in the spectrum is due to ^{40}K. The peak from ^{137}Cs (usually the most searched-for γ-ray emitter) is masked by natural ^{214}Bi and not easily visible. Recently the NEMO observatory was deployed in the NW Irish Sea to search for ^{137}Cs resuspension from sediment (Osvath et al., 2004).

Neutron activation analysis (NAA)

Activation analysis is a sensitive method for determining long-lived radionuclides, although it is more frequently used for the determination of stable isotopes and trace elements (Rosenberg, 1993; Harwey & Chazal, 1993; Hou, 2001; Yonezawa et al., 2001). According to the type of bombarding particles, activation analysis is divided into neutron, photon and charge particle activation analysis. Of these, only NAA has been widely employed in the determination of radionuclides. Neutrons of different energies (cold, thermal or fast neutrons) can be used with different pre-irradiation and post-irradiation chemistry, e.g. instrumental (i.e. non-destructive) NAA, radiochemical NAA, pre-concentration NAA, etc. As it is not possible to discuss NAA in depth in this review, we shall illustrate a few examples for long-lived radionuclides like ^{53}Mn (Bibron et al., 1974), ^{99}Tc (Muramatsu et al., 1988; Ikeda et al., 1989), ^{129}I (Muramatsu et al., 1988; Hou et al., 1999, 2000, 2001), ^{135}Cs (Chao & Tseng, 1996), ^{230}Th, ^{232}Th (Huk, 1988), ^{235}U, ^{238}U (Benedik & Byrne, 1995), ^{237}Np (Byrne, 1986; Germain et al., 1987; and ^{231}Pa (Byrne & Benedik, 1999), where after neutron irradiation, a nuclide with a shorter half-life and better measurement characteristics (for γ-ray spectrometry) can be analysed. As it is increasingly difficult to irradiate samples in research reactors, and usually mass spectrometry methods give better detection limits, of the above list, the two most interesting radionuclides for measurement by NAA are ^{53}Mn (although it can be analysed by a higher energy AMS, too (Knie et al., 2000), and ^{135}Cs (which has only been measured in real samples by NAA and TIMS (Lee et al., 1993; Chao & Tseng, 1996)).

As the reaction ^{53}Mn (n, γ) ^{54}Mn with thermal neutrons can be accompanied by fast neutron reactions on ^{55}Mn (n, 2n) and ^{54}Fe (n, p), it is important to use a well-thermalised neutron channel. The method has been applied to studies of ^{53}Mn ($T_{1/2} = 3.7 \times 10^6$ y) in sediments, in manganese nodules and crust, and in Antarctic ice (Bibron et al., 1974).

As ^{135}Cs is a long-lived ($T_{1/2} = 2.06 \times 10^6$ y) β-ray emitter and its determination by RMT is rather difficult, TIMS and NAA have been applied in its measurement (Lee et al., 1993; Chao & Tseng, 1996), and the AMS method has been under development (Zhao et al., 1999). A detection limit of 10^{-4} Bq (1 pg) has been obtained by NAA (Chao & Tseung, 1996).

As an alternative method, NAA, even though very sensitive for the determination of some of the long-lived radionuclides, is restricted in its use because a nuclear reactor is needed. It could, however, be very useful as an independent method in the certification of reference materials.

4.2. *Developments in mass spectrometry techniques for marine radionuclide measurements*

Several mass spectrometric techniques have been successfully applied in the past for analysis of radionuclides in the marine environment (IAEA, 1993). They have used traditional isotope mass spectrometers, like those used for the analysis of ^{3}He, which is a decay product of ^{3}H. TIMS and ICPMS have been used for the analysis of several long-lived radionuclides, along with RIMS (Michel, 2001). More recently, AMS was introduced to widen the spectrum of radionuclides from ^{14}C to transuranics, and because the background of the system is significantly reduced by accelerating analysed ions to several tens or hundreds of MeV (Kutschera, 1986), this has become the most sensitive technique at present for the analysis of long-lived radionuclides in the environment (for recent works see Kutschera et al., 2000).

^{3}H–^{3}He *mass spectrometry*

Direct analysis of environmental ^{3}H levels by isotope mass spectrometry is not possible because of the high background and relatively short half-life of ^{3}H ($T_{1/2} = 12.3$ y). However, the ^{3}H decay product, ^{3}He is very rare, so its analysis is most suitable for measuring ^{3}H concentrations down to 0.01 TU (Clarke et al., 1976). At present, this is the most sensitive method of isotopic analysis as isotopic ratios of ^{3}H/^{1}H can be measured down to 10^{-20}. The only disadvantage of the ^{3}He method is the long waiting period (about 6 months) for in-growth of ^{3}He in a water sample hermetically sealed in a glass flask. On the other hand, if appropriate sampling techniques and containers are used for water sampling, it is possible to analyse ^{3}He and ^{4}He gas in the water column simultaneously. The helium isotopes can give additional information on the water age (^{3}H–^{3}He method) and on circulation of water masses in the ocean (^{4}He). If a given water parcel in the sea is not subject to mixing with water of different ages, the relation between its ^{3}He content becomes the time (age) elapsed since the last contact with the atmosphere (Jenkins & Clarke, 1976; Torgersen et al., 1979).

As an example, in Fig. 18 we present the distribution of ^{3}H and ^{3}He in the Pacific Ocean (Schlosser et al., 1999) developed in the WOCE programme, the most comprehensive study done till now with radionuclides as oceanic tracers, in which the spatial resolution exceeded that achieved by GEOSECS programme by about a factor of ten. ^{3}H and ^{3}He sections provide valuable information on the spreading of water masses from the surface into the interior of the Pacific Ocean.

The traditional isotopic technique has not been successful in the analysis of other radioactive isotopes in marine samples, mainly because of problems with high background (e.g. in the case of ^{14}C). This technique is nevertheless widely used for the analysis of isotopic ratios of many stable isotopes, e.g. ^{2}H/^{1}H, ^{13}C/^{12}C, ^{15}N/^{14}C, ^{18}O/^{16}O, etc., which give important additional information on the processes in the water column, on sources of contamination, etc. (Roether et al., 1996; Froehlich et al., 1999; Schlosser et al., 1999).

Thermal ionisation mass spectrometry (TIMS)

TIMS is the most sensitive method for absolute isotope abundance measurements and until the introduction of ICPMS, it was the most widely used technique for ultra-low analysis of long-lived radionuclides (especially actinides) in the environment, as well. TIMS is a proven analytical method and is known for providing the most precise isotope ratio determinations with precision down to a few ppm (Halverson, 1984; White & Wood, 1988; Taylor et al., 1998). With TIMS positive ions are produced in the spectrometer's ion source by evaporation of a sample from a heated metal surface. Source filaments are usually chosen from metals with relatively high work functions, high melting points and the necessary purity and mechanical properties (e.g. platinum, rhenium, tantalum or tungsten). After heavy separation chemistry,

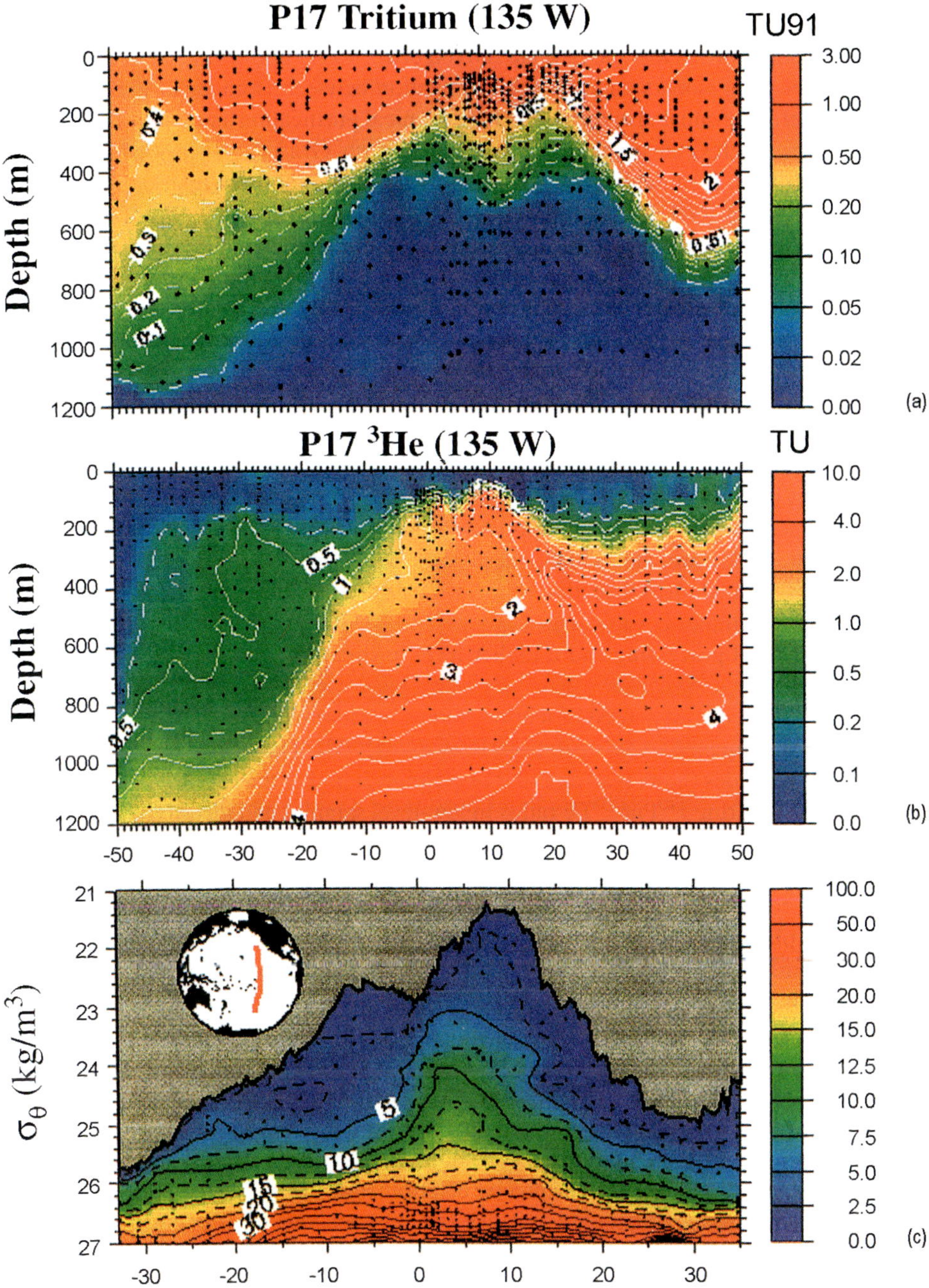

Fig. 18. Distribution of (a) ^3H, (b) $\delta\,^3He$ and (c) the $^3H/^3He$ age along WOCE WHP line P17 (approx. 135° W) in the Pacific Ocean (Schlosser et al., 1999).

samples are loaded as highly purified solutions onto the filament or by electrodeposition. By adding a known quantity of yield tracer to the sample, isotope dilution analysis is used to measure concentrations below the fg scale (Dixon et al., 1997; Inkret et al., 1989).

Interferences from isobars and mass discrimination between isotopes of a given element need to be carefully evaluated and excluded by chemical purification and control of ionisation conditions. Nevertheless the problem of interferences with TIMS is not as great as in ICPMS. Isotope fractionation is more difficult to control, but usually produces smaller systematic errors. However, using total sample evaporation method this problem has been solved as the sample is completely evaporated from the filament. The final results are given in averaged isotopic mean values when measurement has been completed and no measurable amount of sample is left on the filament. Thus, the limiting factor of TIMS in environmental analysis is very often the purity of the sample, as improvements and innovations in mass spectrometer hardware have augmented analytical precision. The new state of the art TIMS machines with multi-collector arrays have considerably improved operation in multi-dynamic mode (e.g. the TRITON from Finnigan), producing a sensitivity of around 10^5 atoms for ^{239}Pu, which is comparable to AMS.

When using measurement techniques with the extreme sensitivity of TIMS at levels below 10^{-15} g per sample, it is very important to minimise the effects of airborne dust particles, reagents, glassware, etc. which can all contribute significantly to the sample blank, and as a result, to analytical detection limits which are actually defined by the contamination of the blank sample. For this reason, successful TIMS requires more careful sampling, very clean chemical processing and a higher level of instrumental expertise than any other method.

TIMS has not been widely used in marine studies because of the high cost of the machine and the expertise required for its operation (Buesseler, 1993). The main advantage is its high sensitivity for actinides (more than an order of magnitude higher than for SAS), as well as the possibility of measuring ^{240}Pu/^{239}Pu ratios in low-level samples and identifying the origin of plutonium in marine samples (Bertine et al., 1986; Buesseler & Halverson, 1987; Buesseler & Scholkovitz, 1987). A wide range of ^{240}Pu/^{239}Pu atom ratios have been found in the marine environment: from 0.05 in Mururoa lagoon sediment, through global fallout ratio (0.186) to discharges of plutonium from reprocessing plants (e.g. 0.24 for Irish Sea water or 0.20 for Irish Sea sediment (Kershaw et al., 1999) to 0.30 for Bikini Lagoon sediment (Muramatsu et al., 2001). It has been found (Buesseler, 1997) that plutonium delivered as stratospheric fallout appears to be relatively soluble and has a longer residence time in seawater than plutonium from close-in fallout which is more rapidly removed from the water column to deep ocean sediments. High sensitive TIMS has recently been used for size-fractionated plutonium isotope studies in the coastal environment (Dai et al., 2001).

Several studies on natural U and Th in the marine environment have also used TIMS. It has proved to be especially efficient for the precise determination of e.g. ^{234}U/^{238}U and ^{236}U/^{238}U ratios on small samples giving highly sensitive and very accurate measurements (Chen et al., 1986). This has lately become of great interest in connection with the use of depleted uranium in military weapons. For Th isotopes (^{228}Th, ^{230}Th, ^{232}Th), TIMS has been even more effective because of their lower concentrations in seawater (in comparison with U isotopes (with the exception of ^{236}U)) when performing analysis with RMT, requiring large volume samples (1000 l). TIMS permits reduction by a factor of 1000 of the size of the sample (Broecker et al., 1973; Bacon & Andersen, 1982; Chen et al., 1986; Dixon et al., 1997). Other long-lived actinides such as ^{231}Pa, ^{237}Np, ^{243}Am, ^{244}Cm and ^{246}Cm as well as some long-lived fission products (e.g. ^{99}Tc, ^{135}Cs) have been analysed by TIMS, as well (Anderson & Walker, 1980; Halverson, 1984; Heumann, 1988; Koppenaal, 1988; Dixon et al., 1997; Kelley et al., 1999; Cooper et al., 2000), yet more work is needed for the full exploitation of these tracers in marine studies. Moreover, TIMS can be effectively used in speciation and size-fractionation studies (Dai et al., 2001) and also for certification of marine reference materials (Povinec & Pham, 2000; Inn et al., 2001a, b; Lewis et al., 2001).

Inductively coupled plasma mass spectrometry (ICPMS)
ICPMS is a powerful technique for the analysis of elements, stable and long-lived radioactive isotopes in the marine environment. With the introduction of the present state of the art high resolution ICPMS machines, this technique competes with TIMS in many respects. The principal advantages of ICPMS are its capability to determine long-lived radioisotopes of metallic elements down to fg levels, to analyse aqueous samples directly and rapidly (in a few minutes), the low cost per analysis and small sample size. However, it is not free of matrix and isotopic effects, and the formation of chlorides and oxides requires extra careful purification procedures, e.g. extraction chromatography. A steady increase over the past decade in radioanalytical applications using ICMPS has resulted in a decrease in both the price of instruments and detection limits. New generation sector field instruments with double-focusing (Sturup et al., 1998) and even multi-collector ICPMS (Taylor et al., 2001) have improved sensitivity (by about an order of magnitude) and precision over traditional quadrupole machines (Wyse et al., 2001).

ICPMS has been used in both higher-resolution and lower-resolution modes (Sturup et al., 1998; Wyse et al., 2001). The higher-resolution mode has the advantage of addressing polyatomic interferences, although it cannot solve all the problems with isobaric interferences, which may be caused by incomplete separation chemistry. On the other hand, maximum sensitivity can be reached in the lower-resolution mode. Thus a combination of the two modes appears to be the best compromise for reaching maximum sensitivity and controlling interferences. The higher count rates under lower-resolution mode (Fig. 19) give better analytical peaks with lower uncertainties and optimal data quality. Analytes with a relatively strong probability of polyatomic interferences on the isotopes of interest (e.g. ^{238}U which produces a hydride peak that would interfere with ^{239}Pu) should always be scanned. Even so there are problems with the relatively poor abundance sensitivity of sector field ICPMS in the measurement of isotopes with one mass below an abundance peak (e.g. ^{237}Np in the presence of high ^{238}U), and two mass units below (e.g. ^{230}Th in the presence of ^{232}Th, or ^{236}U in the presence of high ^{238}U content).

Even when sample matrices are reasonably clean and care has been taken to minimise oxides during tuning, measurements made near the detection limit are sensitive to overestimation due to polyatomic interferences. The use of chromatographic resins (Lee et al., 2001) has been found to be a suitable technique for processing small volumes, removing possible interferences by additional cleaning, as well as for cleaning leached plutonium samples electrodeposited on stainless steel disks, previously analysed by SAS.

Marine radionuclide measurements by ICPMS have been carried out on long-lived radioisotopes of actinides in seawater, sediment and biota, like Th, U, Np and Pu isotopes (Sturup

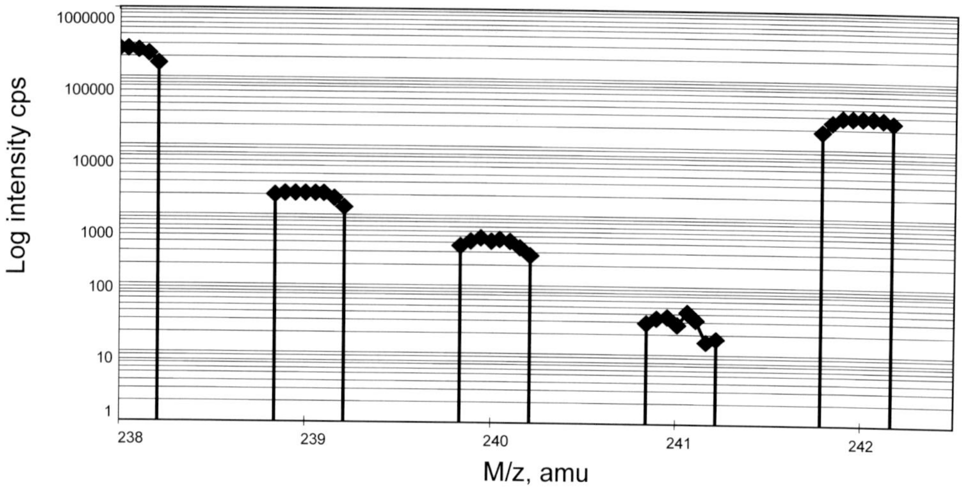

Fig. 19. A typical spectrum of Pu isotopes as obtained by high resolution ICPMS (Wyse et al., 2001).

et al., 1998; Chiappini et al., 1999; Lee et al., 2001; Muramatsu et al., 2001). Long-lived fission radionuclides (like ^{99}Tc) and ^{129}I have been studied in seawater and seaweeds. Usually solution neutralisation is done for liquid samples with detection limits below pg. The ETV technique has also been used for determining U and Pu isotopes in seawater and ^{129}I in seaweeds.

As an example, Figure 20 shows typical ^{239}Pu and ^{240}Pu water profile activity concentrations (together with ^{240}Pu/^{239}Pu mass ratios) taken during the IAEA'97 NW Pacific Ocean cruise close to the Enewetak Atoll (IAEA, 2002). The medium depth peak, located at 500 m water depth, is clearly visible for both radionuclides, as well as the higher concentrations measured in the bottom sample. The ^{240}Pu/^{239}Pu ratio is higher than expected from global fallout (0.186), indicating the influence of close Bikini and Enewetak Atolls on Pu concentrations in the water column (Livingston et al., 2001).

ICPMS has also been used as an effective method for the certification of concentrations of Th, U, Pu isotopes in marine reference materials – seawater, sediment and biota (Lee et al., 2001; Inn et al., 2001a, b). Although molecular, isobaric and isotopic interferences remain crucial for successful operation of ICPMS, this technique has a large potential for automation by direct coupling with new generation of chromatography instruments (Michel, 2001).

Resonance ionisation mass spectrometry (RIMS)

RIMS, which is based on highly efficient and selective ionisation of elements by tuneable lasers (and subsequent analysis of ions by a mass spectrometer) has been found to be a powerful technique whose main advantages are almost complete isotopic suppression, high ionisation efficiency, low background ion detection and high isotopic selectivity, resulting in an overall sensitivity of detection limits down to the fg scale (Wendt et al., 2000). As only atoms of the desired element (or isotope) are excited and ionised by laser (Fig. 21; Wendt et al., 2000), undesired ions in the mass spectrometer can be suppressed to a significant extent.

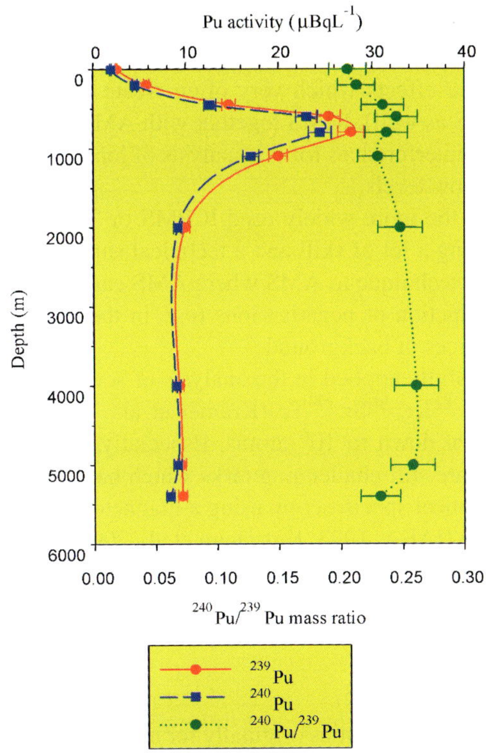

Fig. 20. ^{239}Pu, ^{240}Pu profiles in the water column of the Pacific Ocean (close to Enewetak Atoll) (IAEA, 2002).

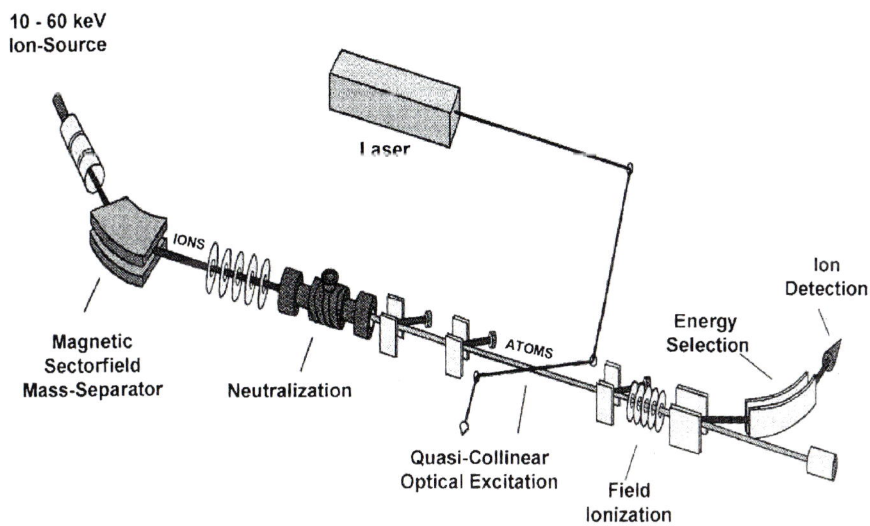

Fig. 21. Experimental set-up for RIMS in collinear geometry (Wendt et al., 2000).

Therefore, of the various applications of RIMS, ultra low-level determination of long-lived radionuclides in the environment has been the most widely acknowledged. As ICPMS is very often limited by molecular effects which very often create isobaric interferences (as occasionally occurs with TIMS as well), RIMS together with AMS is currently the most selective technique with minimum interferences for the analysis of long-lived radionuclides in the marine environment at very low levels.

However, compared to the more widely used ICPMS or TIMS, RIMS is still a too complicated technique requiring a lot of skill and a technical support. It would seem that RIMS will remain an alternative technique to AMS where AMS cannot be used effectively because of insufficient or no production of negative ions (e.g. in the case of ^{81}Kr), strong isobaric interferences or other sources of background.

RIMS has been successfully applied in the analysis of several long-lived radionuclides in the environment such as, ^{81}Kr, ^{90}Sr, ^{99}Tc (Kramer et al., 1986; Wendt et al., 2000) with a detection limit for ^{239}Pu down to 10^6 atoms. Especially, ultratrace isotope detection of noble gases (^{39}Ar, ^{81}Kr) are still challenging tasks which have recently been addressed also by AMS, as well as by optical spectroscopy using a magneto-optical trap and collinear laser photon burst spectroscopy (IAEA, 1993; Kutschera et al., 2000; Bailey et al., 2000).

Although the application of RIMS in marine radionuclide studies has been very limited, its great potential could be in speciation studies (Baxter, 1993), especially when it is coupled to chromatographic systems, as well as in certification of reference materials.

Accelerator mass spectrometry (AMS)

AMS was developed in the late seventies originally for analysis of ^{14}C in environmental samples (for recent works see Kutschera et al., 2000; Fifield, 2000). However, the first successful attempts to analyse isotopes using accelerated ions are much older but the technique was forgotten for almost 40 years. Attempts were made to use cyclotrons, as well as tandem accelerators for acceleration of ions, but the problems with current stability in cyclotrons led later to the wide spread of tandem accelerators in AMS. The real breakthrough was the development of dedicated tandem accelerators for AMS, firstly for ^{14}C and then for other radionuclides as well (^{10}Be, ^{26}Al, ^{129}I). A typical schema for an AMS facility operating in light mass as well as heavy mass ranges is shown in Fig. 22 (Lawson et al., 2000). It consists of three main parts – the ion source, the accelerator and the mass spectrometer. As in tandem accelerators only negative ions can be used for acceleration, the AMS technique can only be applied for those elements (the great majority) forming negative ions. For elements which only form positive ions (e.g. noble gases) cyclotrons can be used, e.g. for the analysis of ^{39}Ar and ^{81}Kr. However, this technique is still in the stages of development.

As AMS has been extensively reviewed recently (Kutchera et al., 2000; Fifield, 2000), in this paper we shall concentrate only on a few radionuclides (^{3}H, ^{14}C, ^{10}Be, ^{26}Al, ^{32}Si, ^{39}Ar, ^{99}Tc, ^{129}I, U, Pu isotopes) with highest application potential in the marine environment. Some other long-lived radionuclides have been successfully analysed by AMS (or are under development) with possible applications in marine sciences, like ^{36}Cl (Hatori et al., 2000; Jakobsen et al., 2000), ^{53}Mn (Knie et al., 2000), ^{135}Cs (Zhao et al., 1999) and ^{237}Np (Fifield, 2000).

Tritium. It was expected that AMS analysis of ^{3}H in environmental samples would have the same advantages as in ^{14}C analysis by AMS i.e. an increase in sensitivity compared to LSS,

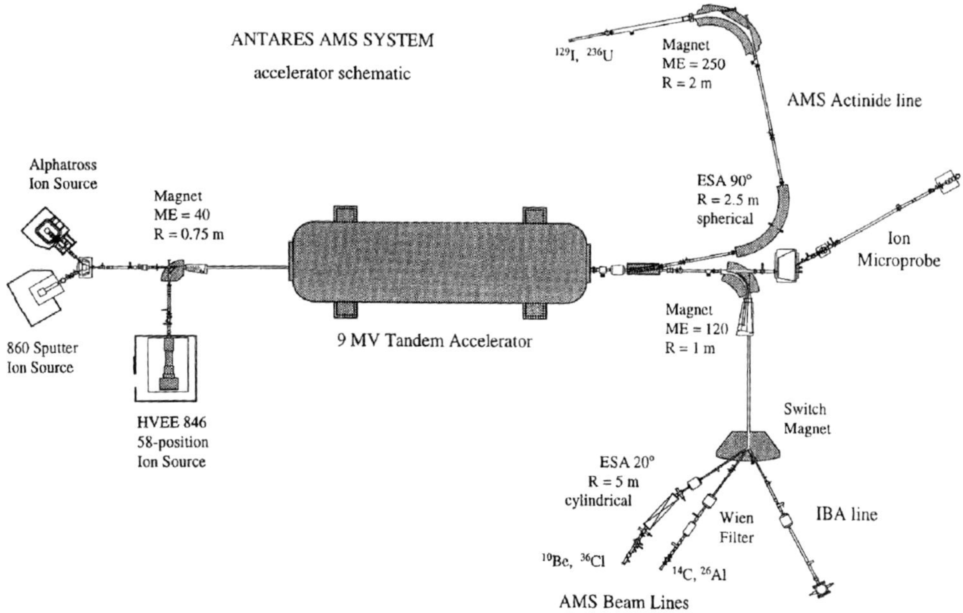

Fig. 22. Schema of the ANTARES AMS facility (Lawson et al., 2000).

permitting the reduction of sample size. Nevertheless, the AMS prototypes (Roberts et al., 2000) developed until now, based on a radio frequency quadruple (RFQ) linac, can measure $^{3}H/^{1}H$ ratios down to 10^{-13} only (and with further improvement probably down to 10^{-15}) in milligram-sized samples, which makes its application possible only in biological research dealing with ^{3}H-labelled compounds.

Another possibility, of course, would be to analyse ^{3}He (the daughter product of ^{3}H decay (as it was done originally by Alvarez group in the late thirties) in the same way as in isotope mass spectrometry, but this would require the use of an ion source with positive ions.

Carbon-14. Carbon-14 ($T_{1/2} = 5730$ y) has been recently analysed by AMS in seawater samples as a world-wide tracer of water mass movement in the major WOCE programme (Key, 1996; McNichol et al., 2000), as well as in several other works (Kumamoto et al., 2000; Povinec et al., 2000, 2001c; Aramaki et al., 2001). It has also been used for the investigation of sediment dynamics (Somayajulu et al., 1999), dating of deep sea sediments (Brown et al., 2000), as well as in studies of ^{14}C in suspended particles (Kawahata & Murayama, 2000), in marine biota (Yoneda et al., 2000) and in corals (also in connection with the El Niño phenomenon) (Guilderson & Schrag, 1998).

We shall illustrate ^{14}C analysis of seawater samples in more detail. A typical schema of a ^{14}C line used at IAEA-MEL for the preparation of samples is shown in Fig. 23. Usually only dissolved inorganic carbon (DIC) is extracted from the water sample (in other applications dissolved organic carbon can be extracted from larger samples using a different technique) by acidification of the water sample to pH 3 and the carbon dioxide released in a flow of high purity oxygen (or nitrogen) is collected in traps cooled with liquid nitrogen. After purification,

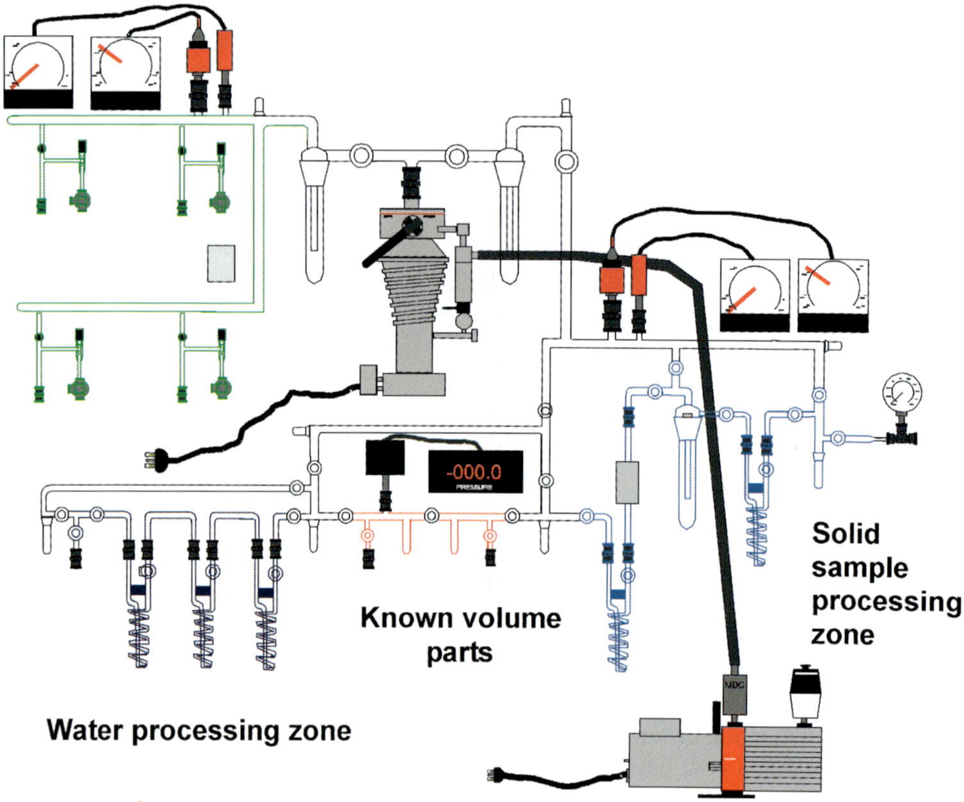

Fig. 23. Line for the preparation of ^{14}C samples for AMS analysis at IAEA-MEL.

the CO_2 is converted to graphite over, e.g. a Fe catalyst. The graphite sample is inserted in a sample holder and fixed in the carousel of the ion source. Carbon ions liberated from the graphite sample are accelerated to energies of a few MeV/nucleon and further analysed by a magnetic spectrometer. The ^{14}C activity in seawater samples is usually expressed by Δ^{14}C defined as

$$\Delta^{14}C = (R - 1)1000 \ (‰),$$

where

$$R = A_S/0.7459A_O e^{\lambda(y-1950)},$$

where

A_S and A_O are equal to the $^{14}C/^{12}C$ ratios of a sample and the NIST ^{14}C standard (oxalic acid) normalised to $\delta^{13}C_{PBD} = -25‰$, $\lambda = 1/8267\ y^{-1}$ and y is equal to a measurement year.

The precision of AMS measurements is typically below ±5‰.

Desk top AMS systems for ^{14}C analysis using both tandem (Suter et al., 2000) and cyclotron (Chen et al., 2000) accelerators have recently been developed and will broaden the applications of AMS in environmental and life sciences.

Berylium-10,^{26}Al. Berylium-10 ($T_{1/2} = 1.5 \times 10^6$ y) and ^{26}Al ($T_{1/2} = 7.1 \times 10^5$ y) measured by RMT were used as useful tracers in marine studies for more than 30 years (Somayajulu, 1967; Lal, 1999). As the AMS technique of ^{10}Be and ^{26}Al determination was developed almost simultaneously with ^{14}C, both radionuclides have been very often used in ocean chemistry (Raisbeck et al., 1996; Ku et al., 1990), dating of manganese nodules and crusts (Somayajulu, 1967; Kobayashi et al., 2000) and dating of sediments (Aldahan & Possnert, 2000; McHargue et al., 2000). The recent studies in ocean waters have revealed their short residence times and appreciable effects of exchange fluxes at the coastal and ocean interfaces, indicating possible variations in the past with temporal changes in climate and biological productivity (Dong et al., 2001).

Silicon-32,^{99}Tc. Silicon-32 is a cosmogenic radionuclide ($T_{1/2} = 150$ y) and ^{99}Tc is an anthropogenic radionuclide ($T_{1/2} = 2.13 \times 10^5$ y) which has been widely distributed in the marine environment through global fallout and releases from nuclear reprocessing plants. Both radionuclides have been determined in the past by RMT as well by TIMS, ICPMS and RIMS (only for ^{99}Tc) and applied in oceanic water studies (Somayajulu et al., 1987, 1991; Peng et al., 1993; Beasley & Lorz, 1986; McCartney & Rajendran, 1999; Kershaw et al., 1999). Silicon-32 (Treacy et al., 2000; Morgenstern et al., 2000) and ^{99}Tc (Fifield, 2000; Berquist et al., 2000), have been recently successfully analysed by AMS, however, their AMS applications in marine studies are still in development stage.

Argon-39,^{81}Kr. Argon-39 being a medium-lived radionuclide ($T_{1/2} = 269$ y), i.e. between ^3H and ^{14}C, it would be an ideal natural tracer for studying water mixing and transport (Loosli et al., 1986; Broecker & Peng, 2000), however, due to requirement of large volume samples for analysis (over 1000 l) it has not been yet widely applied in marine studies. Therefore, AMS and/or RIMS (in the case of ^{81}Kr ($T_{1/2} = 2.1 \times 10^5$ y)) techniques, which are still under development, would help to bring these radionuclides in more frequent use in oceanic research.

Iodine-129. Iodine-129 ($T_{1/2} = 15.7 \times 10^6$ y) thanks to AMS has been widely used as an oceanic tracer to study water transport (Raisbeck & Yiou, 1999; Fehn & Snyder, 2000). Half-litre seawater samples collected and stored without additives are treated in the laboratory following the schema given in Fig. 24. An NaI carrier (10 mg) is added to the sample and after several preparation steps, the AgI sample produced is used as a target for AMS analysis. The results of ^{129}I analysis in seawater samples can be expressed either as activity concentrations (Bq m^{-3}), as a concentration (^{129}I atoms l^{-1}), or as a $^{129}I/^{127}I$ ratio. AMS measurements should be normalised with respect to a reference material. Normally, no background subtraction is necessary as the AMS background is very low. Nonetheless, subtraction of a blank sample, processed in a similar way to the real samples is usually necessary (Povinec et al., 2000).

Uranium isotopes. In the uranium isotopic sector the most interesting development was the AMS ability to analyse ^{236}U in environmental samples (Hotchkis et al., 2000; Berkovits

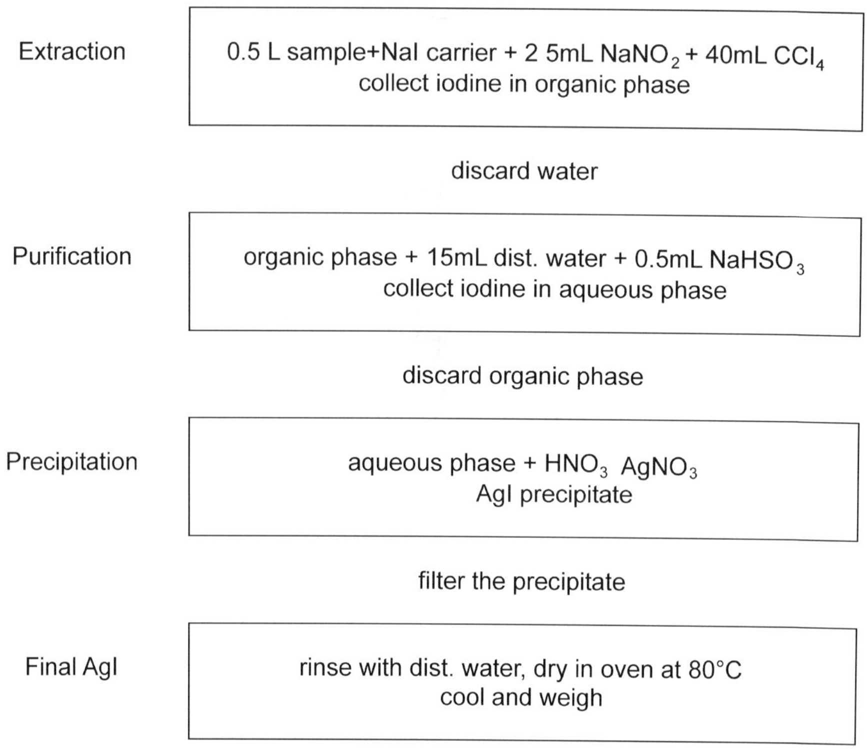

Fig. 24. Flow chart for the preparation of AgI from water samples for ^{129}I analysis by AMS.

et al., 2000). This isotope has been a challenge for AMS as it is embedded in a matrix of ^{235}U and ^{238}U which are potentially sources of background. The principal applications could be in the area of safeguards, as the ^{236}U/^{238}U ratio clearly shows the reactor origin of uranium. Recently, the ^{236}U/^{238}U ratio has been used as an indicator of use of depleted uranium in military munitions.

Plutonium isotopes. AMS has also been used recently in the analysis of Pu isotopes (^{239}Pu, ^{240}Pu and ^{242}Pu) in marine samples. The samples were prepared either by leaching electrodeposited plutonium from previously analysed stainless steel disks by SAS, or by direct preparation of small volume samples using EiChrom resins (Fig. 4). In the final stage of sample preparation for AMS, the plutonium is dispersed in an iron oxide matrix, adjusted to the atom ratios of ^{242}Pu: Fe $= 10^{-10}$ to 10^{-9}, evaporated to dryness, and baked at 550°C to obtain Fe_2O_3. The resulting sample is then mixed with Al powder (in the proportion of 4:1 by weight), which is serving as an electrical and thermal conductor. Finally, the sample is pressed into the sample holders, fixed in the carousel and analysed by AMS. The activities of Pu isotopes in a sample are calculated on the basis of the isotopic dilution technique using either ^{236}Pu (if ^{242}Pu is to be analysed) or ^{242}Pu as tracers.

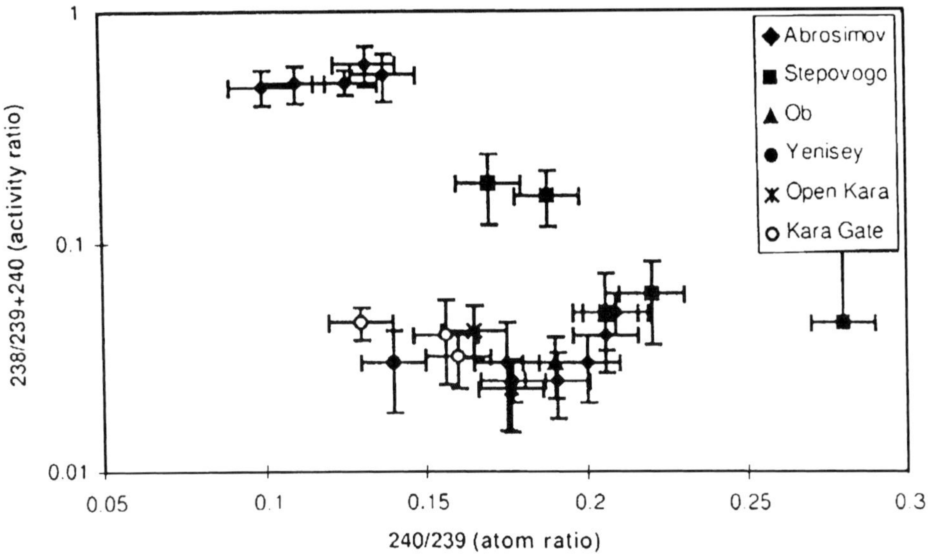

Fig. 25. $^{240}Pu/^{239}Pu$ atom ratios and $^{238}Pu/^{239,240}Pu$ activity ratios in sediment samples from the Kara Sea (Oughton et al., 2001).

We shall illustrate Pu AMS studies on Kara Sea sediment samples (Oughton et al., 1999, 2001) where deviations from global fallout $^{240}Pu/^{239}Pu$ ratios were identified in samples collected at radioactive waste dumping sites in Novaya Zemlya (Fig. 25). The sediments collected at the Kara Gate, Yenisey Estuary and Abrosimov Bay were influenced by low burn-up or military sources, while the ratios seen in contaminated sediment in Stepovoy Bay are concordant with high burn-up, non-military sources.

AMS has been tested, together with sample preparation techniques, on sediment, biota and seawater samples which are in various stages of preparation as IAEA reference materials (Lee et al., 2001). It was possible to compare the results obtained by SAS, LSS and ICPMS at the same time. The results are presented in Table 1 for ^{238}Pu, ^{239}Pu, ^{240}Pu, $^{239,240}Pu$, ^{241}Pu and ^{242}Pu in IAEA-134 (Irish Sea cockle flesh), IAEA-135 (Irish Sea sediment), IAEA-381 (Irish Sea water), IAEA-384 (Fangataufa Lagoon sediment) and IAEA-414 (Irish and North Sea fish). In general, a good agreement was found between the results obtained using the different techniques. It is interesting to note that the ICPMS results were very often slightly higher (up to 10%) than the SAS results (for combined $^{239,240}Pu$ data) and the AMS results. These discrepancies may have been caused by experimental errors contributing to the total uncertainty. It has often been seen with ICPMS analyses of Pu isotopes, that some matrix or interference effects (such as a contribution of ^{238}U, polyatomic chlorides and oxides and different forms of hydrates) could contribute to the final results. However, for the $^{240}Pu/^{239}Pu$ atom ratios, AMS and ICPMS results were in very good agreement. The relative precision of $^{239,240}Pu$ results obtained to date by SAS, AMS and ICPMS were around 5%.

Table 1
A comparison of results of analysis of Pu isotopes in IAEA Reference Materials

Method	^{238}Pu (Bq kg^{-1})	^{239}Pu (Bq kg^{-1})	^{240}Pu (Bq kg^{-1})	^{239,240}Pu (Bq kg^{-1})	^{241}Pu (Bq kg^{-1})	^{242}Pu (Bq kg^{-1})	^{238}Pu/^{239,240}Pu$^{\#}$ (Bq kg^{-1})	^{240}Pu/^{239}Pu$^{\$}$ (Bq kg^{-1})	^{242}Pu/^{239}Pu (Bq kg^{-1})
IAEA-134 Irish Sea cockle flesh									
SAS	3.0 ± 0.2			16.3 ± 0.8			0.184 ± 0.015		
AMS									
ICPMS		9.8 ± 0.8	7.7 ± 0.6	17.5 ± 1.0				0.212 ± 0.008	
LSS									
IAEA-135 Irish Sea sediment									
SAS	42 ± 1			222 ± 8			0.189 ± 0.008		
AMS		129 ± 13	92 ± 9	221 ± 16				0.192 ± 0.027	0.0062 ± 0.0002
ICPMS		127 ± 10	98 ± 8	221 ± 13	3970 ± 200	0.051 ± 0.001		0.210 ± 0.010	0.0068 ± 0.0008
LSS					3000 ± 650	0.056 ± 0.006			
IAEA-381 Irish Sea water									
SAS	3.3 ± 2			14.0 ± 1.0			0.236 ± 0.020		
AMS		8.2 ± 0.3	7.0 ± 0.2	15.7 ± 0.4		0.0046 ± 0.0005		0.242 ± 0.008	0.0084 ± 0.0005
ICPMS		8.1 ± 0.8	7.0 ± 0.7	15.1 ± 1.1		0.0047 ± 0.0002		0.240 ± 0.010	0.0089 ± 0.0006
LSS					200 ± 60				
IAEA-384 Fangataufa Lagoon sediment									
SAS	40 ± 2			111 ± 3			0.360 ± 0.020		
AMS		109 ± 11	14 ± 1	123 ± 11		2.5 ± 0.4		0.048 ± 0.002	0.0004 ± 0.0001
ICPMS		102 ± 8	18 ± 2	120 ± 8	240 ± 20	1.9 ± 0.5		0.051 ± 0.002	0.0003 ± 0.0001
LSS					100 ± 70				
IAEA-414 Irish and North Sea fish									
SAS	0.025 ± 0.002			0.120 ± 0.005		0.044 ± 0.004	0.208 ± 0.019		
AMS		0.087 ± 0.003	0.053 ± 0.005	0.14 ± 0.01		0.068 ± 0.010		0.180 ± 0.009	0.012 ± 0.002
ICPMS		0.063 ± 0.007	0.046 ± 0.005	0.11 ± 0.02	2.7 ± 0.6			0.189 ± 0.010	0.008 ± 0.001
LSS									

$^{\#}$Activity ratio.
$^{\$}$Mass ratio.

5. Quality assurance and quality control of marine radionuclide measurements

Accurate and precise determinations of radionuclide concentrations in marine samples are important aspects of the assessment of the marine environment and the use of radionuclides in studies of oceanographic processes. Data credibility is becoming essential in regional and world-wide programmes to which laboratories with different practises contribute, especially if the resulting data are input into regional and global radionuclide databases. Quality assurance (QA) as a system of activities and actions (Garfield, 1992) has two main components:

(i) *Quality assessment* – a mechanism to verify that the system is operating within acceptable limits;
(ii) *Quality control* (QC) – a mechanism to control data quality (errors).

The objectives of the QA programme include maintaining a continuous assessment and control of data quality, identifying proper analytical methods, providing permanent records of the performance of instruments, standardising analytical procedures, ensuring sample integrity, improving record-keeping and identifying training needs. The common aim of all these objectives is to provide high quality data.

The QA functions include the development (or selection) of proper methods of sampling, sample preservation, sample pre-treatment, sample analysis and methods for evaluation and reporting of results. Further, they include intralaboratory and interlaboratory methods of validation and evaluation, establishing quality control guidelines and maintaining quality control sample programmes. A laboratory without a proper QA programme cannot operate successfully. This is especially important when the laboratory produces series of data that are of interest in regional or world-wide programmes (Michel, 2001).

A QA programme (including good laboratory practises) should be described in one of the most important laboratory documents – the QA manual. It should contain QA policy and objectives, staff responsibilities, analytical methods (including sampling, field measurements, sample handling, protocols, data reduction and evaluation (including uncertainty budget (Dovlete & Povinec, 1999)), materials and standards used, QA procedures, results of interlaboratory comparisons, the recording system and database. The QA manual should be regularly up-dated with any new developments in the laboratory.

The International Atomic Energy Agency's Marine Environment Laboratory in Monaco has been assisting marine laboratories in quality assurance and control for over 30 years (Parr et al., 1998; Povinec et al., 1999a, b). This has been done through the IAEA's Analytical Quality Control Services (AQCS) by organising intercomparison exercises and proficiency tests for the analysis of radionuclides in the marine environment, by establishing QA programmes in the participating laboratories, by providing examples of QA manuals, by QA training and QA missions.

5.1. *Intercomparison exercises*

Both world-wide and regional intercomparison exercises are the important part of a laboratory's QA programme. They have considerably helped laboratories to assess their performance, supplied them with reference materials and enabled them to gather information on their performance. Although there has been remarkable growth both in the number of participating laboratories (from 40 to more than 150) and in the quality of data, the performance of all participating laboratories is not yet satisfactory.

We shall illustrate recent developments in the performance of laboratories in three intercomparison exercises organised by IAEA-MEL, IAEA-135 (Irish Sea sediment), IAEA-315 (Arabian Sea sediment) and IAEA-384 (Fangataufa Lagoon sediment), by comparing the results obtained for the analysis of ^{210}Pb. ^{210}Pb is a radionuclide frequently analysed in sediments as in optimal cases it is used for dating surface sediment and for estimating sedimentation rates. Figure 26 compares the results obtained for ^{210}Pb analysis in IAEA-135, IAEA-315 and IAEA-384. It can be seen that about 20–25 laboratories out of the 35 which submitted their results (the total number of participating laboratories was much higher – about 140 laboratories/exercise), performed very well in these intercomparison exercises. Generally, there was an improvement in the laboratories' performances as documented by the associated uncertainties – around 12% for IAEA-135 and IAEA-315, but 7% for IAEA-384. However, in this last run, there were still 8 laboratories (23%) which reported outlying results. As γ-ray spectrometry was the main technique used for ^{210}Pb analysis, these results show that laboratories have problems with the measurement of low-energy γ-rays.

Following the IUPAC (ISO, 1997) recommendations for assessment of laboratory performance, Z-score methodology has been introduced in the evaluation of intercomparison results,

$$Z = (x_{\mathrm{i}} - x_{\mathrm{a}})/s_{\mathrm{b}},$$

where:

x_{i} is the robust mean of the reported values of massic activity in the sample,
x_{a} is the assigned value (a mean value of accepted results),
s_{b} is the target standard deviation.

The performance of a laboratory is considered acceptable if the difference between the robust mean of the laboratory and the assigned value (in s_{b} units) is less than or equal to two. The analysis is regarded as being out of control when $|Z| > 3$. The Z-score evaluation represents a simple method which gives participating laboratories a normalised performance score for bias. As an example, Figure 27 shows Z-score distribution for ^{60}Co in IAEA-384 (Fangataufa Lagoon sediment). The performance of laboratories in this instance was very good.

5.2. *Reference materials*

Reference Materials (RMs) represent samples of well established properties used for the assessment of analytical methods. More rigorous materials – Certified Reference Materials (CRMs) or Standard Reference Materials (SRM, issued by the National Institute of Standards and Technology – NIST, USA) have property values certified by technically valid procedures (at least with two independent methods) traceable to SI units (ISO, 1996).

RMs and CRMs have great impact on the development of methods of known accuracy. They represent important benchmarks in QA, identifying weak methodologies, detecting training

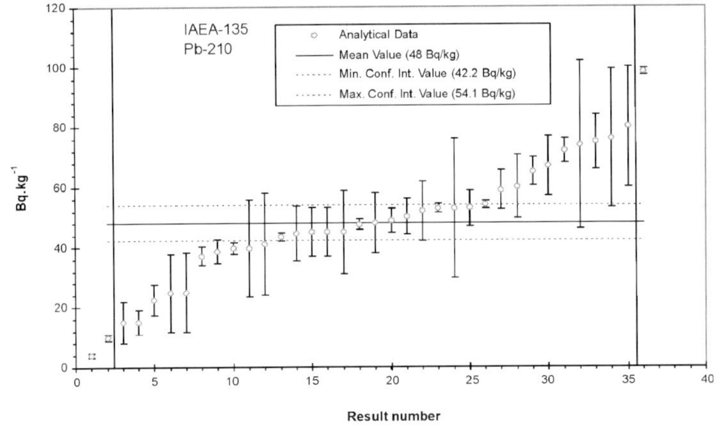

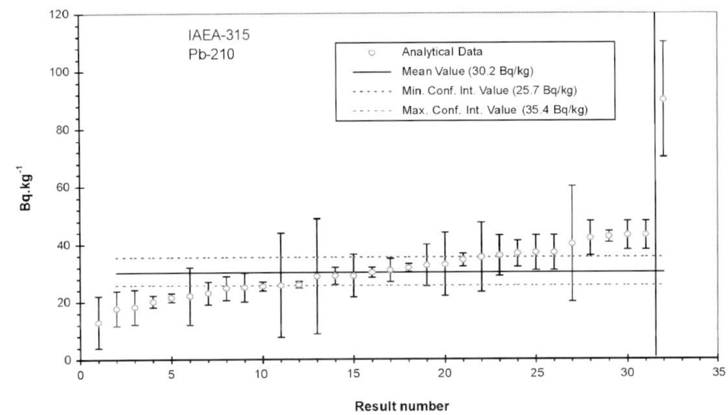

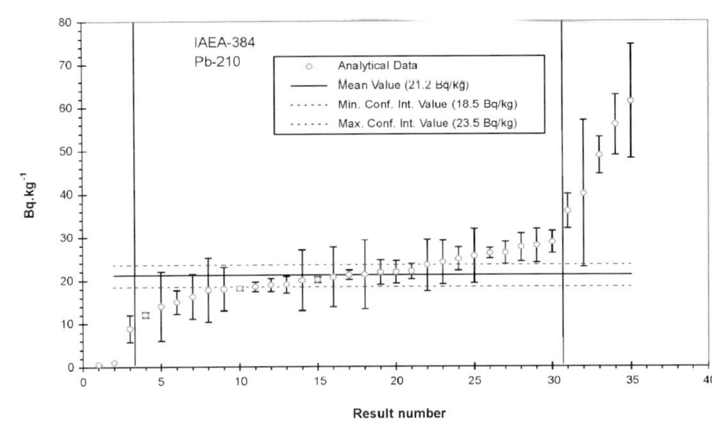

Fig. 26. Data evaluation for ^{210}Pb in Irish Sea sediment IAEA-135 (top, 1992), Arabian Sea sediment IAEA-315 (middle, 1993) and Fangataufa Lagoon sediment IAEA-384 (bottom, 1999).

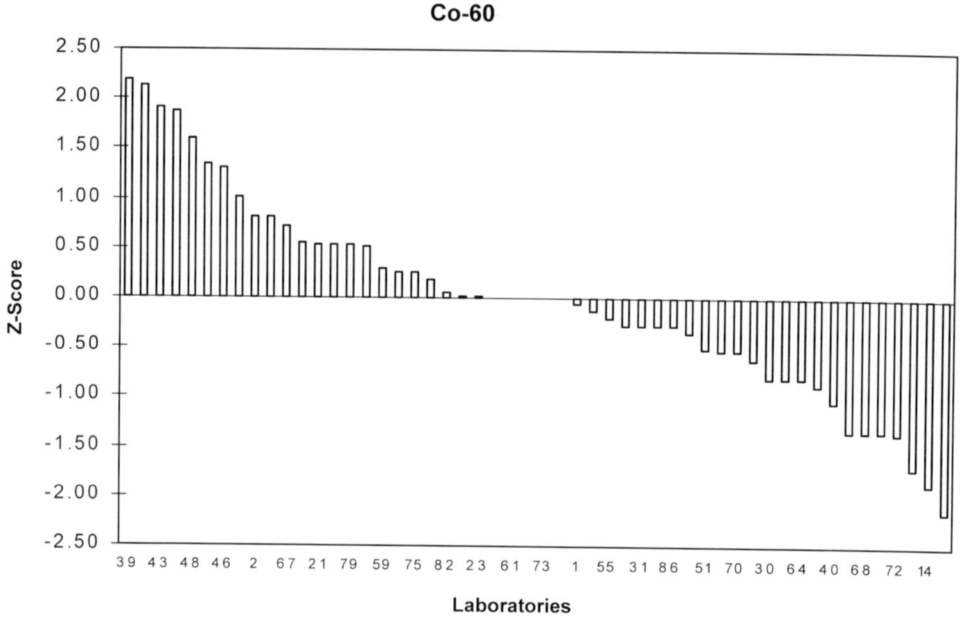

Fig. 27. Z-score evaluation for ^{60}Co in Fangataufa Lagoon sediment IAEA-384.

needs, up-grading the quality of laboratories' performances and assessing the validity of analytical methods. The reference methods can only be accepted on the basis of interlaboratory tests performed on selected CRMs.

As a typical example of the use of RMs, we present in Fig. 28 the quality control chart for analysis of ^{239,240}Pu in Irish Sea sediment RM (IAEA-135), which is frequently used as a reference material at IAEA-MEL. The obtained data are mainly within a confidence interval of $\alpha = 0.05$, although a few exceptions were observed. This documents the necessity of regular use of RMs, especially when new staff carry out analyses, or new methods are incorporated into the existing established procedures.

In order to improve data quality, to provide the required traceability to SI standards and to improve the accuracy and precision of laboratories' measurements, priority should be given to the production of CRMs. The CRMs should be available for the different environmental matrices e.g. sediment, water and biota. The required long-term availability of CRMs (over 10 years) necessitates their long-term stability and the collection and preparation of large volume samples (over 100 kg). The relative precision of all reported data should be better than 5%. This would require highly homogenised samples thoroughly tested for any inhomogeneities of major elements (inhomogeneities should be below 1%). The principal analyses should be accompanied by supporting characterisation e.g. for multielemental composition, mineralogy, particle size distribution, radionuclide speciation studies etc. At least two independent analytical methods should be used for reporting certified values. This may not be a major problem in the case of radionuclides, where different methods are available (e.g. RMT, ICPMS, TIMS, AMS, NAA, etc.).

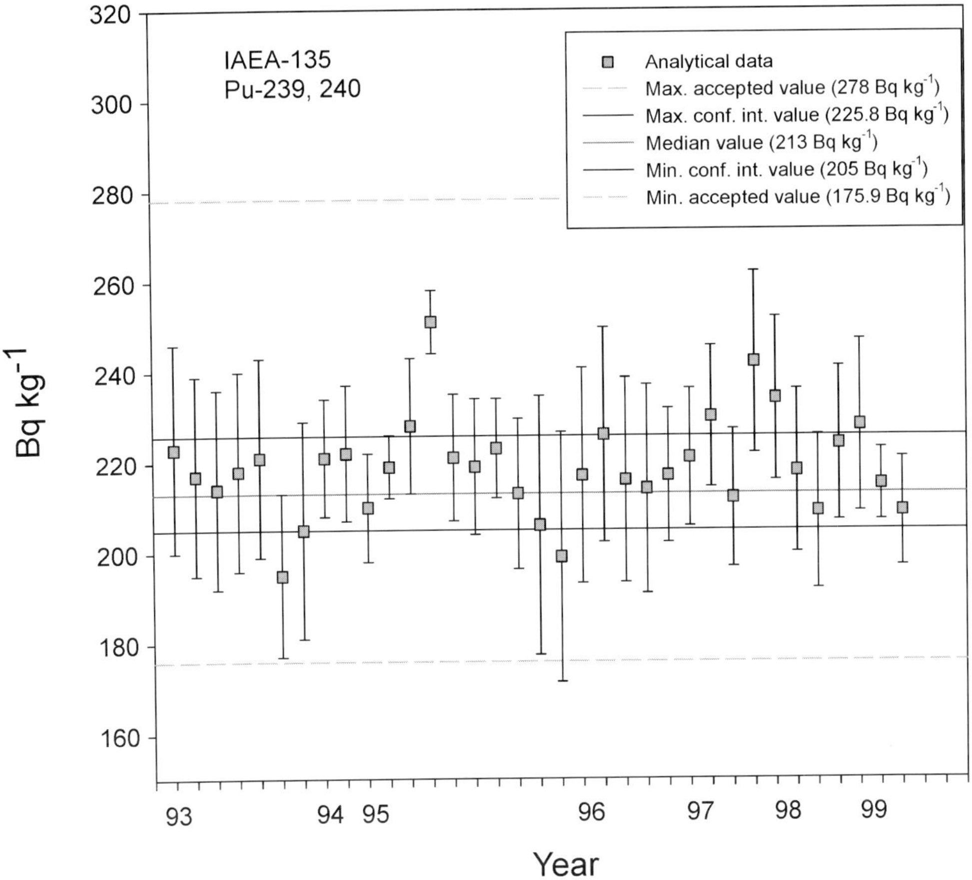

Fig. 28. Quality control chart from IAEA-MEL's Radiometrics Laboratory for the analysis of ^{239,240}Pu in IAEA-315 (Irish Sea sediment).

The specific needs for the production of CRMs for mass spectrometric methods (e.g. AMS, ICPMS, TIMS), where usually small samples with very low radionuclide concentrations are necessary and which require special treatment, should be addressed. Tables 2 and 3 list RMs and CRMs (or SRMs produced by NIST) for the analysis of anthropogenic and natural radionuclides in marine sediments, seawater and biota, which are either available or under production.

5.3. *Reference methods*

Reference methods (IAEA, 1970, 1975) represent a wide-ranging series of methods and guidelines used in marine radioactivity studies. They have considerable influence on the accuracy, precision and general reliability of data and represent a dynamic system based on both current methods and new methods still under development. Periodic evaluation and modifica-

Table 2
Available Reference Materials for anthropogenic and natural radionuclides in the marine environment

Code	Type	Matrix	Place of origin	Reference
IAEA-MA-B3/RN	RM	Fish flesh	Baltic Sea	
IAEA-135	RM	Sediment	Irish Sea	Ballestra et al. (1993)
IAEA-300	RM	Sediment	Baltic Sea	Ballestra et al. (1994)
IAEA-315	RM	Sediment	Arabian Sea	Ballestra et al. (1997)
IAEA-352	RM	Tuna fish flesh	Mediterranean Sea	Ballestra et al. (1990)
IAEA-381	CRM	Seawater	Irish Sea	Povinec et al. (2002)
IAEA-384	CRM	Sediment	Fangataufa Lagoon	Povinec & Pham (2000)
IAEA-368	RM	Sediment	Pacific Ocean	Ballestra et al. (1991)
IAEA-414	CRM	Fish	Irish and North Seas	
NIST-SRM 4357	SRM	Sediment		

Table 3
Reference Materials for radionuclides in the marine environment – under preparation

Code	Type	Matrix	Place of origin	Expected date available
IAEA-418	CRM	Water	Mediterranean Sea	2004
IAEA-385	CRM	Sediment	Irish Sea	2004
IAEA-415	CRM	Fish flesh	Atlantic Ocean	2005
IAEA-412	CRM	Sediment	Pacific Ocean	2006
IAEA-410	CRM	Sediment	Bikini Atoll	2007
NIST-SRM 4358	SRM	Shellfish	Barents Sea	2004
NIST-SRM 4359	SRM	Seaweed		
NIST-SRM 4360	SRM	Fish		

tion of reference methods should be made to up-date the list with recent trends in analytical measurement techniques. Reference methods should be available for different sample matrices and elements. Their provision involves an extensive and time-consuming effort, including global cooperation and testing in expert laboratories. A new IAEA document on reference methods is under preparation, and will reflect recent developments in both radiometrics as well as mass spectrometric methods of low-level radionuclide analyses of marine samples.

6. Conclusions and outlook

6.1. *Conclusions*

It has been demonstrated that there have recently been several important innovations in sampling, shipboard measurements and laboratory analysis of marine samples. Innovations in RMT include applications of high efficiency HPGe detectors used both in the laboratory and on shipboard. Stationary γ-ray monitors based on NaI(Tl) detectors with satellite data transmission and towing systems based on NaI(Tl) and HPGe detectors have been used for monitoring radionuclides in seawater and on the seabed. LSS with anticosmic and/or pulse shape

Table 4
A comparison of detection limits for frequently analysed long-lived radionuclides in the marine environment (Bq)

	^{3}H	^{14}C	^{99}Tc	^{129}I	^{236}U	^{237}Np	^{240}Pu
RMT	10^{-3}	10^{-4}	1×10^{-2}	10^{-2}	1×10^{-3}	1×10^{-4}	$1 \times 10^{-5\#}$
NAA			1	10^{-7}		5×10^{-4}	
ICPMS	10^{-3*}		1×10^{-5}	10^{-3}	5×10^{-9}	5×10^{-6}	5×10^{-6}
TIMS			8×10^{-6}	10^{-8}	1×10^{-10}	1×10^{-9}	0.5×10^{-6}
RIMS			1×10^{-5}				$\sim 10^{-6}$
AMS	10	10^{-7}	6×10^{-6}	10^{-10}	1×10^{-10}	1×10^{-10}	0.4×10^{-6}

$\#^{239,240}$Pu.
*He in-growth MS.

discrimination has been the most important development for analysis of soft β-ray emitters like ^{3}H and ^{241}Pu.

SAS remains a powerful technique for the analysis of α-ray emitters because of the simplicity of measurement, reasonable resolution and sensitivity for radionuclides with shorter half-lives. However, low activity samples require long counting times and the separation of ^{239}Pu and ^{240}Pu is difficult. A new state of the art TIMS operating in multi-dynamic mode with a multi-collector detection system has produced better precision and accuracy in isotopic ratio and concentration measurements. TIMS is now the most sensitive technique, relatively free of interferences, employed for low energy range mass spectrometry of long-lived actinides. ICMPS has proved to be powerful tool for the analysis of long-lived radionuclides because of its high sensitivity, rapid analysis, multi-isotopic composition and the low cost per analysis. Nevertheless, there could be problems with molecular, isobaric and isotopic interferences even if careful purification procedures are used.

The most exciting breakthrough in the analysis of long-lived radionuclides in the marine environment has been made in the AMS sector. AMS operates at the highest level of sensitivity, requires minimum sample size and minimises matrix and interference effects. However, its operation is complex and experienced operators are needed, which means that analysis is more expensive.

A comparison of detection limits for a number of long-lived radionuclides measured by RMT, ICPMS, TIMS and AMS is given in Table 4. It can be seen that the most sensitive technique is AMS which gives the lowest detection limits (with the exception of ^{3}H), three to eight orders of magnitude lower than RMT.

6.2. *Outlook*

It is always difficult to predict new and important developments in any branch of science. As concerns the field of analytical techniques over the next 10–20 years, e.g. it is clear that due to decreasing concentrations of anthropogenic radionuclides in the marine environment, more sensitive methods will be needed. They will include:

(i) Quicker and simpler methods of chemical separation enabling multi-radionuclide separations from one sample and a decrease in manpower by the use of new generation of chromatographs and robotic systems, directly coupled with spectrometers.

(ii) Modern, state of the art coincidence–anticoincidence γ-ray spectrometers with anti-cosmic–anticompton shielding, operating underground for the analysis of γ-ray emitters with relatively short half-lives.
(iii) Further development and more frequent use of mass spectrometry techniques (TIMS, ICPMS, AMS, RIMS) for ultra low-level analysis of long-lived radionuclides. It is expected that AMS will be more widely using 'desk-top' machines. Accelerators working with positive ions (e.g. for ^{39}Ar and ^{81}Kr analysis) will be further developed and applied in marine studies. Direct coupling of ICPMS (liquid sources) and AMS (gas sources) with new generation of chromatographs will be further developed and applied in marine studies. Other long-lived radionuclides, like ^{135}Cs, will be mastered as well, in order to enlarge the group of long-lived radionuclides used as oceanic tracers.

These new techniques will not only greatly improve detection limits for some radionuclides, but will require smaller sample volumes so that sampling, e.g. of the water column, will be much easier. Further,

(i) New techniques will be developed for speciation studies of radionuclides associated with colloid and particles for better understanding of water processes, water–sediment interaction and biological uptake of radionuclides.
(ii) Separate radionuclide analyses of group of particles, biota species and organic compounds in suspended matter will be carried out (thanks to a sub-milligram sample size requirements for the analysis) to better understand processes in the water column and bioaccumulation of radionuclides.
(iii) Bulk sample radionuclide analysis of sediments will be replaced by radionuclide analysis of different minerals and organic compounds present in sediment, thus allowing better understanding of biogeochemistry of different elements in the marine environment.
(iv) More attention will be focused on the study of natural radionuclides in the marine environment – both for the investigation of marine processes and for the protection of man and the marine environment.
(v) Quality assurance and quality control including the certification of laboratories, will further progress and data quality will improve to the extent that national, regional and global marine information systems will be built.
(vi) High quality CRMs covering all marine matrixes and radionuclides of interest will play an important role in the marine radioactivity studies.

All these we would call 'expected predictions' which are foreseeable in the near future. However, any breakthrough in new technology is far more difficult to predict. For example, twenty years back it was thought that new 'laser-based technologies' would be the major advance in analytical techniques, but this has not happened. RIMS (with its sensitivity similar to ICPMS) is still not a 'desk-top' technique that would permit widespread use. Laser Microprobe Mass Analysis (LAMMA), as well as other mass spectrometry techniques such as Secondary Ionization Mass Spectrometry (SIMS) and Fourier Transform–Ion Cyclotron Resonance (FT–ICR), all have detection limits several orders of magnitude below those discussed in this review.

The techniques discussed in this paper, together with new techniques, which should be available in the near future, will further enhance applications of radionuclides as tracers in marine studies. Therefore, thanks to new tracers, better detection limits and smaller samples, new exciting investigations are expected to be carried out, which will further improve our knowledge of oceanic processes, and they will contribute significantly to better understanding and protecting the marine environment.

Acknowledgements

The author would like to express his appreciation for the support he received from the IAEA-MEL directors M. S. Baxter, H. D. Livingston and R. F. C. Mantoura during his R&D work at IAEA-MEL. The co-operation by the staff of IAEA-MEL in the development of radiometrics and mass spectrometry methods for analysis of radionuclides in the marine environment, namely by J. La Rosa, S.-H. Lee, I. Osvath, E. Wyse, J.-F. Comanducci, J. Gastaud, I. Levy, L. Liong Wee Kwong and M.-K. Pham and, as well as by C. Gustavsen for her help in the preparation of the manuscript is highly acknowledged. The collaboration with W. Burnett (State University of Florida, Tallahassee), K. Fifield (Australian National University, Canberra), M. Hotchkis, D. Fink and C. Tuniz (ANSTO, Sydney), T. Jull and G. Burr (University of Arizona, Tucson), L. Keiser (University of Toronto), H. H. Loosli (University of Bern), U. Morgenstern (Institute of Geological and Nuclear Sciences, Lower Hutt, New Zealand), P. Vojtyla (CERN, Geneva) and Z. Top (University of Florida, Miami) is highly acknowledged. IAEA-MEL operates under a bilateral agreement between the IAEA and the Government of the Principality of Monaco.

References

Aldahan, A. & Possnert, G. (2000). The ^{10}Be marine record of the last 3.5 Ma. *Nuclear Instruments and Methods in Physics Research, B 172*, 513–517.

Anderson, T. J. & Walker, R. L. (1980). Determination of picogram amounts of technetium-99 by resin bead mass spectrometric isotope dilution. *Analytical Chemistry, 52*, 709.

Aramaki, T., Mizushima, T., Kuji, T., Povinec, P. P. & Togawa, O. (2001). Distribution of radiocarbon in the southwestern North Pacific. *Radiocarbon, 43*, 857–867.

Bacon, M. P. & Anderson, R. F. (1982). Distribution of thorium between dissolved and particulate forms in the deep sea. *Journal of Geophysical Research, 87*, 2045.

Bailey, K., Chen, C. Y., Du, X., Li, Y. M., Lu, Z.-T., O'Connor, T. P. & Young, L. (2000). ATTA – A new method of ultrasensitive isotope trace analysis. *Nuclear Instruments and Methods in Physics Research, B 172*, 224–227.

Ballestra, S., Vas, D., Lopez, J. J. & Noshkin, V. (1990). Report on the Intercomparison Run IAEA-352. Radionuclides in tuna fish flesh. IAEA/AL/032. Vienna: IAEA.

Ballestra, S., Lopez, J. J., Gastaud, J., Parsi, P., Vas, D. & Noshkin, V. (1991). Report on the Intercomparison Run IAEA-368. Radionuclides in Pacific Ocean sediment. IAEA/AL/047, IAEA/MEL/47. Vienna: IAEA.

Ballestra, S., Gastaud, J., Lopez, J. J., Parsi, P. & Vas, D. (1993). Report on the Intercomparison Run IAEA-135. Radionuclides in Irish Sea sediment. IAEA/AL/063. Vienna: IAEA.

Ballestra, S., Gastaud, J., Lopez, J. J., Parsi, P. & Vas, D. (1994). Report on the Intercomparison Run IAEA-300. Radionuclides in marine sediment. IAEA/AL/046. Vienna: IAEA.

Ballestra, S., Pettersson, H., Gastaud, J., Parsi, P. & Vas, D. (1997). Report on the Intercomparison Run IAEA-315. Radionuclides in marine sediment. IAEA/AL/65. Vienna: IAEA.

Baxter, M. S. (1993). Resonance ionization mass spectrometry. In *Development and Evaluation of Alternative Radioanalytical Methods, Including Mass Spectrometry for Marine Materials* (pp. 53–64). IAEA-TECDOC-683. Vienna: IAEA.

Beasley, T. M. & Lorz, H. V. (1986). A review of the biological and geochemical behaviour of technetium in the marine environment. *Journal of Environmental Radioactivity, 3*, 1–22.

Benedik, L. & Byrne, A. R. (1995). Simultaneous determination of trace uranium and thorium by radiochemical neutron activation analysis. *Journal of Radioanalytical and Nuclear Chemistry Articles, 189* (2), 325–331.

Bergquist, B. A., Marchetti, A. A., Martinelli, R. E., McAninch, J. E., Nimz, G. J., Proctor, I. D., Southon, J. R. & Vogel, J. S. (2000). Technetium measurements by accelerator mass spectrometry at LLNL. *Nuclear Instruments and Methods in Physics Research, B 172*, 328–332.

Berkovits, D., Feldstein, H., Ghelberg, S., Hershkowitz, A., Navon, E. & Paul, M. (2000). ^{236}U in uranium minerals and standards. *Nuclear Instruments and Methods in Physics Research, B 172*, 372–376.

Bertine, K. K., Chow, T. J., Koide, M. & Goldberg, E. D. (1986). Plutonium isotopes in the environment: Some existing problems and some new ocean results. *Journal of Environmental Radioactivity, 3*, 189.

Bibron, R., Chesselet, R., Crozaz, G., Leger, G., Mennessier, J.-P. & Picciotto, E. (1974). Extra-terrestrial ^{53}Mn in Antarctic ice. *Earth and Planetary Science Letters, 21*, 109.

Brodzinski, R. L. (1991). Low-level gamma-ray spectrometry. *Journal of Physics G: Nuclear Particle Physics, 17*, 5403–5413.

Broecker, W. S. & Peng, T.-H. (2000). Comparison of ^{39}Ar and ^{14}C ages for waters in the deep ocean. *Nuclear Instruments and Methods in Physics Research, B 172*, 473–478.

Broecker, W. S., Kaufman, A. & Trier, R. M. (1973). The residence time of thorium in surface sea water and its implication regarding the reactive pollutants. *Earth and Planetary Science Letters, 20*, 35.

Brown, L., Cook, G. T., Mackenzie, A. B., Naysmith, P., Anderson, R., Thomson, J. & Nixon, S. (2000). Characterisation of differential mixing of foraminifera and bulk carbonate in NE Atlantic sediments. *Nuclear Instruments and Methods in Physics Research, B 172*, 490–494.

Buesseler, K. O. (1993). Thermal ionization mass spectrometry. In *Development and Evaluation of Alternative Radioanalytical Methods, Including Mass Spectrometry for Marine Materials* (pp. 45–52). IAEA-TECDOC-683. Vienna: IAEA.

Buesseler, K. O. (1997). The isotopic signature of fallout plutonium in the North Pacific. *Journal of Environmental Radioactivity, 36*, 69–83.

Buesseler, K. O. & Halverson, J. E. (1987). The mass spectrometric determination of fallout ^{239}Pu and ^{240}Pu in marine samples. *Journal of Environmental Radioactivity, 5*, 425.

Buesseler, K. O. & Scholkovitz, E. R. (1987). The geochemistry of fallout plutonium in the North Atlantic: I. A pore water study in shelf, slope and deep-sea sediments. *Geochimica et Cosmochimica Acta, 51*, 2605.

Buesseler, K. O., Cochran, J. K., Bacon, M. P., Livingston, H. D., Casso, D., Hirschberg, M. C. & Hartman, A. (1992). Determination of thorium isotopes in seawater by non-destructive and radiochemical procedures. *Deep-Sea Research, 39*, 1103–1114.

Burnettt, W. C., Corbet, D. R., Schultz, M., Horwitz, E. P., Chiarizia, R., Dietz, M., Thakkar, A. & Fern, M. (1997). Preconcentration of actinides elements from soils and large volume water samples using extraction chromatography. *Journal of Radioanalycal and Nuclear Chemistry, 226*, 121–127.

Byrne, A. R. (1986). Determination of ^{237}Np in Cumbrian (UK) sediments by neutron activation analysis: Preliminary results. *Journal of Environmental Radioactivity, 4*, 133–144.

Byrne, A. R. & Benedik, L. (1999). Application of neutron activation analysis in determination of natural and man-made radionuclides, including Pa-231. *Czechoslovak Journal of Physics, 49, SI*, 265–271.

Chao, J. H. & Tseng, C. L. (1996). Determination of ^{135}Cs by neutron activation analysis. *Nuclear Instruments and Methods in Physics Research, A 272*, 275–279.

Chen, J. H., Edwards, R. L. & Wasserburg, G. J. (1986). ^{238}U, ^{234}U and ^{232}Th in seawater. *Earth and Planetary Science Letters, 80*, 241.

Chen, M., Xiangshun, L., Deming, L., Yonghao, L., Zhou, W., Chen, G., Shen, L., Xu, S. & Zhang, Y. (2000). Minicyclotron (SMCAMS)-based accelerator mass spectrometry and real ^{14}C measurements. *Nuclear Instruments and Methods in Physics Research, B 172*, 193–200.

Chernayev, A. M., Gapanov, I. A. & Lapushkina, L. V. (1999). Development of an underwater Cherenkov detector to reveal sources of technogenic radionuclides. IAEA-SM-354/93P. In *Proceedings of the International Symposium on Marine Pollution* (pp. 417–419). Monaco, October 5–9, 1998. IAEA-TECDOC-1094. Vienna, Austria.

Chiappini, R., Pointurier, F., Millies-Lacroix, J.-C., Lepetit, G. & Hemet, P. (1999). $^{240}Pu/^{239}Pu$ isotopic ratios and $^{239+240}Pu$ total measurements in surface and deep waters around Mururoa and Fangataufa atolls compared with Rangiroa atoll (French Polynesia). *Science of the Total Environment, 237/238*, 269–276.

Clarke, W. B., Jenkins, W. J. & Top, Z. (1976). Determination of tritium by mass spectrometric measurement of ^{3}He. *International Journal of Applied Radiation Isotopes*, *27*, 515–522.

Cochran, J. K., Buesseler, K. O., Bacon, M. P. & Livingston, H. D. (1993). Thorium isotopes as indicators of particle dynamics in the upper ocean: Results from the JGOFS North Atlantic Bloom Experiment. *Deep-Sea Research I*, *40*, 1569–1595.

Cook, G. T. & Anderson, R. (1991). The determination of ^{241}Pu by liquid scintillation spectrometry using the Packard 2250CA. *Journal of Radioanalytical and Nuclear Chemistry. Letter Section*, *154*, 319–330.

Cook, G. T., Harkness, D. B., MacKenzie, A., Miller, B. F. & Scott, E. M. (Eds) (1996). Liquid Scintillation Spectrometry 1994. In *Proceedings of the International Conference*. Glasgow, 1994. University of Arizona, Tucson: Radiocarbon.

Cooper, J. A. & Perkins, R. W. (1971). An anticoincidence-shielded dual Ge(Li) gamma-ray spectrometer for low-level environmental radionuclide analysis and gamma–gamma coincidence studies. *Nuclear Instruments and Methods in Physics Research*, *94*, 29–38.

Cooper, J. A. & Perkins, R. W. (1972). A versatile Ge(Li)–NaI(Tl) coincidence–anticoincidence gamma-ray spectrometer for environmental and biological problems. *Nuclear Instruments and Methods in Physics Research*, *99*, 125–146.

Cooper, L. W., Kelley, J. M., Bond, L. A., Orlandini, K. A. & Grebmeier, J. M. (2000). Sources of the transuranic elements plutonium and neptunium in Arctic marine sediments. *Marine Chemistry*, *69*, 253–276.

Coursey, B. M. & Taylor, D. M. (2000). Proceedings of the ICRM Low-level Radioactivity Measurements Techniques Conference. Mol, 1999. *Applied Radiation and Isotopes*, *53* (1/2), 410.

Dai, M. H., Buesseler, K. O., Kelley, J. M., Andrews, J. E., Pike, S. & Wacker, J. F. (2001). Size-fractionated plutonium isotopes in a coastal environment. *Journal of Environmental Radioactivity*, *53*, 9–25.

Debertin, K. & Helmer, R. G. (1998). *Gamma and X-Ray Spectrometry with Semiconductor Detectors*. Amsterdam: Elsevier.

Deuser, W. G. (1986). Seasonal and interannual variations in deep-water particle fluxes in the Sargasso Sea and their relation to surface hydrography. *Deep-Sea Research*, *33*, 225–246.

Dixon, P., Curtis, D. B., Musgrave, J., Roesnch, F., Roach, J. & Rokop, D. (1997). Analysis of naturally produced technetium and plutonium in geologic materials. *Analytical Chemistry*, *69*, 1692–1699.

Dong, W., Lal, D., Ransom, B., Berger, W. & Caffee, M. W. (2001). Marine biogeochemistries of Be and Al: A study based on cosmogenic ^{10}Be, Be and Al in marine calcite, aragonite, and opal. Proceedings, Indian Academy of Sciences. *Earth and Planetary Science Letters*, *110* (2), 95–102.

Dovlete, C. & Povinec, P. P. (1999). Quantification of uncertainty in gamma-spectrometry – a Monte Carlo approach. *Acta Physica Universitatis Comenianae*, *XL*, 43–55.

Fehn, U. & Snyder, G. (2000). ^{129}I in the Southern Hemisphere: Global redistribution of an anthropogenic isotope. *Nuclear Instruments and Methods in Physics Research*, *B 172*, 366–371.

Fifield, L. K. (2000). Advances in accelerator mass spectrometry. *Nuclear Instruments Methods Physics Research*, *B 172*, 134–143.

Florkowski, T., Morawska, L. & Rozanski, K. (1988). Natural production of radionuclides in geological formulations. *Nuclear Geophysics*, *2*, 1.

Fowler, S. W., Ballestra, S., La Rosa, J. & Fukai, R. (1983). Vertical transport of particulate-associated plutonium and americium in the upper water column of the Northeast Pacific. *Deep-Sea Research*, *30*, 1221–1233.

Fowler, S. W., Noshkin, V. E., La Rosa, J. & Gastaud, J. (2000). Temporal variations in plutonium and americium inventories and their relation to vertical transport in the northwestern Mediterranean Sea. *Limnology and Oceanography*, *45* (2), 446–458.

Froehlich, K., Rozanski, K., Povinec, P., Oregioni, B. & Gastaud, J. (1999). Isotope studies in the Caspian Sea. *Science of the Total Environment*, *237/238*, 419–427.

Garcia-Leon, M. & Madura, G. (Eds) (1991). *Low-Level Measurements of Man-Made Radionuclides in the Environment. Proceedings of the Second International Summer School* (468 pp.). Huelva, 1990. Singapore: World Scientific.

Garcia-Leon, M. & Garcia-Tenorio, R. (Eds) (1994). *Low-Level Measurements of Radioactivity in the Environment. Proceedings of the Third International Summer School* (563 pp.). Huelva, 1993. Singapore: World Scientific.

Garfield, F. M. (1992). *Quality Assurance Principles for Analytical Laboratories*. New York: AOAC Int.

Germain, P., Guéguéniat, P., May, S. & Pinte, G. (1987). Measurement of transuranic elements, chiefly ^{237}Np (by neutron activation analysis), in the physical and biological compartments of the French shore of the English Channel. *Journal of Environmental Radioactivity*, *5*, 319.

Guilderson, T. P. & Schrag, D. P. (1998). Abrupt shift in subsurface temperatures in the Tropical Pacific associated with changes in El Niño. *Science*, *281*, 240–243.

Halverson, J. E. (1984). A review of applications of mass spectrometry to low level radionuclide metrology. *Nuclear Instruments and Methods in Physics Research*, *223*, 349.

Harms, I. & Povinec, P. P. (1999). The outflow of radionuclides from Novaya Zemlya bays – modeling and monitoring strategies. *Science of the Total Environment*, *237/238*, 193–201.

Harvey, B. R. & Chazal, A. (1993). Conventional radiometric methods of analysis including neutron activation. In *Development and Evaluation of Alternative Radioanalytical Methods, Including Mass Spectrometry for Marine Materials* (pp. 7–24). IAEA-TECDOC-683. Vienna: IAEA.

Hatori, S., Nawata, H., Ohseki, M., Matsuzaki, H., Misawa, T. & Kobayashi, K. (2000). Gas-filled magnet at MALT – Measurement of ^{36}Cl. *Nuclear Instruments and Methods in Physics Research*, *B 172*, 211–217.

Heumann, K. G. (1988). Isotope dilution mass spectrometry. In F. Adams, R. Gijbels & R. Van Grieken (Eds), *Organic Mass Spectrometry* (pp. 301–376). New York: Wiley-Interscience.

Heusser, G. (1992). In P. Povinec (Ed.), *Organic Mass Spectrometry Proceedings of 14th Europhysics Conference on Nuclear Physics* (p. 247). Singapore: World Scientific.

Heusser, G. (1994). Background in ionizing radiation detection – illustrated by Ge-spectrometry. In M. Garcia-Leon & R. Garcia-Tenorio (Eds), *Proceedings of the Third International Summer School* (pp. 69–112). Huelva, 1993. Singapore: World Scientific.

Heusser, G. (1995). Low-radioactivity background techniques. *Annual Review of Nuclear and Particle Science*, *45*, 543–590.

Holm, E., Garcia-Leon, M. & Jerome, S. M. (Eds) (1996). Proceedings of the International Committee for Radionuclide Metrology Conference on Low-Level Measurement Techniques. *Applied Radiation and Isotopes*, *47*, 827–1149.

Holm, E., Jerome, S. & Hurtgen, C. (Eds) (2000). Proceedings of the ICRM Low-level Radioactivity Measurement Techniques Conference. *Applied Radiation and Isotopes*, *53*, 3–410.

Hotchkis, M. A. C., Child, D., Fink, D., Jacobsen, G. E., Lee, P. J., Mino, N., Smith, A. M. & Tuniz, C. (2000). Measurement of ^{236}U in environmental media. *Nuclear Instruments and Methods in Physics Research*, *B 172*, 659–665.

Hou, X. (2001). Activation analysis for the determination of long-lived radionuclides. In *Proceedings of the NKS Seminar*. Lund, May 3–4, 2001.

Hou, X. L., Dahlgaard, H., Rietz, B. Jacobsen, U., Nielsen, S. P. & Aarkrog, A. (1999). Determination of ^{129}I in seawater and some environmental materials by neutron activation analysis. *Analyst*, *124*, 1109–1114.

Hou, X. L., Dahlgaard, H. & Nielsen, S. P. (2000). Iodine-129 time series in Danish, Norwegian and northwest Greenland coast and the Baltic Sea by Seaweed. *Estuarine, Coastal and Shelf Science*, *51*, 571–584.

Hou, X. L., Dahlgaard, H. & Nielsen, S. P. (2001). Chemical speciation analysis of ^{129}I in seawater and a preliminary investigation to use it as a tracer for geochemical cycle study of stable iodine. *Marine Chemistry*, *74* (2/3), 145–155.

Huh, C. A. (1988). Determination of thorium in seawater by neutron activation analysis and mass spectrometry. In J. C. Guary, P. Guéguéniat & R. J. Pentreath (Eds), *Radionuclides: A Tool for Oceanography* (pp. 84–91). Amsterdam: Elsevier.

IAEA (1970). *Reference Methods for Marine Radioactivity Studies I*. Technical Report Series No. 118. Vienna: IAEA.

IAEA (1975). *Reference Methods for Marine Radioactivity Studies II*. Technical Report Series No. 169. Vienna: IAEA.

IAEA (1980). *Proceeding of Methods of Low-Level Counting and Spectrometry*. Vienna: IAEA.

IAEA (1993). Development and evaluation of alternative radioanalytical methods, including mass spectrometry for marine materials. IAEA-TECDOC-683. Vienna: IAEA.

IAEA (1999). Marine pollution. *Proceedings of a Symposium Held in Monaco* (694 pp.). October, 1998.

IAEA (2002). *Marine Radioactivity Studies in the World Oceans* (MARS). Monaco: IAEA-MEL.

Ikeda, N., Seki, R., Kamemoto, M. & Otsuji, M. (1989). Activation analysis for technetium-99 by the use of a neutron excitation reaction. *Journal of Radioanalytical and Nuclear Chemistry Articles*, *131* (1), 65–71.

Inkret, W. C., Efurd, D. W., Miller, G., Rokop, D. J. & Benjamin, T. M. (1998). Applications of thermal ionization mass spectrometry to the detection of ^{239}Pu and ^{240}Pu intakes. *International Journal of Mass Spectrometry, 178*, 113–120.

Inn, K. G. W., McCurdy, D., Bell III, T., Loesch, R., Morton, J. S., Povinec, P., Burns, K., Henry, R. & Barss, N. M. (2001a). Standard, intercomparisons and performance evaluations for low-level and environmental radionuclide mass spectrometry and atom counting. *Journal of Radioanalytical and Nuclear Chemistry, 249* (1), 109–113.

Inn, K. G. W., McCurdy, D., Kuruvilla, L., Barss, N. M., Pietrzak, R., Kaplan, E., Inkret, W., Efurd, W., Rokop, D., Lewis, D., Gautier, P. & Bell III, R. T. (2001b). Intercomparison study of inductively coupled plasma mass spectrometry, thermal ionization mass spectrometry and fission track analysis of μBq quantities of ^{239}Pu in synthetic urine. *Journal of Radioanalytical and Nuclear Chemistry, 249* (1), 121–131.

ISO (1996). *Quality System Guidelines for the Production of Reference Materials.* Guide 34. Geneva: ISO/IEC.

ISO (1997). *Proficiency Testing and Interlaboratory Comparisons.* Guide 43 1. Geneva: ISO/IEC.

Jacobsen, G. E., Hotchkis, M. A. C., Fink, D., Child, D. P., Tuniz, C., Sacchi, E., Levins, D. M., Povinec, P. P. & Mulsow, S. (2000). AMS measurement of ^{129}I, ^{36}Cl and ^{14}C in underground waters from Mururoa and Fangataufa atolls. *Nuclear Instruments and Methods in Physics Research, B 172*, 666–671.

Jenkins, W. J. & Clarke, W. B. (1976). The distribution of ^{3}He in the western Atlantic Ocean. *Deep-Sea Research, 23*, 481–494.

Johnson, O., Osborne, D., Wildgoose, C., Worstell, W. & Loebel, P. (1997). Development and testing of a large-area underwater survey device. *IEEE Transactions on Nuclear Science, 44*, 792–798.

Jones, D. G. (2001). Development and applications of offshore gamma-ray spectrometry measurements: A review. *Journal of Environmental Radioactivity, 53*, 313–333.

Jones, D. G., Miller, J. M. & Roberts, P. D. (1984). The distribution of Cs-137 in surface intertidal sediments from the Solway Firth. *Marine Pollution Bulletin, 15*, 187–194.

Jones, D. G., Roberts, P. D. & Miller, J. M. (1988). The distribution of gamma-emitting radionuclides in surface subtidal sediments near the Sellafield Plant. *Estuarine Coastal and Shelf Science, 27*, 143–161.

Jones, D. G., Roberts, P. D., Strutt, M. H., Higgo, J. J. & Davis, J. R. (1999). Distribution of Cs-137 and inventories of Pu-238, Pu-239/240, Am-241 and Cs-137 in Irish Sea intertidal sediments. *Journal of Environmental Radioactivity, 44*, 159–189.

Katagiri, M., Kobayashi, Y., Taguchi, Y. & Uchida, T. (1995). General-purpose germanium gamma-ray detector cooled with two Stirling refrigerators (pp. 80–81). KEK-PROC-95-1. Tsukuba.

Kawahata, H. & Murayama, M. (2000). Radiocarbon of settling particles from the hemipelagic region of the Western Pacific Warm Pool. *Nuclear Instruments and Methods in Physics Research, B 172*, 485–489.

Kelley, J. M., Bond, L. A. & Beasley, T. M. (1999). Global distribution of Pu isotopes and ^{237}Np. *Science of the Total Environment, 237/238*, 483–500.

Kershaw, P. J., McCubbin, D. & Leonard, K. S. (1999). Continuing contamination of North Atlantic and Arctic waters by Sellafield radionuclides. *Science of the Total Environment, 237/238*, 119–132.

Key, R. M. (1996). WOCE Pacific Ocean Radiocarbon Program. *Radiocarbon, 38* (3), 415–423.

King, S. E., Aakenes, U. R., McCoy, K. & Phillips, G. W. (1997). Development of the Arctic radionuclide monitoring station. *IEEE Transactions on Nuclear Science, 44*, 777–782.

Knie, K., Faestermann, T., Korschinek, G., Rugel, G., Ruhm, W. & Wallner, C. (2000). High-sensitivity AMS for heavy nuclides at the Munich Tandem accelerator. *Nuclear Instruments and Methods in Physics Research, B 172*, 717–720.

Kobayashi, Y., Takahashi, R., Shima, S., Katagiri, M. & Takahashi, K. (1999). Development of a submersible Ge detector system (pp. 501–503). IAEA-TECDOC-1094. Vienna, Austria: IAEA.

Kobayashi, T., Nagai, H. & Kobayashi, K. (2000). Concentration profiles of ^{10}Be in large manganese crusts. *Nuclear Instruments and Methods in Physics Research, B 172*, 579–582.

Koczy, F. F. (1958). Natural radium as a tracer in the ocean. In *Proceedings of the 2nd International Conference of Peaceful Uses of Atomic Energy* (pp. 351–357). Geneva.

Koppenaal, D. W. (1988). Atomic mass spectrometry. *Analytical Chemistry, 60*, 113.

Kramer, S. D., Hurst, G. S., Chen, C. H., Payne, M. G., Allman, S. L., Philips, R. C., Lehmann, B. E., Oescher, H., Loosli, H. H., Willis, R. D. & Thonnard, N. (1986). Analysis of ^{81}Kr in groundwater using laser resonance ionization spectroscopy. *Nuclear Instruments and Methods in Physics Research, B 17*, 395–401.

Ku, T. L., Kusakabe, M., Measures, C. I., Southon, J. R., Cusimano, G., Vogel, J. S., Nelson, D. E. & Nakaya, S. (1990). Beryllium isotope distribution in the western North Atlantic: A comparison to the Pacific. *Deep-Sea Research, 37*, 795–808.

Kumamoto, Y., Honda, M. C., Murata, A., Harada, N., Kusakabe, M., Hayashi, K., Kison, N., Katagiri, M., Nakao, K. & Southon, J. R. (2000). Distribution of radiocarbon in the western North Pacific: Preliminary results from MR97-02 cruise in 1997. *Nuclear Instruments and Methods in Physics Research, B 172*, 495–500.

Kutschera, W. (1986). Accelerator mass spectrometry and nuclear physics. *Nuclear Instruments and Methods in Physics Research, B 17*, 377–384.

Kutschera, W., Golser, R., Priller, A. & Strohmeier, B. (Eds) (2000). Accelerator mass spectrometry. *Proceedings of the 8th International Conference on Accelerator Mass Spectrometry. Nuclear Instruments and Methods in Physics Research, B 172* (997 pp.).

Lal, D. (1999). An overview of five decades of studies in cosmic ray produced nuclides in oceans. *Science of the Total Environment, 237/238*, 3–13.

La Rosa, J. J., Burnett, W., Lee, S.-H., Levy, I., Gastaud, J. & Povinec, P. P. (2001). Separation of actinides, cesium and strontium from marine samples using extraction chromatography and sorbents. *Journal of Radionalytical and Nuclear Chemistry, 248* (3), 765–770.

Lawson, E. M., Elliott, G., Fallon, J., Fink, D., Hotchkis, M. A. C., Hua, Q., Jacobsen, G. E., Lee, P., Smith, A. M., Tuniz, C. & Zoppi, U. (2000). AMS at ANTARES – The first 10 years. *Nuclear Instruments and Methods in Physics Research, B 172*, 95–99.

Lee, S.-H., Gastaud, J., La Rosa, J. J., Liong Wee Kwong, L., Povinec, P. P., Wyse, E., Fifield, L. K., Hausladen, P. A., De Tada, L. M. & Santos, G. M. (2001). Analysis of plutonium isotopes in marine samples by radiometric, ICP-MS and AMS techniques. *Journal of Radioanalytical and Nuclear Chemistry, 248* (3), 757–764.

Lee, T., Ku, T.-L., Hu, H.-L. & Chen, J.-C. (1993). First detection of fallout Cs-135 and potential applications of $^{137}Cs/^{135}Cs$ ratio. *Geochimical et Cosmochimical Acta, 57*, 3493–3497.

Leon Vintro, L., Mitchell, P. I., Condren, O. M., Moran, M., Vives, I. Battle, J. & Sanchez-Cabeza, J. A. (1996). Determination of the $^{240}Pu/^{239}Pu$ atom ratio in low activity environmental samples by alpha spectrometry and spectral deconvolution. *Nuclear Instruments and Methods in Physics Research, A 369*, 597–602.

Lewis, D., Miller, G., Duffy, C. J., Efurd, D. W., Inkret, W. C. & Wagner, S. E. (2001). Los Alamos National Laboratory thermal ionisation mass spectrometry results from intercomparison study of inductively coupled plasma mass spectrometry, thermal ionization mass spectrometry, and fission track analysis of μBq quantities of ^{239}Pu in synthetic urine (LA-UR-001698). *Journal of Radioanalytical and Nuclear Chemistry, 249*, 115–120.

Liong Wee Kwong, L., La Rosa, J. J., Lee, S.-H. & Povinec, P. P. (2001). Liquid scintillation spectrometry of beta-emitters in marine samples. *Journal of Radioanalytical and Nuclear Chemistry, 248* (3), 751–755.

Livingston, H. D. & Povinec, P. P. (2000). Anthropogenic marine radioactivity. *Ocean and Coastal Management, 43*, 689–712.

Livingston, H. D. & Povinec, P. P. (2002). A Millennium perspective on the contribution of global fallout radionuclides to ocean science. *Health Physics, 82*, 656–668.

Livingston, H. D., Povinec, P. P., Ito, T. & Togawa, O. (2001). The behaviour of plutonium in the Pacific Ocean. In A. Kudo (Ed.), *Plutonium in the Environment* (pp. 267–291). Amsterdam: Elsevier.

Loosli, H. H., Moll, M., Oeschger, H. & Schotterer, U. (1986). Ten years low-level counting in the underground laboratory in Bern, Switzerland. *Nuclear Instruments and Methods in Physics Research, B 17*, 402–405.

MacCartney, M. & Rajendran, K. (1999). Historical levels of ^{99}Tc in the Irish Sea. IAEA-SM-354/44. In *Proceedings of the International Symposium on Marine Pollution*. (pp. 105–110). Monaco, October 5–9, 1998. IAEA-TECDOC-1094 Vienna: Austria, IAEA.

McHargue, L. R., Donahue, D., Damon, P. E., Sonett, C. P., Biddulph, D. & Burr, G. (2000). Geomagnetic modulation of the late Pleistocene cosmic-ray flux as determined by ^{10}Be from Blake Outer Ridge marine sediments. *Nuclear Instruments and Methods in Physics Research, B 172*, 555–561.

McNichol, A. P., Schneider, R. J., von Reden, K. F., Gagnon, A. R., Elder, K. L., NOSAMS, Key, R. M. & Quay, P. D. (2000). Ten years after – The WOCE AMS Radiocarbon Program. *Nuclear Instruments and Methods in Physics Research, B 172*, 479–484.

Michel, R. (2001). Environmental radioactivity measuring methods. In F. Brechignac & B. J. Hoeard (Eds), *Radioactive Pollutants* (pp. 27–62). Les Ulis, France: EDP Sciences.

Miller, J. M., Thomas, B. W., Roberts, P. D. & Creamer, S. C. (1982). Measurement of marine radionuclide distributions using a towed sea-bed spectrometer. *Marine Pollution Bulletin, 13* (9), 315–319.

Moore, W. (1996). Large groundwater input to coastal waters revealed by ^{226}Ra enrichments. *Nature*, *380*, 612–614.
Moore, W. (1999). The subterranean estuary: A reaction zone of ground water and sea water. *Marine Chemistry*, *65*, 111–125.
Moore, W. S. & Arnold, R. (1996). Measurement of ^{223}Ra and ^{224}Ra in coastal waters using a delayed coincidence counter. *Journal of Geophysical Research*, *101* (C1), 1321–1329.
Morgenstern, W., Fifield, L. K. & Zondervan, A. (2000). New frontiers in glacier ice dating: Measurement of natural ^{32}Si by AMS. *Nuclear Instruments and Methods in Physics Research*, *B 172*, 605–609.
Muramatsu, Y., Ohmomo, Y. & Sumiya, M. (1988). Determination of iodine-129 and iodine-127 in environmental samples collected in Japan. *Journal of Radioanalytical and Nuclear Chemistry Articles*, *123*, 181.
Muramatsu, Y., Hamilton, T., Uchida, S., Tagami, K., Yoshida, S. & Robison, W. (2001). Measurement of ^{240}Pu/^{239}Pu isotopic ratios in soils from the Marshall Islands using ICP-MS. *Science of the Total Environment*, *278*, 151–159.
Noakes, J. E., Noakes, S. E., Dvoracek, D. K., Culp, R. A. & Bush, B. (1999). Rapid coastal survey of anthropogenic radionuclides, metals, and organic compounds in surficial marine sediments. *Science of the Total Environment*, *237/238*, 449–458.
Ostlund, H. G. & Dorsey, H. G. (1975). Rapid electrolytic enrichment and hydrogen gas proportional counting of tritium. In *Low-Radioactivity Measurements and Applications* (pp. 55–60). Bratislava: SPN.
Osvath, I. & Povinec, P. P. (2001). Seabed γ-ray spectrometry: Applications at IAEA-MEL. *Journal of Environmental Radioactivity*, *53*, 335–349.
Osvath, I., Povinec, P. P., Huynh-Ngoc, L. & Comanducci, J.-F. (1999). Underwater gamma-surveys of Mururoa and Fangataufa lagoons. *Science of the Total Environment*, *237/238*, 277–286.
Osvath, I., Povinec, P. P., Livingston, H. D., Ryan, T., Comanducci, J.-F. & Mulsow, S. (2004). In situ stationary monitoring of ^{137}Cs in seawater of the NW Irish Sea. *Journal of Radioanalytical and Nuclear Chemistry* (in print).
Oughton, D. H., Skipperud, L., Salbu, B., Fifield, L. K., Cresswell, R. C. & Day, J. P. (1999). Determination of ^{240}Pu/^{239}Pu isotope ratios in Kara Sea and Novaya Zemlya sediments using accelerator mass spectrometry. IAEA-SM-354/59. In *Proceedings of the International Symposium on Marine Pollution* (pp. 123–128). Monaco, October 5–9, 1998. IAEA-TECDOC-1094. Vienna, Austria.
Oughton, D., Day, P. & Fifield, K. (2001). Plutonium measurement using accelerator mass spectrometry: Methodology and applications. In A. Kudo (Ed.), *Plutonium in the Environment* (pp. 47–62). Amsterdam: Elsevier.
Parr, R. M., Fajgelj, A., Dekner, R., Vera Ruiz, H., Carvalho, F. P. & Povinec, P. P. (1998). Analytical Quality Assurance Programmes to meet the Present and Future Needs of Developing Countries. *Journal of Analytical Chemistry*, *360*, 287–290.
Peng, T.-H., Maier-Reimer, E. & Broecker, W. S. (1993). Distribution of ^{32}Si in the world ocean: Model compared to observation. *Global Biogeochemical Cycles*, *7*, 463–474.
Povinec, P. (1981). Dual parameter gamma-ray spectrometer for low-level counting. *Isotopenpraxis*, *18*, 92–97.
Povinec, P. (Ed.) (1982). Low-level counting. In *Proceedings of the Second International Conference, Low Radioactivities'80* (434 pp.). Bratislava: VEDA.
Povinec, P. (Ed.) (1986). Low-level counting. *Proceedings of the Third International Conference, Low Radioactivities '85. Nuclear Instruments and Methods in Physics Research*, *B 17*, 377–588.
Povinec, P. (Ed.) (1987). Low-level counting and spectrometry. In *Proceedings of the Third International Conference, Low Radioactivities'85* (308 pp.). Bratislava: VEDA.
Povinec, P. & Usacev, S. (Eds) (1977). Low-radioactivity Measurements and Applications. In *Proceedings of the International Conference* (542 pp.). Bratislava: SPN.
Povinec, P. P. (Ed.) (1991a). Rare nuclear decays and fundamental physics. *Proceedings of the 14th Europhysics Conference on Nuclear Physics. Journal of Physics*, *G 17* (543 pp.).
Povinec, P. P. (1991b). Low-level gas counting. In M. Garcia-Leon & G. Madurga (Eds), *Proceedings of the Second International Summer School* (pp. 38–72). Singapore: World Scientific.
Povinec, P. P. (1992). Rare nuclear processes. In *Proc. 14th Europhysics Conference on Nuclear Physics* (441 pp.). Singapore: World Scientific.
Povinec, P. P. (1994). Underground low-Level counting. In M. Garcia-Leon & R. Garcia-Tenorio (Eds), *Low-Level Measurements of Radioactivity in the Environment* (pp. 113–140). Singapore: World Scientific.
Povinec, P. P. (Ed.) (1999). Marine environment – understanding and protecting for the future. *Science of the Total Environment*, *237/238*, 1–526.

Povinec, P. P. & Pham, M. K. (2000). Report on the Intercomparison Run IAEA-384. Radionuclides in Fangataufa Lagoon sediment. IAEA/AL/126. Vienna: IAEA.

Povinec, P. P., Osvath, I., Baxter, M. S. & Harms, I. (1995). *In-situ* monitoring of gamma radioactivity in the Kara Sea. *Proceedings of the Workshop on Monitoring of Nuclear Contamination in Arctic Seas*. NRL/MR/6610-95-7674 (pp. IV 2–17). Washington, DC.

Povinec, P. P., Baxter, M. S., Gastaud, J., Liong Wee Kwong, L. & Oregioni, B. (1996a). Low-level liquid scintillation spectrometry of soft beta emitters. In G. T. Cook, D. D. Harkness, A. B. Mackenzie, B. F. Miller & E. M. Scott (Eds), *Liquid Scintillation Spectrometry 1994* (pp. 149–155). Tucson: Radiocarbon.

Povinec, P. P., Osvath, I. & Baxter, M. S. (1996b). Underwater gamma-spectrometry with HP Ge and NaI(Tl) detectors. *Applied Radiation and Isotopes*, *47* (9/10), 1127–1133.

Povinec, P. P., Osvath, I., Baxter, M. S., Harms, I., Huynh-Ngoc, L., Liong Wee Kwong, L. & Pettersson, H. B. L. (1997). IAEA-MEL's contribution to the investigation of Kara Sea radioactivity and radiological assessment. *Marine Pollution Bulletin*, *35*, 235–241.

Povinec, P. P., Gastaud, J. & Pham, M. K. (1999a). IAEA-MEL's AQCS programme for marine radioactivity measurements (pp. 87–93). IAEA-TECDOC-1094. Vienna, Austria.

Povinec, P. P., Pham, M. K. & Ballestra, S. (1999b). Report on the Intercomparison Run and Certified Reference Material IAEA-381. Radionuclides in Irish Sea water. IAEA/AL/118. Vienna: IAEA.

Povinec, P. P., Oregioni, B., Jull, A. J. T., Kieser, W. E. & Zhao, X.-L. (2000). AMS measurements of ^{14}C and ^{129}I in seawater around radioactive waste dump sites. *Nuclear Instruments and Methods in Physics Research, B 172*, 672–678.

Povinec, P. P., La Rosa, J. J., Lee, S.-H., Mulsow, S., Osvath, I. & Wyse, E. (2001a). Recent developments in radiometric and mass spectrometric methods for marine radioactivity measurements. *Journal of Radioanalytical and Nuclear Chemistry*, *248* (3), 713–718.

Povinec, P. P., Osvath, I., Mulsow, S. & Comanducci, J.-F. (2001b). La surveillance in-situ de la radioactivite marine au large de Monaco. *Proceedings of the 36th CIESM Congress*. Monaco. 155.

Povinec, P. P., Jull, A. J. T. & Burr, G. S. (2001c). Radiocarbon in seawater at radioactive waste dumping sites in the Northeast Atlantic and Northwest Pacific. *Radiocarbon*, *43*, 879–886.

Povinec, P. P., Badie, C., Baeza, A., Barci-Funel, G., Bergan, T. D., Bojanowski, R., Burnett, W., Eikenberg, J., Fifield, L. K., Serradell, V., Gastaud, J., Goroncy, I., Herrmann, J., Hotchkis, M. A. C., Ikaheimonen, T. K., Jakobson, E., Kalimbadjan, J., La Rosa, J. J., Lee, S.-H., Liong Wee Kwong, L., Lueng, W. M., Nielsen, S. P., Noureddine, A., Pham, M. K., Rohou, J.-N., Sanchez-Cabeza, J. A., Suomela, J., Suplinska, M. & Wyse, E. (2002). Certified Reference Material for radionuclides in seawater IAEA-381 (Irish Sea water). *Journal of Radioanalytical and Nuclear Chemistry*, *251*, 369–374.

Povinec, P. P. et al. (2003). IAEA'97 expedition to the NW Pacific Ocean – results of oceanographic and radionuclide investigations of the water column. *Deep-Sea Research II*, *50*, 2607–2638.

Povinec, P. P., Comanducci, J.-F. & Levy, I. (2004). IAEA-MEL's underground laboratory for the analysis of radionuclides in the environment at very low levels (in print).

Raisbeck, G. M. & Yiou, F. (1979). ^{10}Be concentrations and residence time in the ocean surface layer. *Earth and Planetary Science Letters*, *43*, 237–240.

Raisbeck, G. M. & Yiou, F. (1999). I-129 in the ocean: Origins and applications. *Science of the Total Environment*, *237/238*, 31–41.

Reidel, T., Hentig, R. V. & Feilitzsch, F. V. (2000). Anticompton spectrometer element studies. *Applied Radiation and Isotopes*, *53*, 231–236.

Roberts, M. L., Hamm, R. W., Dingley, K. H., Chiarappa-Zucca, M. L. & Love, A. H. (2000). A compact tritium AMS system. *Nuclear Instruments and Methods in Physics Research*, *B 172*, 262–267.

Roether, W., Manca, B. B., Klein, B., Bregant, D., Georgopolous, D., Beitzel, V., Kovacevic, V. & Luchetta, A. (1996). Recent changes in eastern Mediterranean deep waters. *Science*, *271*, 333–335.

Rosenberg, R. J. (1993). Non-conventional measurement techniques for the determination of some long-lived radionuclides produced in nuclear fuel. A literature survey. *Journal of Radioanalytical and Nuclear Chemistry*, *171*, 465–482.

Ryan, T. P., Mitchell, P. I., Vives I Battle, J., Sanchez-Cabeza, J. A., McGarry, A. T. & Schell, W. R. (1993). Low-level ^{241}Pu analysis by supported-disk liquid scintillation counting. In J. E. Noakes, F. Schönhofer & H. A. Polach (Eds), *Liquid Scintillation Spectrometry 1992* (pp. 75–82). Tucson: Radiocarbon.

Saito, M., Iwamoto, S. & Sumino, T. (2000). Tritium concentration in Aquatic Environment determined by SPE tritium enrichment apparatus. In J. Inaba, S. Hisamatsu & Y. Ohtouka (Eds), *Proceedings: Distribution and Speciation of Radionuclides in the Environment* (pp. 230–132). Rokkasho: IES.

Salbu, B., Lind, O. C., Borretzen, P. & Oughton, D. H. (2001). Advanced speciation techniques for radionuclides associated with colloids and particles. In F. Brechignac & B. J. Howard (Eds), *Radioactive Pollutants* (pp. 243–260). Les Ulis: EDP Sciences.

Schlosser, P., Bayer, R., Boenisch, G., Cooper, L. W., Ekwurzel, B., Jenkins, W. J., Khatiwala, S., Pfirman, S. & Smethie, W. M. (1999). Pathways and residence times of dissolved pollutants in the ocean derived from transient tracers and stable isotopes. *Science of the Total Environment, 237/238*, 15–30.

Sokolov, A. D., Danengirsh, S. G., Popov, S. V., Pchelintsev, A. B., Gostilo, V. V., Druzhinin, A. A., Maksimov, M. Y., Lebedev, N. M. & Shapovalov, V. I. (1999). A deep-water gamma-ray spectrometer on the basis of a high-purity germanium detector. *Instruments and Experimental Techniques, 42*, 563–568.

Somayajulu, B. L. K. (1967). Beryllium-10 in a manganese nodule. *Science, 156*, 1219–1220.

Somayajulu, B. L. K., Rengarajan, R., Lal, D., Weiss, R. F. & Craig, H. (1987). GEOSECS: Atlantic ^{32}Si profiles. *Earth and Planetary Science Letters, 85*, 329–342.

Somayajulu, B. L. K., Rengarajan, R., Lal, D. & Craig, H. (1991). GEOSECS Pacific and Indian Ocean ^{32}Si profiles. *Earth and Planetary Science Letters, 107*, 197–216.

Somayajulu, B. L. K., Bhushan, R., Sarkar, A., Burr, G. S. & Jull, A. J. T. (1999). Sediment deposition rates on the continental margins of the eastern Arabian Sea using ^{210}Pb, ^{137}Cs and ^{14}C. *Science of the Total Environment, 237/238*, 429–439.

Soukissian, T. H., Chronis, G. T. & Nittis, K. (1999). POSEIDON: Operational marine monitoring system for Greek seas. *Sea Technology, 40*, 31–37.

Sturup, S., Dahlgaard, H. & Nielsen, S. C. (1998). High resolution inductively coupled plasma mass spectrometry for the trace determination of plutonium isotopes and isotope ratios in environmental samples. *Journal of Analytical Atomic Spectrometry, 13*, 1321–1326.

Suter, M., Jacob, S. W. A. & Synal, H.-A. (2000). Tandem AMS at sub-MeV energies – Status and prospects. *Nuclear Instruments and Methods in Physics Research, B 172*, 144–151.

Taylor, C. B. (1977). Tritium enrichment of environmental waters by electrolysis: Development of cathodes exhibiting high isotopic separation and precise measurement of tritium enrichment factors. In P. Povinec (Ed.), *Low-Radioactivity Measurements and Applications* (pp. 131–140). Bratislava: SPN.

Taylor, R. N., Croudace, I. W., Warwick, P. E. & Dee, S. J. (1998). Precise and rapid determination of ^{238}U/^{235}U and uranium concentration in soil samples using thermal ionisation mass spectrometry. *Chemical Geology, 144*, 73.

Taylor, R. N., Warneke, T., Milton, J. A., Croudace, I. W., Warwick, P. E. & Nesbitt, R. W. (2001). Plutonium isotope ratio analysis at femtogram to nanogram levels by multicollector ICP-MS. *Journal of Analytical Atomic Spectrometry, 16*, 279.

Theodorsson, P. (1996). *Measurement of Weak Radioactivity* (333 pp.). Singapore: World Scientific.

Thomas, B. W., Clayton, C. G., Ranasinghe, V. V. C. & Blair, I. M. (1983). Mineral exploration of the sea bed by towed sea bed spectrometers. *Applied Radiation and Isotopes, 34*, 437–449.

Torgersen, T., Clarke, W. B. & Jenkins, W. J. (1979). The tritium/helium-3 method in hydrology. In *Proceedings of the Symposium on Isotope Hydrology* (pp. 917–930). Vienna 1979. Vienna: IAEA.

Treacy Jr., D. J., Knies, D. L., Grabowski, K. S., Hubler, G. K., DeTurck, T. M. & Mignerey, A. C. (2000). Determination of ^{32}Si by AMS at the US Naval Research Laboratory. *Nuclear Instruments and Methods in Physics Research, B 172*, 321–327.

Tyler, A. N., Sanderson, D. C. W. & Scott, E. M. (1996). Estimating and accounting for Cs-137 source burial through in-situ gamma spectrometry in salt marsh environments. *Journal of Environmental Radioactivity, 33*, 195–212.

Van der Loef, M. R., Friedrich, J. & Bathmann, U. V. (1997). Carbon export during the Spring Bloom at the Antarctic Polar Front, determined with the natural tracer ^{234}Th. *Deep-Sea Research II, 44*, 457.

Vojtyla, P., Beer, J. & Stavina, P. (1994). Experimental and simulated cosmic muon induced background of a Ge spectrometer equipped with a top side anticoincidence proportional chamber. *Nuclear Instruments and Methods in Physics Research, B 86*, 380–386.

Vojtyla, P. & Povinec, P. P. (2000). A Monte Carlo simulation of background characteristics of low-level HP Ge detectors. *Applied Radiation and Isotopes, 53*, 185–190.

Wendt, K., Trautmann, N. & Bushaw, B. A. (2000). Resonant laser ionization mass spectrometry: An alternative to AMS? *Nuclear Instruments and Methods in Physics Research, B 172*, 162–169.

White, F. A. & Wood, G. M. (1988). *Mass Spectrometry – Applications in Science and Engineering*. New York: Wiley-Interscience.

Wyse, E., Lee, S.-H., La Rosa, J. J., Povinec, P. & de Mora, S. J. (2001). ICP-sector field mass spectrometry analysis of plutonium isotopes: Recognizing and resolving potential interferences. *Journal of Analytical Atomic Spectrometry*, *16*, 1107–1111.

Yoneda, M., Kitagawa, H., van der Plicht, J., Uchida, M., Tanaka, A., Uehiro, T., Shibata, Y., Morita, M. & Ohno, T. (2000). Pre-bomb marine reservoir ages in the western North Pacific: Preliminary result on Kyoto University collection. *Nuclear Instruments and Methods in Physics Research*, *B 172*, 377–381.

Yonezawa, C., Matsue, H., McKay, K. & Povinec, P. (2001). Analysis of marine samples by neutron-induced prompt gamma-ray technique and ICP-MS. *Journal of Radioanalytical and Nuclear Chemistry*, *248*, 719–725.

Zhao, X.-L., Kieser, W. E., Litherland, A. E., Smith, J. N., Fowler, S. W. & Miquel, J.-C. (1999). The AMS analysis of ^{129}I and ^{135}Cs: Development of tools for oceanographic tracing and pollution source identification (pp. 243–248). IAEA-TECDOC-1094. Vienna, Austria: IAEA.

Zvara, I., Povinec, P. & Sykora, I. (1994). Determination of very low levels of radioactivity. *Pure and Applied Chemistry*, *66* (12), 2537–2586.

Index of Authors

Subject Index